S0-ARK-456

$24.50

# PROPERTIES
# AND
# MANAGEMENT
# OF
# FOREST
# SOILS

# PROPERTIES AND MANAGEMENT OF FOREST SOILS

William L. Pritchett

Professor of Forest Soils
Soil Science Department
University of Florida
Gainesville

JOHN WILEY & SONS

New York   Chichester   Brisbane   Toronto

Copyright © 1979, by John Wiley & Sons, Inc.

All rights reserved. Published simultaneously in Canada.

Reproduction or translation of any part of
this work beyond that permitted by Sections
107 and 108 of the 1976 United States Copyright
Act without the permission of the copyright
owner is unlawful. Requests for permission
or further information should be addressed to
the Permissions Department, John Wiley & Sons.

Library of Congress Cataloging in Publication Data

Pritchett, William L
  Properties and management of forest soils.

  Bibliography: p.
  Includes index.
  1. Forest soils.  2. Soil management.  3. Forest
management.  I. Title.
SD390.P74      634.9     78-23196
ISBN 0-471-03718-4
Printed in the United States
10 9 8 7 6 5 4 3 2 1

# PREFACE

Forest land management has undergone some major changes during the past decade. Mounting pressure for more intensive and diverse use of forest land has been generated by worldwide population increases, rises in per capita consumption of wood products, reduction in the forest land base, and increases in the cost of acquiring and owning land. Significant advances have been made in silviculture, especially in reforestation technology and soil management of short rotation forests for fiber production. The use of genetically superior planting stock, intensive seedbed preparation, chemical pest control, and correction of nutrient deficiencies is common practice in pine forests of the southern United States, in coniferous forests of the Pacific Northwest, central Europe and Scandinavia, and in exotic forests of Australia-New Zealand, South America, and southern Africa. Although these practices have been successfully used to increase tree growth and shorten the rotation period of the forest crop, there is a general lack of understanding of their effects on basic soil processes, long-term site productivity, and the total environment.

Concurrent with advances in forest technology has been an awakening of the public to the value of forests for recreation, wildlife, watershed protection, and in preservation of a quality environment. These public pressures also have exerted an influence on forest land management. With the growing awareness of the integral role of forest soils in the total forest ecosystem, it becomes increasingly necessary to understand forest soil properties as they relate to modern management practices. Foresters need to know what happens to the soil and to the environment when stands are clear-cut harvested; when fires sweep through our forests; when sites are intensively prepared by blading, discing, or bedding; and when chemicals are used to control pests or to correct soil nutrient deficiencies. It is obvious that the changing techniques in forest management and the public's attitude toward silvicultural operations are rapidly converting foresters from generalists to specialists. It appears equally obvious that forest soils is one of the areas in which there will be an increasing demand for specialists.

While most of the basic principles of soil science apply to forest soils as well as to agricultural soils, there are a number of fundamental differences between silviculture and agriculture that result in striking differences in soil properties and in soil management practices. For example, the continual deposition of leaves and other debris creates a kind of organic mulch on the surface of forest soils, which results in a more stable soil microclimate and provides conditions for a wider spectrum of soil animals and microorganisms than are found in most cultivated soils or grasslands. The activity of these organisms and the leaching of organic acids and other decomposition products result in substantial differences in chemical, physical, and biological properties of the two groups of soils. These differences are sufficiently great to warrant separate study and treatment.

This book is for the advanced student and for the practicing forester as a source of current information on the fundamental properties and processes of forest soils, with particular reference to the application of soil science to silviculture. Material covered herein is presented on the assumption that the reader has completed basic courses in soils and silviculture, and possesses some understanding of the nature and function of forest ecosystems. Selected references cited in the text are listed in the Literature Cited section for further reading. Special attention is given to recent advances in intensive management of soils of short-rotation plantations and to the effect of these management practices on the environment. It is primarily in this latter respect that this text differs from the classical works of the past by Lutz and Chandler (1946) and Wilde (1958).

Note that common names are used for tree species of the United States, while the scientific names of other species are generally used in the text. Scientific names of United States species are given in Appendix A. The I.S.U. system is used throughout the text in recognition of the trend toward worldwide acceptance of this simplified method of measurement. However, a conversion table is provided in Appendix B for the more common units.

I express my appreciation to Walter Lyford and others at Harvard Forest where I spent a year of sabbatical leave writing the first draft of this manuscript. I also thank my colleagues of the Soil Science Department and the School of Forest Resources and Conservation at the University of Florida, especially R. F. Fisher, C. A. Hollis, and W. H. Smith, for assistance in developing and organizing the materials; to C. B. Davey, G. L. Switzer, and A. G. Wollum for their helpful reviews of the text; and to my wife, Peggy, for proofreading and assisting in countless other tasks.

W.L.P.

# CONTENTS

# PROPERTIES
# AND
# MANAGEMENT
# OF
# FOREST
# SOILS

# Part 1
## PROPERTIES AND DYNAMIC PROCESSES

# 1

---

# INTRODUCTION

Soil is often defined as the outer, mostly unconsolidated, layer of the earth's crust, ranging from a few centimeters to more than 3 m in thickness. It is further described as a natural body composed of a mixture of organic and mineral materials in which plants grow. Of course, this definition does not hold for all soils, and, furthermore, some plants growing in the air and sea do not have direct contact with soil. However, the description does emphasize the close relationship between soil and plants, which is the basis for the study of forest soils.

Soil, however, has different meaning for different people. To the civil engineer and builder, it is a material on which to construct roads or lay foundations for building; to the homeowner and sanitary engineer it may be a recipient for domestic and municipal wastes; to the hydrologist it is viewed as a vegetated water-transmitting mantle; and to the ecologist it is an ecosystem or an ecosystem component involving a vast number of organisms and individual chemical processes (Stone, 1975). The forester also looks upon the soil as more than a medium for tree growth. Because forests are expected to serve the multiple uses of recreation, aesthetics, wildlife habitat, and watershed protection, the soil that supports these forests must also serve many purposes.

## HISTORICAL PERSPECTIVE

Soil science had its real beginning in ancient times, but the accumulation of knowledge about soils, their properties and management, has been very slow. It is said that the early Chinese used a schematic soil map as a basis for taxation; there are biblical references to the benefits of the use of dung around plants; and early Greeks and Romans apparently employed some sound soil management practices including the use of legumes, ashes, and other soil amendments (Brady, 1974). But the origin of scientific study of soil can probably be attributed to the

German chemist Julius von Liebig who developed the concept of the essentiality of mineral elements in the soil and added manures for plant growth. Lawes and Gilbert put Liebig's theories to test at the now-famous Rothamsted Experiment Station in England and found them to be generally sound, particularly the concept which came to be known as the "Law of the Minimum."

Following these beginnings, scientists in many different fields contributed to the development of soil science, including chemists, physicists, microbiologists, hydrologists, and especially geologists. Names such as Dokuchaiev, Davey, Darwin, Gedroiz, Glinka, Hartig, Heyer, Hilgard, Hopkins, King, Marbut, Ramann, Sprengel, and Thaer are prominent among the early investigators, but a complete listing would be almost impossible. Most of the early research on soils was directed toward its use for agricultural purposes, because agricultural problems were recognized first and given a higher priority than those of forestry. After all, there were seemingly endless supplies of forest for game, fuel, and building materials, and the use of forests was generally exploitative in nature.

The importance of soils to the natural forest ecosystem was recognized by some early scientists, but these pioneers were apparently few in number. In the last half of the nineteenth century as the practice of forestry developed in Europe, the need for information on forest soils became more apparent. In Germany, Heyer, Ebermayer, and others made significant contributions to the understanding of the physical properties of forest soils and of the importance of the litter layer to the stability of the forest. The classical studies of Müller (1879), a Danish scientist who described the forest humus layers and the influence of biological activity on their development, and of Ramann (1893) who applied soil chemical, physical, and biological information to certain forest practices, were among the earliest books that dealt primarily with forest soils. By the start of the twentieth century a number of students of forest soils had made significant contributions to the knowledge of this component of the forest ecosystem.

The science of forest soils was slow to develop in North America because of the general lack of a compelling need for this kind of information concerning our exploited forests. Only after World War I did the idea of managing selected forests for sustained yields, reforestation of abandoned farm lands, and the establishment of shelterbelts in the Midwest begin to take hold. Perhaps the greatest impetus to the scientific study of forest soils followed the publication of the texts on the subject by Lutz and Chandler (1946) and Wilde (1958). These two books were widely used by students throughout the United States for many years.

## WHAT IS A FOREST SOIL?

The need for a separate study of forest soils is sometimes questioned on the assumption that a forest soil is no different from a soil supporting other tree crops, such as citrus or pecans, or, for that matter, a soil devoted to grassland or cultivated annual crops. This assumption is generally made by persons who are not well acquainted with forest ecosystems and have not noted even the most obvious properties of soils associated with forests. The forest cover and its resultant forest floor provides a microclimate and a spectrum of microorganisms different from those associated with most other soils. Such dynamic processes as cycling of nutrients among components of the forest stand and the formation of organic acids from decaying debris and the subsequent leaching of bases give a distinctive character to soils with a forest cover.

In contrast, grassland soils generally possess thicker and darker-colored A1 horizons, with higher contents of bases and organic matter than those of forest soils. They occur over a range of temperature zones, but are generally associated with moderately low rainfall conditions. The concentration of soil organic matter resulting from the continuous regeneration of fiberous grass roots plus the low level of effective leaching results in soils that are generally more fertile than forest soils. Bacteria are the predominant microflora, and earthworms are probably the most important mesofauna of grasslands.

In the broadest sense, a forest soil is considered any soil that has developed under the influence of a forest cover. This view recognizes the unique effects of deep rooting by trees, specific organisms associated with forest vegetation, and the litter layer and leaching promoted by the products of its decomposition on soil genesis. By this definition, forest soils can be considered to cover approximately one-half of the earth's land surface. As a matter of fact, essentially all soils except those of the tundra, marshes, grasslands, and deserts were developed under a forest cover and have acquired some distinctive properties as a result of that association. Of course, not all of these soils are in forests today. Perhaps as much as one-third of the former forest soils are now devoted to agricultural use. Kellogg (1964) estimated that about 30 percent of the world's land surface is presently covered by forests of various types.

When European settlers arrived in the United States, forests covered about half of the land area; another two-fifths grew grass and herbaceous plants, while the remainder was mostly arid and barren. The eastern seaboard was entirely forested, and, largely as a consequence of the difficulties of clearing new land, agricultural settlement was mostly confined to the Atlantic slope until the end of the eighteenth century. Slowly settlement began expanding west of the Appalachian Mountains and extensive land clearing was underway. By the middle of the nineteenth century more than $1\frac{1}{4}$ million hectares of virgin forests had been cleared and the land converted to agricultural purposes. This area of cleared forest land gradually increased until well into the twentieth century. By this time improvements in agricultural technology permitted the release of millions of hectares of former croplands for other uses. Large areas of degraded farmland, especially in the eastern regions, reverted to forests, and by 1950 forests covered about a third of the total land area of the United States. This forest land base is again slowly diminishing under the relentless pressures of increasing population and industrial development.

Few truly virgin forests exist today in populated regions of the globe. The conversion of forestlands to croplands and back to forests has gone through more cycles in sections of central Europe and Asia than in the United States. Many European forestlands have been managed rather intensively for centuries. At the other extreme, an example of relatively short-term shifts in land use is found in the tropics where "swidden" agriculture, or shifting cultivation, is a form of crop rotation involving 2 or 3 years of cultivated crops alternating with 10 to 20 years of regenerated forests. Such practices alter many properties of the original forest soil, as do many other artificially created disturbances, as well as some acts of nature, exemplified by storms and lightning fires.

In recent years, intensively managed plantation forests have been created in several countries of the world. Among these man-made forests are several million hectares of exotic pine forests in the Southern Hemisphere and an even larger area of plantations employing native species in the Northern Hemisphere. The latter includes some 8 million hectares of pine plantations in the coastal plains of the southern United States, and large areas of Douglas-fir in the Pacific Northwest.

Because of the alteration in certain properties of forest soils as a result of intensive management, it is obvious that the distinction between forest soils and agricultural soils has become progressively less evident. Although some properties acquired by soil during its development persist long after the forest cover has been removed and the soil subjected to cultivation, other characteristics are drastically modified by practices associated with agricultural use. For this reason we will treat forest soils in this text in the more narrow sense as *soils that presently support a forest cover*. This is particularly necessary because only cursory attention will be given to genesis and classification of soils. Emphasis will be placed on understanding the various physical, chemical, and biological properties of forest soils, and how these properties influence site management for timber production and the other uses to which they are subjected.

## FOREST SOILS COMPARED TO CULTIVATED SOILS

The soil is more than just a medium for the growth of land plants, and a provider of physical support, moisture, and nutrients. The soil is a dynamic system that serves as a home for a myriad of organisms, a disposal area for nature's "wastes," a filter of toxic substances, and a storehouse for nutrients. The soil is a product of its environment, the quality of which is, in large part, a function of the forest stand it supports. At the same time the soil is a major feature of the habitat influencing plant growth. Although it is only one of several environmental factors controlling distribution of vegetation types, it can be the most important one under some conditions. For example, the further removed a tree is from the region of its climatic optimum, the more discriminating it becomes in respect to the soil. This means that the range of soil conditions favorable to the growth of a species narrows under unfavorable climatic conditions for that species.

Some of the physical, chemical, and biological properties that are characteristic of forest soils will be discussed in detail in later chapters. It will suffice at this time to point out certain properties of forest soils that differ from those of cultivated soils. These differences derive, in part, from the fact that the most "desirable" soils have often been selected for agricultural use, and the remainder left for native vegetation, such as forests and grasses. Fortunately, soil requirements for agricultural crops often differ from requirements for forest crops. It is not unusual to find that some highly productive forest sites are exceedingly poor for agricultural use. Nevertheless, they may be used for agricultural crops because of their location in reference to markets or centers of population. Poor drainage, steep slopes, or the presence of large stones are examples of soil conditions that favor forestry over agricultural use. However, the choice of land use often results from differences in requirements of the crops. Good examples are the wet flatlands of many coastal areas around the world. These important forest soils cannot be effectively used for agricultural purposes without considerable investments in water control, lime, and fertilizers.

Physical properties have long been regarded as of outstanding importance in forest soils research, while chemical properties were generally given the most attention in studies of agricultural soils. For example, coastal marine deposits of silicious sands are often very low in nutrient reserves, but variations in tree growth on these sands appear to be primarily related to the small differences in elevation, which affect soil moisture conditions. The fact that pines grow rather well on some almost-pure quartz sands in southeastern United States probably led to the oft-heard conclusion among early foresters of the region that "given sufficient moisture, southern pine can grow in a bed of marbles." The coastal soils of the area are certainly more than a bed of marbles, of course, for even the youngest of these soils have had several millennia to collect atmospheric dust and aerosols and rainfall containing minerals and small amounts of nitrogen from lightning discharges. In time, these atmospheric deposits led to colonization by lower forms of plant and animal life, incuding dinitrogen-fixing blue-green algae and lichens. Simpler forms of higher plant life followed, and finally southern pines and other associates became established.

Forest trees customarily occupy a site for many years, sending their roots well into the subsoil. During that period considerable amounts of organic materials are returned to the soil in the form of leaf and litter fall, and decaying roots. This litter layer exerts a profound influence on the physical, chemical, and biological properties of the soil. The tree canopy also shades the soil and keeps it several degrees cooler than cultivated soils. The presence of forest vegetation and litter results in more uniform moisture and temperature conditions that produce a soil climate that is rather oceanic in nature. The more favorable soil climate, plus the acid condition resulting from the decomposition and leaching from the litter layer, promote a more diverse and active microflora and fauna population than that found in agricultural soils.

Recent research has revealed that tree growth rates can be substantially increased by intensive management, including the addition of plant nutrients to deficient soils. Nevertheless, the net demands for nutrients by forest trees are considerably less than those made by most agricultural crops. Agricultural crops, many of which are exotic to a given area, are generally grown on the more productive soils that have been drastically modified by cultivation, liming, and fertilization. They are generally short-lived crops, and the whole plant, minus the roots, is usually harvested annually.

Agricultural soils, therefore, may be described as artificial products of man's activities, while forest soils are natural bodies that exhibit a well-defined succession of natural horizons. This contrast was certainly true a few decades ago, and it continues to be true in most areas today. But the contrast has diminished greatly in the exotic forests of the Southern Hemisphere, in the short-rotation forests of the southeastern United States, and in other intensively managed forests of the North American Pacific Northwest, and central and northern Europe. Clear-cut harvesting of trees disturbs the surface litter and results in short-term increases in temperature and moisture of the surface soil. Seedbed preparation by root raking, discing or plowing, and sometimes bedding, incorporates the litter layer with the mineral soil, which favors microbial activity. Fertilizing increases the nutrient level of the surface soil, but may also affect the rate of breakdown of the organic layer. At any rate, all of these practices exert a short-term influence on the characteristics of surface soils and render them more and more like agricultural soils. Fortunately, most of these changes are relatively temporary, existing only until a forest cover again becomes well established. With the development of the forest canopy and a humus layer on the forest floor, the soil again acquires many of the properties that distinguish it from cultivated soils.

## SOILS AND MODERN FORESTRY

It should be well appreciated that soils of natural forest ecosystems present few problems to the forest manager. These soils possess certain chemical, physical, and biological properties unique to the conditions under which they developed. They are normally stable and resilient bodies that are only temporarily altered by fire, windthrow, and other acts of nature. However, the effects of perturbations associated with intensive management of forest soils on their physical properties, organic matter, and nutrient contents, microbial population, and long-term productivity are relatively recent problems that have not been well researched.

The fact that many management practices tend to alter soil properties to approach those of cultivated soils, at least until the new forests become well established, does not necessarily permit managers to apply results of agricultural research to forest soil problems. Many of the problems remain peculiar to the forest environment and require special treatment. It is for these reasons that the interest in forest soil science has reached new heights in concert with the widespread adoption of modern forest management techniques in attempts to meet world demands for wood products.

The first 13 chapters of this text deal with some basic properties of forest soils and the dynamic processes important to forest ecosystems. Later chapters concern forest soil management and the consequences of forest management on the environment.

# 2
# FOREST SOILS AND VEGETATION DEVELOPMENT

In the earth's beginning there was only the atmosphere, the seas, and the bare rocks. When the sun shone and warmed the earth it evaporated water from the seas and deposited it on the land as rain. In time the parent rocks weathered under the relentless forces of nature, and plants and animals gradually evolved, multiplied, and occupied the land. Plants took root in the weathered rock, and upon dying, contributed organic matter in the age-old process of soil formation. Thus soils and vegetation developed in concert—neither independent of the other.

However, the development of soil and associated forest vegetation has not been a simple and straightforward process. It has taken place over many thousands of years through a sequence of complex, interrelated events. While a number of relatively independent factors are involved in the development of both soil and vegetation, perhaps none is more important than climate. Climate, vegetation, and soil form an interrelated dynamic complex, and when one member of this complex is altered the others similarly change and a new equilibrium is established.

Soil, like climate, plays an essential role in the growth and development of forests. It provides the water, nutrients, and medium of support for trees and other forest vegetation. Soils are derived from parent material of different mineral composition and these materials result in soil properties that influence both the composition of forest vegetation and the rate of tree growth. As the parent rock weathers, a soil profile develops, influenced not only by the physical factors of the environment but by the biota of the area that contribute to mineral weathering and organic matter content. Eventually four horizons comprise a typical, well-drained forest soil in the temperate regions: an organic layer (O); a leached zone (A); an area of accumulation (B); and the unweathered parent material (C). The formation of these distinctive soil horizons results from a series of complicated reactions and comparatively simple rearrangements of matter, termed *pedogenic processes* (Buol, Hole, and McCracken, 1973).

## PEDOGENIC PROCESSES
Perhaps all pedogenic processes operate to some degree in all soils, but some of them are not particularly important in the genesis of forest soils and, consequently, will not be discussed. They all involve some phase of (1) addition of organic and mineral materials to the soil as solids, liquids, and gases; (2) losses of these materials from the soil; (3) translocation of materials from one point to another in soils; or (4) transformations of mineral and organic matter within the soil.

Additions to the soil are in both organic and inorganic forms. Wind and rain act as agents of transport for dust particles, aerosols, and organic compounds washed from the forest canopy. These additions are generally not large in the interior of extensive forests, but can be substantial where open fields are adjacent to forests or as a result of catastrophic storms or volcanic action. The most significant addition to forest soil is organic matter. The surface litter layer contributes to the accumulation of soil organic matter and exerts considerable influence on the underlying mineral soil, and the associated population of microorganisms and soil animals.

Processes involving soil losses are leaching of mobile ions such as calcium, magnesium, and potassium and surface erosion by water or wind, solifluction, creep, and other means of mass wasting. The latter are generally not problems in forested areas except in some disturbed steep and mountainous areas. The important processes in forest soils are those that relate to translocation within the soil body, characterized as *eluviation* and *illuviation*.

Eluviation, involving mobilization and translocation, is analogous to leaching by solution, except that leaching generally implies removal from the entire solum. Eluviation is the transfer of materials, often colloidal, from a zone of impoverishment to a zone of enrichment, while illuviation is the reverse of this process. Some of the processes involved in one or the other of these two phases of translocation are: *decalcification* or the eluviation of carbonates, a process operating in humid areas; and *calcification,* the accumulation of carbonates, commonly observed in more arid regions. *Desalinization,* refers to the leaching of soluble salts from saline soils, while *salinization* implies the accumulation of these salts, usually in arid or coastal depressions. If sodium is the dominant cation, the processes of removal or accumulation of salt are termed *dealkalization* or *alkalization*.

Probably the two most important soil-forming processes in forested areas are *podzolization* and *laterization*. Although these two processes are rather ill defined, the terms have been widely used in soils literature. Podzolization is the process by which organic materials and sesquioxides are translocated from the upper soil horizon and subsequently deposited in the B horizon. It is the dominant process under forests in cool, moist climates. The process is intensified under vegetation that produce acid litter layers, as is the case with certain conifers, especially hemlocks, spruces, and pines in northern latitudes, kauri forest *(Agathis australis)* in New Zealand, and with heath *(Calluna vulgaris)* of northern Europe. In the coastal plain of southeastern United States extensive areas of Aquods (Groundwater Podzols) are developed under southern pines and associated oaks and understory plants (Buol et al., 1973).

Most, but not all, soils formerly named Podzols are Spodosols. The latter are widely known as acid, ashy gray sands over dark sandy loams. The characteristic contrasting horizons (Figure 2.1) are apparently developed by water percolating through quartose sand in which soluble organic compounds from the

**Figure 2.1** A podzolized profile of a Spodosol in a spruce-fir stand in a boreal forest in Quebec.

acid litter layer clean the sand grain in the first horizon and coat the grains in the second horizon with a dark humus-iron oxide mixture. Romans (1970) considered the formation of a Spodosol to begin with the accumulation of a surface raw humus mat, with constituents from the humus layers weathering the upper mineral horizon and forming soluble complexes with the iron and aluminum ions, rendering them mobile.

Laterization is found in intertropical zones with high temperatures and extreme leaching that favor rapid desilication and accumulation of ferric oxides under oxidizing conditions. The process produces an oxic horizon within 2 m of the surface or plinthite that forms a continuous phase within 30 cm of the mineral surface and with no spodic or argillic horizon overlying the oxic horizon (Soil Survey Staff, 1975). Oxisols, which include most soils previously classed as Latosols or Lateric, have a high degree of weathering and a low cation exchange capacity. They are the soils most often associated with tropical rain forests, but it should be pointed out that not all soils in the tropics are Oxisols. Oxisols and their associated Ultisols and Inceptisols support forests except in more populated areas where they are used for *shifting cultivation*. This is a practice whereby small patches of forest are cleared and cultivated for a few years until nutrients are exhausted. The clearings are then abandoned and forest vegetation is allowed to reclaim the areas.

## FACTORS OF SOIL FORMATION

The pedogenic processes involved in the genesis of soils are conditioned by a number of external factors. Simply stated, these factors are the forces of weather and of living organisms, modified by topography, acting over time upon the parent rock. These factors were first outlined by Dokuchaev, the Russian scientist who laid the foundation for soil genesis in 1883.

### Climate

Climate is generally considered the most important single factor influencing soil formation, and it tends to dominate the soil-forming processes in most forested areas. It directly affects weathering through the influence of temperature and moisture on the rate of physical and chemical processes. Physical weathering plays an especially important role during the early stages of soil development— particularly in the degradation of parent materials. In glaciated areas, soils may have been largely formed, transporated, and deposited by ice or meltwater (Figure 2.2). These drift materials had generally been subjected to only minimal chemical weathering at time of deposit, but thereafter chemical weathering played a major role. In a similar fashion, sands of the coastal plain were deposited, and the topography greatly influenced, by the alternate withdrawal and release of water during glacial and interglacial periods. Fluctuations in sea levels during these periods sorted the textural separates, as layer upon layer of sands were deposited during interglacial periods. Sinkholes and caverns often

**Figure 2.2**  Cobbles in a Eutrochrept (Brown Forest) soil in Finnish esker. (Note only slight soil podzolization in this low-rainfall area.)

developed by solution erosion in regions underlain by limestone, and windblown sand dunes formed when the areas were above water.

*Chemical weathering* can manifest itself in many ways. Initially the principal agent is percolating rainwater charged with carbon dioxide dissolved from the atmosphere. Calcium and magnesium carbonates and other rock minerals such as felspars and micas, consisting principally of aluminum silicates containing such cations as Na, K, Ca, Mg, and Fe, are affected. These latter minerals are hydrolyzed by the acid solution to produce clay minerals and to release metal ions, some of which become attached to the clay particles. Resistant materials, such as quartz, are little affected and tend to accumulate in the soil as sand and silt particles, while soluble carbonates tend to be dissolved and removed.

*Leaching* of the readily soluble components continually depletes the surface soil of the basic cations as well as chlorides, sulphates, and carbonates, producing acidic soil conditions. The degree of this impoverishment is controlled by the intensity of leaching from water passing through the soil, the speed of the return of materials to the soil surface via organic remains, and the weathering of rock fragments. In most forested regions there is a positive rainfall balance (after subtraction of transpiration, evaporation, and runoff losses) resulting in percolation, the formation of acid surface soils, and the absence of any accumulated carbonates in a subsoil horizon.

*Temperature* controls the rate of many soil processes, including the activity of organisms concerned with the breakdown of organic material, as well as root growth and water absorption. Rapid changes in temperature and frost action play important roles in fracturing bedrock and reducing the particle size of parent materials.

Climate also exerts indirect effects through its profound influence on the distribution of plants and their associated fauna and microflora. As a consequence of the overriding influence of climate and vegetation on development, soils tend to be distributed geographically in broad zones which correspond roughly to the vegetation zones of the world. In fact, the soil orders (zonal, intrazonal, and azonal) recognized in the classification scheme used in the United States prior to the adoption of the present system of soil taxonomy (Soil Survey Staff, 1975) were determined primarily on the basis of the climate and vegetation under which the soils developed. Because this concept aids in the understanding of processes in vegetation development and because it is still used as a basis for soil classification in a number of countries, it is briefly discussed here.

The *zonal* concept of soil classification is based on the premise that a certain set of climatic conditions ultimately produces a given type of soil regardless of the original parent material. The obvious importance of weathering, leaching, and organic matter decay, which are largely regulated by climate, cannot be denied. However, it is equally obvious that not all soils are *mature*; that is, they have not reached a "climatic climax." In many places poor drainage results in swamps and marshes and prevents soils from reaching equilibrium with the climate. Bogs are sometimes formed by mosses in very humid areas on soils that would otherwise be considered well drained. At the other moisture extreme, sodium or calcium salts may accumulate in surface soils of arid regions and render these dry soils even less hospitable to plants. Soils that reflect the influence of local conditions over that of climate and vegetation are called *intrazonal* soils.

Another group of soils is characterized by the nature of its parent material. These are generally immature soils, such as shallow soils over bedrock, unconsolidated sand dunes and loess deposits, and recent alluvial soils. They are without significant horizon differentiation and are in the *azonal* order in the older classification scheme. Therefore, within a given area it can be said that climate exerts a prevailing tendency toward the development of a certain type of soil, but it doesn't necessarily dictate the dominant properties at any given time.

The zonal concept envisions the gradual development of soil characteristics so that ultimately an equilibrium is reached in response to climatic conditions. However, major climatic changes have taken place in the middle and high latitudes during the past several thousand years, and the profile features of many soils today may not be the product of present climatic conditions. The zonal method of classifying soils is an attractive one, but, because of its limitations, it has been replaced in the United States and many other countries by a basic system of soil taxonomy (Soil Survey Staff, 1975). This latter system is based on measurable soil properties as found in the field at the present time, and it employs a new nomenclature relating to the important characteristics of the soil components.

## Topography, Time, and Vegetation

As noted above, *topography* can have a profound local influence on soil development, overshadowing that of climate. Soil relief affects development mainly through its influence on soil moisture and temperature and on leaching and erosion, as well as plant cover. For example, in most coastal plains small variations in elevation can have a pronounced effect on drainage. Water in soils with restricted drainage often becomes stagnant because microorganisms and plant roots use dissolved oxygen faster than it can be restored. Under anaerobic conditions, Fe, Al, Mn, and some other heavy metals are reduced to a more soluble condition. Under reduced conditions these metals move in the soil solution until they eventually oxidize and precipitate. Such precipitation produces gleyed conditions that are often characterized by a gray to grayish brown matrix sprinkled with yellow, brown, or red mottles or concretions in the B horizon. Subsoil colors are generally reliable guides to drainage conditions, ranging from bluish-gray for very poorly drained, reduced condition to yellows and reds for better-drained areas.

The genesis of a soil begins when a catastrophic event initiates a new cycle of soil development. The length of *time* required for a soil to develop, or to reach equilibrium with its environment, depends upon its parent material, climatic conditions, living organisms, and topography. Weakly podzolized layers have developed under hemlock within 100 years or so after some New England fields were removed from cultivation. These layers have also formed in relatively short periods in local depressions resulting from tree uprootings in hemlock forests. Recognizable mineral soil development under black spruce, and perhaps red spruce and balsam fir, also can be fairly rapid (Lyford, 1952). Buol et al. (1973) pointed out instances where forest soils on glacial moraines in Alaska developed a litter layer in 15 years, a brownish A horizon in silt loam by 250 years, and a thick Spodosol (Podzol) profile in 1000 years. Soil development from bare rock to muskeg took about 2000 years at one spot in Alaska. The time required to develop modal Hapludalf (Gray-Brown Podzolic) soil was estimated to be greater than 1000 years in northeastern United States. Development of distinctive profile characteristics can be extremely slow in arid regions and in some localized areas where drainage or other conditions may slow the weathering processes.

The important roles of *vegetation* in soil development can hardly be over-emphasized. Tree roots grow into fissures and aid in the breakdown of bedrock. They penetrate some compacted layers and improve aeration, soil structure, water infiltration and retention, and nutrient-supplying capacity. In addition to tree roots, living organisms in the soil are responsible for organic matter additions and decomposition, accretion of nitrogen, and structural stability. Furthermore, protection of the soil from erosion may be afforded by a vegetative ground cover. A forest cover can significantly modify the temperature and moisture condition of the soil below by its influence on the amount of water that reaches the soil surface, by a reduction in runoff and an increase in percolation, and by an increase in water loss as a result of evapotranspiration. Roots of trees reach deep into the profile, extract bases, and return them to the surface in litter fall. The litter of many deciduous species decomposes rather rapidly and results in base-enriched humus. On the other hand, the litter of most conifers is rather resistant to decay and it is strongly acid, which encourages the process of podzolization. The composition of leachates produced by various plant covers has a profound effect on the type and speed of soil-forming processes. Indeed, the weathering of rocks that takes place under coniferous trees has been called "acid weathering." The presence of a thin, continuous Bh horizon below the A2 horizon on some windthrow mounds, but not on other mounds in a disturbed forest, may depend on the particular species of tree or vegetation that grew on or near some mounds (Lyford and MacLean, 1966).

Pyatt (1970) outlined examples of changes in soil processes that appear to have resulted from change in vegetation brought about by human activity. One example was the increase in soil wetness resulting from deforestation during Bronze Age times in areas of Great Britain. Increased soil moisture apparently was responsible for the development of gleying within formerly well drained soils on the north York moors. A similar process commenced during medieval times with the spread of heathy vegetation at the expense of woodland on some sites in south Wales.

Because of the close relationship between soils and plant communities and the mutual influence of one on the formation of the other, an examination of the processes of vegetation development should help in the understanding of the influence of the soil on these processes.

## VEGETATION DEVELOPMENT

Simultaneously with the development of soils, the vegetation of earth evolved slowly and differently in various regions in response to soil and climatic variables. It is assumed that from primitive marine organisms evolved the present-day members of the plant kingdom called *Thallophyta*: the bacteria, fungi, and algae.

These microorganisms probably invaded the land during the Cambrian period and were ancestors to clubmosses, horsetails, and ferns. Thereafter, these spore-bearing members of the *Pteridophyta* phylum apparently dominated the landscape until the end of the Carboniferous period, when seed-bearing plants began to exert their influence. Two major classes of seed plants evolved from the pteridophytes during this period. The primitive *Gymnospermae* of which the only large surviving group are the conifers, made their appearance before the end of the Carboniferous period. The conifers were predominant during the Mesozoic era, when the more advanced *Angiospermae* came on the scene. During the Cretaceous period the angiosperms rapidly gained ground, mainly at the expense of the gymnosperms, and by Eocene times they had become the dominant group of land plants (Eyre, 1963). Thus, it is possible that angiosperm and coniferous trees, similar to today's species, have dominated the earth's vegetation for the past 50 to 100 million years.

It is generally believed that the earth was formed more than 4.5 billion years ago. Furthermore, it is theorized that until some 200 million years ago, the land areas formed a single supercontinent. Then it broke into fragments that largely define today's continents, and the fragments began slow and ponderous voyages across the planet (Matthews, 1973). Although North America and Europe are reported to be moving apart by a distance equal to a man's height during his lifetime, the two continents may have been joined in the north as recently as 65 million years ago, with the Appalacian and Scandia mountains forming a continuous range.

It seems likely that all land surfaces began as bare areas, such as uplifted sea floors, lava flows, and former glacier fields. The northern part of North America and the whole of northwest Europe were scoured by giant glaciers as recently as 10 to 20 thousand years ago. While the vegetation of many of the warmer areas of the world has apparently had a rather constant composition for millennia, reinvasion of the abandoned ice fields and coastal areas inundated during intraglacial periods has taken place relatively recently.

Tree species that evolved and simultaneously colonized the European and North American continents while these two land masses were joined in the pre-Cretaceous period were largely eliminated by the repeated advances of ice sheets during the Ice Age. It appears that while most species could continue to retreat to the south in the advance of the North American glaciers, they were not so fortunate in Europe and Asia. Trapped between the arctic ice mass in the north and the glaciers advancing from the several mountain ranges in the south, many species were apparently eliminated. These events help explain why the flora of much of Europe and Asia is simpler in species composition than that of North America and why Eurasia and North America now have few indigenous plant species in common.

Plant communities are dynamic entities, and changes in microsite flora are continuous, but once established the composition of a particular region tends to remain static for very long periods of time. Only major changes in climate, or catastrophes such as volcanic eruptions, fires, and severe storms cause significant shifts in vegetations of a permanent, or even temporary, nature. Of course, such events do happen and in some regions wildfires and storms occur with some frequency. Because of these events and the natural ageing of ecosystems, many abrupt as well as gradual changes can be expected to take place in soil properties and in plant species composition before a constant vegetation composition, associated with a mature soil is reached.

## SOIL PROPERTIES AND VEGETATION DEVELOPMENT

It has been pointed out that the environmental factors that affect soil development also influence the type of plant community that develops in a particular area. More specifically, the properties of soils developed under a particular regime have a profound influence on the growth and development of plants. Properties such as texture, temperature, pH and nutrient status, moisture relations, and those related to parent materials are particularly important.

The type of parent material from which a soil is derived can influence its base status and nutrient level. Changes in vegetation types have been noted with changes in underlying parent rock. For example, soils derived from sandstone are generally acid and coarse textured, producing conditions under which pines and some other conifers have a great competitive advantage. On the other hand, limestone may weather under the same climatic conditions into finer textured, fertile soils that support demanding hardwoods and such conifers as red cedar. In the Appalachian Mountains it is not unusual to find a band of pine growing on sandstone-derived soils parallel to a strip of hardwoods on limestone soils. Although not all limestone derived soils are productive, this parent material exerts an important influence on the distribution and growth of forest trees. Lutz and Chandler (1946) described forest vegetation in Sweden associated with four groups of parent materials, ranging from low to high percentages of calcium. Soils derived from rocks lowest in calcium support a poor forest of Scots pine. Rocks in the second group are somewhat higher in calcium and produce good soils for pine and mixed conifers. The basic igneous rocks in the third group and the calcareous sedimentary rocks in the fourth have the highest calcium content and give rise to the most productive soils of spruce or hardwood forests.

Differential chemical weathering of the same parent material can also influence the distribution and growth of forest vegetation because of changes in acidity, base status, and nutrient availability associated with intensity of weathering.

Vegetation development on soils developed from transported materials may differ from that on adjacent soils developed from bedrock. In fact, the mode of transport may greatly influence particle-size distribution (texture) of the soil. Wind-deposited soils are likely to be of finer texture than those deposited by water. Outwash sands laid down by water from melting glaciers are generally coarser textured than glacial till soils, formed by the grinding action of advancing glaciers. The latter soils are generally more fertile and desirable for hardwoods. Soils with high silt and clay content may offer more resistance to root penetration than coarser soils and, thus, exclude trees not tolerant of dry conditions in regions with extended drought period. On the other hand, sandy soils can hold less water and minerals but more air than clays. The rate of movement through a soil varies inversely with soil texture, because the minute interstitial spaces in fine-textured soils offer resistance to water movement. While finer textured soils hold more water per unit volume of soil, they also lose more water by direct evaporation than sandy soils.

These factors all relate to the water storage and water-supplying capacity of a soil and the occurrence of forest types. In semiarid regions the amount of available soil moisture frequently determines whether forest or grassland vegetation dominates. Usually site quality improves with increasing amounts of soil moisture—up to a point. An excessive amount of water may be as unfavorable to tree growth as a deficiency. In humid areas, sandy soils with deep water tables are generally regarded as poorer sites for most species than fine-textured soils; but the reverse may be true in arid regions. On the other hand, the presence of a water table near the surface of a sandy soil can compensate for its low water-holding capacity and make it quite productive (White and Pritchett, 1970).

Soil temperature can also have a significant effect on vegetation development. The principal source of heat is solar radiation, and soil temperature at any given time is the balance between heat gained and heat lost. Soil temperatures generally decrease with increasing distances from the equator and with altitude. However, many soil and environmental factors can exert local influences that override the general trend. Aspect and degree of slope influence soil temperature because they determine to an extent the amount of radiation reaching the soil surface. In the Northern Hemisphere south and west slopes are warmer than north and east slopes. Soil texture, color, and content of water have effects on heat absorption and heat radiation. A cover on the soil composed of either living or dead materials, serves as an insulating layer and tends to equalize soil temperature.

Soil temperature influences seed germination and seedling survival, as well as plant growth and development. Poor germination and development of reproduction in dense shade may be partly explained by low soil temperature. On the other hand, full solar radiation can raise the surface soil temperature high enough to be fatal to young seedlings (Lutz and Chandler, 1946). While top growth of many species is related to the sum of day and night temperatures, other species are responsive to various combinations of day and night temperature, termed

thermoperiod. However, root growth may be more dependent on soil temperature than on air temperature. Optimum root growth of ponderosa pine occurred
at 15°C air and 23°C soil temperatures. Optimum root growth for red oak,
basswood, and ash also occurs at relatively high soil temperature (Spurr and
Barnes, 1973).

## SUCCESSION IN FOREST DEVELOPMENT

*Forest succession* refers to the various steps in the development of a community
of forest vegetation toward a stable composition. This theoretically stable community in which the vegetation is in equilibrium with climate and soils is
considered the natural vegetation for an area, or the *climax* vegetation. The basic
premise is that if a naturally well-drained surface is left completely undisturbed
for a protracted period with no major climatic change or other natural cataclysm,
a whole series of plant communities will occupy it but, ultimately, a community
will establish itself and persist, unchanged, quite indefinitely. In the march
toward climax, pioneering communities alter the chemical, physical, and biological properties of the soil through their occupance of the area, thereby preparing
the way for their displacement by a succeeding association of more demanding
species. The development of a plant community that begins with an unoccupied
site and proceeds in the absence of a catastrophic disturbance is termed *primary
succession,* while succession after a disturbance that disrupts rather than
destroys an existing biotic community is termed *secondary.*

Certain developmental trends that characterize forest successions are outlined by Whittaker and Woodwell (1972) as (1) progressive increases in height of
the dominant tree species; (2) increasing diversity of growth forms and stratal
differentiation of communities; (3) increasing species diversity; (4) progressive
soil development, with increasing depth, organic accumulation, and horizon
differentiation; (5) increasing community productivity and respiration, and progress toward balance of these; (6) increasing biomass, and an increase in the stock
of nutrients held in the biomass; and (7) increasing relative stability of species
population. The authors point out that the *climax* is not to be defined by
maximum productivity and species diversity, but will usually be characterized by
maximum biomass accumulation and by a steady state of species populations,
productivity (approximate balance of gross photosynthesis and total respiration),
and nutrient circulation.

Succession toward a climax community is initiated by disturbances that
drastically alter or eliminate existing vegetation. Primary succession may begin
with water or mineral soil under a wide variety of climates, but development on
mineral soil is of particular interest to foresters. Mineral soil may be exposed by
glacial retreat, volcanic ash deposition, avalanches and landslides, devastating
forest fires, sand dune formation, spoil bank development by strip mining, and

others. The specific stages in successional development will vary with local climate, soil materials exposed, and the variety and abundance of plants accessible to the site.

Spurr and Barnes (1973) detailed some interesting examples of types of succession in various parts of the world. Mangrove succession in the tropics, for example, is initiated by red mangrove *(Rhizophora mangle)* becoming established in submerged soil on coastal shoals. A mature mangrove forest develops in time to a height of 10 m or so, accumulating substantial amounts of debris and soil. *Avicennia* replaces *Rhizophora Conocarpus* and fresh water species dominate the site. Under tropical conditions the water table may be sufficiently lowered by transpiration to allow invasion by high forest species with the passage of time.

Succession beginning with bare granite rock surfaces in the temperate zone of the southeastern United States is initiated by mat-forming mosses upon which a lichen (*Cladonia leporina*) becomes established. As the mat thickens, herbs come in, with eventual dominance by bunch grasses (*Andropogon* spp.). Shrubs form the next successional stage, followed over the years by the development of an oak-hickory forest.

Successional stages following glacial retreat in southeastern Alaska were somewhat similar to granite rock colonizations (Spurr and Barnes, 1973). The stages ran from pioneers through the establishment of alder thickets to a spruce-fir forest, followed by forest deterioration leading to muskeg and pit-pond formation. Maximum forest development was reached in about two centuries with a dense spruce-hemlock forest carpeted with a thick layer of mosses and litter. On near-level terrain water retention was increased by the carpet and sphagnum mosses invaded. As aeration diminished the older trees gradually died, resulting in an even more luxuriant growth of sphagnum and the formation of muskeg bogs.

A climax community will be dominated by species which, of all those having access to the site, can compete most successfully in the existing physical conditions (Eyre, 1963). Thus, if these conditions permit the growth of both tall trees and light-loving herbaceous plants in an undisturbed area, the outcome can be predicted; the herbaceous plants will be shaded out and eliminated. The tall trees will become dominants and the resulting climax vegetation will be a forest. Climate is only a permissive factor in this case, because competition may eliminate whole communities that would otherwise flourish. Low-growing shrubs and herbaceous plants can be the most important elements in the climax vegetation only in regions where they are not shaded out by a continuous cover of trees, such as in deserts or arctic fringes.

Forest succession, in the limited sense, begins with the establishment of the pioneer forest trees and proceeds with their replacement by successor trees which profit by the changing environment. Generally, pioneer trees are shade-intolerant species that have low nutrient requirements and can withstand sun scald and frost. Typical pioneer forest species in northern temperate forests are scrub oaks, aspen, birches, pin cherry, alder, and jack pine, while maple, beech, and basswood are tolerant species in the same deciduous zone.

In the Pacific Northwest, Douglas-fir is intermediate in tolerance and pure stands of this species are gradually changed to communities of more tolerant western redcedar, western hemlock, or the true firs, depending on soil and site. In a similar fashion, southern pine eventually will be replaced by more tolerant hardwoods, if left undisturbed. However, the invasion of pioneer species and the stages of succession are governed largely by local soil conditions. Wilde (1958) noted that in the northcentral states cut-over calcareous soils are usually occupied by juniper or redcedar, whereas coarse sands are invaded by pine or scrub oak. Birch, aspen, and cherry require soils of a somewhat higher water-holding capacity, while black spruce is a common pioneer on boreal Spodosols and alder on swampy soils.

The forest ecosystem is a dynamic one, and no stand is really in a stable equilibrium. Trees growing at close spacing are in a constant struggle for light, moisture, and nutrients. Even after the composition of forest stands has gone through progressive or successional change to climax formation, made up of one or a few shade-tolerant species that are best adapted to the soil and climate of the region, there continues to be competition and replacement of individuals within these species.

It is generally conceded that true stability is seldom if ever attained in the forest. Diseases have eliminated some species, such as the American chestnut and American elm, and other diseases and insects have altered the competitive ability of many other species. However, fire and land clearing play the most important role in initiating secondary succession. Most pine species in the United States are pioneers in secondary succession. Even-aged stands of white pine in the northeast, as well as stands of red pine, jack pine, and pitch pine in the lake states are largely the results of past forest fires. In the interior of the western forest, lodgepole pine seeds into recent burns, largely from seed stored in serotinous cones of burned trees. Southern pines are all pioneer species that become established after destructive forest fires and give way to tolerant hardwoods in the continued absence of fires. On the west side of the Cascades and over much of the Coastal Range, Douglas-fir is the pioneer species on burned areas, provided there is a nearby seed source (Spurr and Barnes, 1973).

Numerous examples of secondary succession in abandoned agricultural fields exist in the eastern part of the United States from New England to the southern states, in southwestern France, and in areas of shifting cultivation in the tropics. An interesting case history is that of New England where agriculture flourished until the middle of the nineteenth century. Agricultural areas began to decline with the opening of farmlands of the Midwest via the Erie Canal.

Abandoned upland fields and pastures were invaded by red spruce and white pine mixed with such hardwood pioneers as birch, cherry, and aspen. Once the conifer forest was well established, it was itself invaded by more tolerant hardwoods: white ash, red oak, sugar maple, and black birch. Because the overstory conifers seldom reproduce themselves under their own canopy, they were eventually displaced by these hardwoods, plus the more tolerant hemlock, beech, and basswood (Figure 2.3). Griffith, Hartwell, and Shaw (1930) described the soil profile beneath the virgin white pine-hemlock forest as "podzolic" (Spodosol). Cultivation of the cleared land removed all traces of horizons, thus the invading old-field pines took root in undifferentiated soil profiles. The authors investigated the changes that were brought about in the soil by the white pine and the succeeding hardwood. The pine developed a layer of felted needles on the forest floor. The arrested decomposition of organic debris limited the thickness of the dark brown mineral horizon because of lack of infiltration of humus colloids. They concluded that this degenerating process was reversed with the invasion of hardwoods. The rapid decomposition of the hardwood forest litter increased the thickness of dark brown horizon mineral soil, apparently resulting in a Cryandept (Brown Podzolic soil).

In spite of the many interruptions in succession toward climax, it is apparent that plants with similar climatic requirements tend to form belts or zones around

**Figure 2.3** An example of vegetation succession in New England. Note relic stone fences, invasion of junipers in abandoned fields and pastures, and climax mixed hardwoods in background.

1. Typic Ochraquults found in coastal wetlands support excellent pine growth after fertilization. The grayish sandy clay loam Bt horizons contain yellowish-brown to strong brown mottles. The soils are strongly acid with a seasonably high water table.

2. This Ultic Haplaquod is a principal forest soil of the Southeastern flatwoods. It has a dark gray, sandy surface (A1) layer over a distinctive albic horizon, resting on a spodic horizon at a depth of about 50 cm.

3. A deep yellowish-red Oxisol supporting a tropical rain forest in French Guiana.

4. Typical of the more or less freely drained Spodosols (Podzols) of the boreal and tiaga, this Humod soil developed under spruce-fir forests in Quebec.

5. This Ultisol (Red-yellow Podzolic) in New Zealand probably supported a podocarp forest before it was converted to pasture.

**6**

**7**

**8**

**9**

**10**

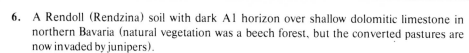

6. A Rendoll (Rendzina) soil with dark A1 horizon over shallow dolomitic limestone in northern Bavaria (natural vegetation was a beech forest, but the converted pastures are now invaded by junipers).

7. A Inceptisol (Brown Podzolic) of the northern mixed hardwood region of New England.

8. An Andept in New Zealand, formed from volcanic ash over an older landform, is planted to radiata pines, which respond to phosphorus fertilization.

9. Deep peat (Histosol); drained and supporting a 10-year-old stand of Scots pine in Finland.

10. Alfisol (Solonetz) profile, high in $CaCO_3$ in the top 50 cm over a natric horizon, in New South Wales. This soil is presently in pasture, but originally supported a *Eucalyptus* forest.

the earth. Plants in these zones are similar in general appearance, even though entirely different species may be found in widely separated segments of a vegetation zone. This type of plant community is termed a *climatic climax* community. It is convenient to group vegetation of alike requirements, form and function of its dominant plants into ecosystems. When done on worldwide scale, variations in composition due to local conditions of soil, drainage, or relief are often ignored. Furthermore, the placing of boundaries between regions is, necessarily, rather arbitrary. Because all areas of a major ecosystem have dominants characterized by the same lifeform they are classed together as a single plant formation type or biome type. The formation type may then be subdivided into distinct plant formations. For example, the original vegetation of western Europe is usually placed in the same formation type as that of the eastern United States, because broad-leaved, deciduous summer forest are dominant in both areas (formations). Whittaker and Woodwell (1972) discussed the appearance of communities of similar physiognomy in the same climates on different continents. They used the example of the giant temperate rain forest found in the highly humid, temperate climates of the northwestern United States and of southeastern Australia. But they point out that the convergence of community structure on different continents is imperfect because of the different evolutionary stock available. For example, eucalypts form the broadleaf rain forest in Australia in contrast with needleleaf rain forest of the United States. It's also interesting that drought-adapted eucalypts form woodlands in Australia in climates that on other continents would probably support grasslands (Whittaker and Woodwell, 1972).

Some well-known examples of climax vegetation are spruce-fir species of the boreal forests of northern Canada and Siberia; hard maple, basswood, yellow birch, and hemlock in the fine-textured soils of the United States northeast; exacting hardwoods, such as beech, maple and oak in central Europe and central United States. Other examples of climax vegetation are given in Chapter 3 dealing with soils of the major forest ecosystems.

# 3

# SOILS ASSOCIATED WITH THE MAJOR FOREST BIOMES

Forests cover about one-third of the earth's land surface and trees play a major role in all ecosystems, except the tundra, deserts, grasslands, and some wetlands. In this context, a *forest ecosystem* can be considered a biotic community in which trees, the dominant feature, interact with the physical environment of that community. Trees have a profound effect upon the soils on which they grow, and to a lesser extent upon the local climate and microclimate. Trees, their growth and survival, are in turn affected by the soil, climate, and other factors of the environment.

Inherent differences in temperature, precipitation, and physiography result in a great diversity in productivity and habitability of various parts of the earth's land areas. Adequate nutrients, along with a proper balance between temperature and precipitation, are of prime importance in determining the suitability of an area for living organisms. Temperature influences the rate of evaporation and, as a consequence, the amount of moisture that will remain available for plant use for a significant period. It also determines whether water will be in a solid or liquid state. For example, the Antarctic continent does not lack for water, but it is relatively lifeless because of the cold and the resulting cover of thick sheets of glacial ice. At the other extreme, deserts are almost uninhabitable because they are too hot and dry for most plant and animal life. Rain that does fall in the deserts is rapidly evaporated. Between these areas of extreme climatic or physiographic features are a tremendous variety of natural areas with distinct biotic communities and associated soils.

Species respond individually to environmental variations and those with similar environmental demands tend to reoccur in community assemblages. The groups are generally identified by key or indicator species common in one floristic group but absent in others. However, Walter (1973) pointed out that the influence of the physical conditions prevailing in the habitat only indirectly affect plant species distribution. The natural limit of distribution of a particular species is reached when, as a result of changing physical environmental factors, its ability to compete is so reduced that it is eliminated by other species. In other words, the distribution of many species would extend much beyond their present natural range if they did not have to compete with better adapted species.

The structure of a plant community is primarily determined by climate and soil; hence, maps of the major natural climax vegetation also serve as a general outline of the climatic and soil regions of the world. For example, boreal forests of North America have types of vegetation, soils, and climate similar to the boreal (taiga) forests of northern Europe and Asia. On the basis of common characteristics, broad areas have been mapped into major ecosystems. The interrelationship between soils and associated vegetation in those ecosystems dominated by forests are of prime concern to forest soil scientists. The groupings herein of similar forest vegetation types (ecosystems) into major biomes and associated soils was done rather arbitrarily and does not necessarily conform to groupings in textbooks on ecology. It generally follows those outlined in Figure 3.1 (Dasmann, 1968).

## BOREAL FOREST

This vast northern evergreen forest, composed largely of conifers, extends southward from the tundra "timberline" across much of Canada and across northern Europe and Asia where it is known as the taiga. Like the tundra, it is largely absent in the Southern Hemisphere. Evergreen trees are apparently well adapted to these areas of cold winters and short growing seasons. There is a minimum of transpiration from their needle-type leaves during the winter when water is unavailable from the frozen ground. Furthermore, they have an advantage over deciduous trees that lose their leaves in the winter and must spend a significant part of the short growing season producing new photosynthetic tissue.

Although the boreal forest may appear to be homogeneous in species composition, it has a varied flora. In North America there is a gradual transition in species across the continent (Figure 3.2). In southern and eastern Canada, red pine, hemlock, yellow birch, and maple are gradually replaced by a predominantly forested belt across the middle of Canada composed chiefly of white spruce, black spruce, balsam fir, jack pine, white birch, and trembling aspen (Figure 3.3). These species give way to white spruce, black spruce, and tamarack as one approaches the tundra. In a somewhat analogous fashion Scots pine*(Pinus sylvestris)* and Norway spruce *(Picea abies)* are predominant in the European boreal (taiga), but as one travels eastward, Siberian species, such as *Abies siberica, Larix siberica,* and *Picea siberica* increase in frequency. There are no sharp boundaries between the coniferous forest of the boreal and the

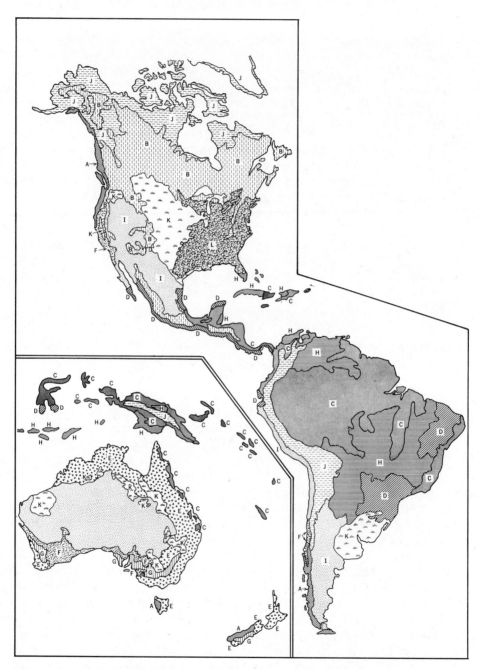

**Figure 3.1**  Major ecosystems of the world (Dasmann, 1968). Used with permission.

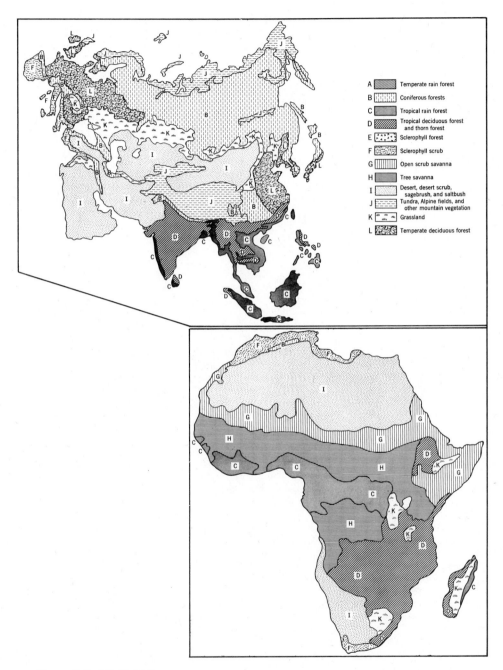

**Figure 3.1** (Continued)

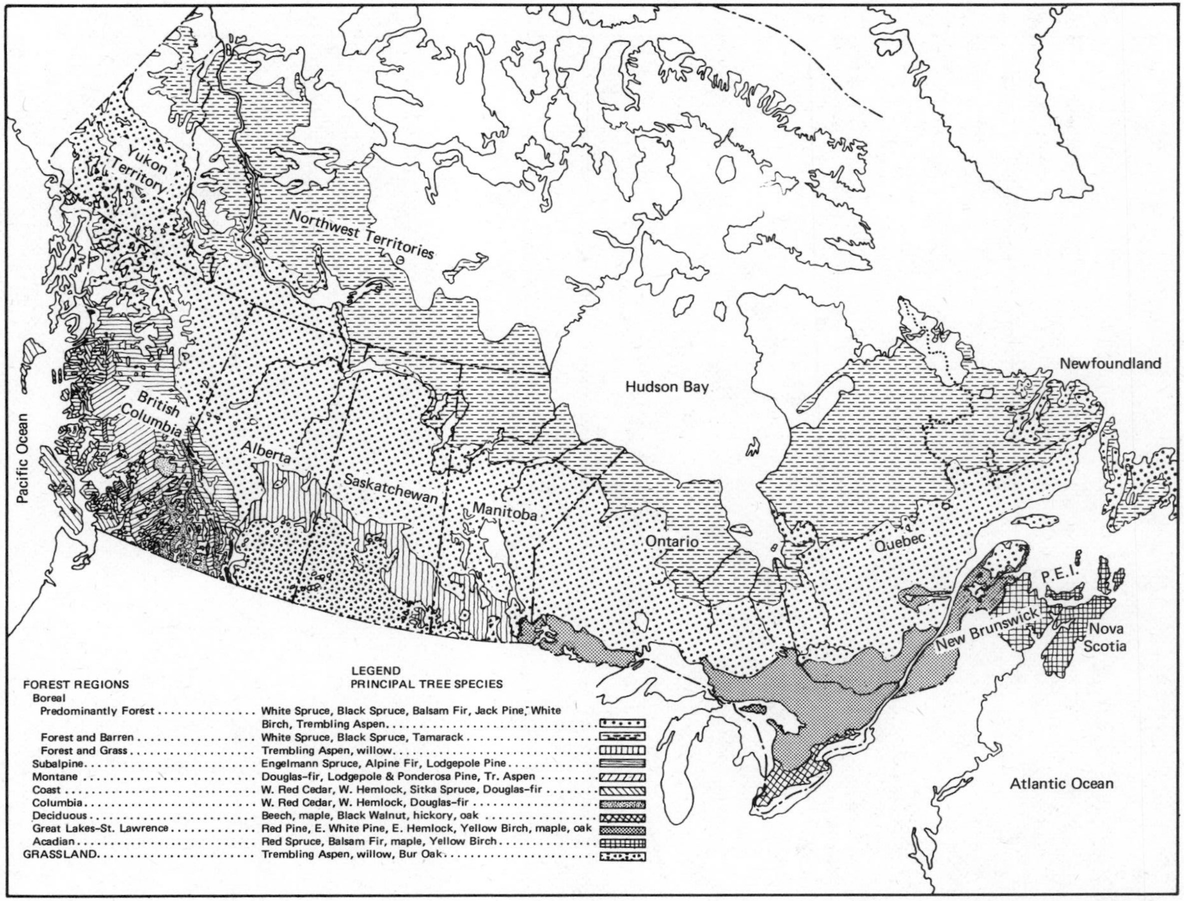
Yukon Territory
Northwest Territories
British Columbia
Alberta
Saskatchewan
Manitoba
Ontario
Quebec
Hudson Bay
Newfoundland
P.E.I.
New Brunswick
Nova Scotia
Pacific Ocean
Atlantic Ocean
LEGEND
FOREST REGIONS
PRINCIPAL TREE SPECIES
Boreal
Predominantly Forest ............... White Spruce, Black Spruce, Balsam Fir, Jack Pine, White Birch, Trembling Aspen
Forest and Barren ............... White Spruce, Black Spruce, Tamarack
Forest and Grass ............... Trembling Aspen, willow
Subalpine ............... Engelmann Spruce, Alpine Fir, Lodgepole Pine
Montane ............... Douglas-fir, Lodgepole & Ponderosa Pine, Tr. Aspen
Coast ............... W. Red Cedar, W. Hemlock, Sitka Spruce, Douglas-fir
Columbia ............... W. Red Cedar, W. Hemlock, Douglas-fir
Deciduous ............... Beech, maple, Black Walnut, hickory, oak
Great Lakes-St. Lawrence ............... Red Pine, E. White Pine, E. Hemlock, Yellow Birch, maple, oak
Acadian ............... Red Spruce, Balsam Fir, maple, Yellow Birch
GRASSLAND ............... Trembling Aspen, willow, Bur Oak

**Figure 3.2** Forest regions of Canada (Rowe, 1972). Reproduced by Permission of the Minister of Supply and Services Canada.

deciduous forests usually found in adjoining more temperate zones. The transition zone consists of either mixed stands of a few conifer species (mainly pine) and a few hardwood species, or of a mosaiclike arrangement of pure deciduous forests on the better soils and pure coniferous forests on poor soils (Walter, 1973). Subalpine and montane species replace the boreal forest in the Rocky Mountains. In Europe the boundary coincides with the northern distribution limit of the oak.

The mean annual rainfall of nearly all of the boreal forests is less than 90 cm and, in fact, is less than 45 cm in most places. However, the evapotranspiration rate is relatively low, even in the summer, and the surface soil beneath the dense tree canopy is perpetually moist. The summers are warm enough so that permafrost is eliminated and the soils tend to be highly leached. The debris returned to the surface under conifers is usually low in nitrogen and the essential bases. It decomposes very slowly and, as a consequence, the ground is generally covered with a thick blanket of mosses growing on needles and decaying wood over a very acid mor humus. The slowly decaying litter forms acid products that dissolve and aid in leaching minerals important for plant growth out of the upper soil (A2) horizon. The sesquioxides of iron and aluminum are removed from this eluviated horizon during the podzolization process and the resultant sandy A2 horizon is light gray in color, as shown in Figure 2.1 and graphically in Figure 3.4. The iron and aluminum compounds that are removed from the upper horizons are mostly redeposited in the illuviated horizons lower in the profile. Soils of the boreal forest were predominantly classed as Spodosols (Podzols) with Histosol (Bog) soils found in bogs or fens (Soil Survey Staff, 1975).

The positive water balance of the boreal results in a high water table in large expanses of nearly level terrain. Where the water table remains at less than 50 cm below the soil surface for a large part of the year, trees grow with great difficulty and forests are replaced by mires. Consequently, extensive areas of the boreal zone are covered by organic soils that do not support the zonal coniferous vegetation. Various mosses (*Sphagnum* and *Polytrichum* species most prominent), a few flowering plants (*Eriophorum* and *Trichophorum* species) and dwarf shrubs (*Vaccinium* and *Calluna* species) are types of vegetation usually found on the wet organic soils. Various names are used for these mires in different countries: bogs, muskegs, and fens to mention a few. Walter (1973) recognized three types of mires, distinguished according to (the origin of) their water source. (1) *Topogenous* mires or fens occupy the lowest part of the terrain and are associated with very high water tables. Although the ground water of the boreal is generally acid and nutrient impoverished, the water contains sufficient minerals to produce eutrophic fens (Figure 3.5). (2) *Ombrogenous* mires or raised bogs are higher than their surroundings and are watered by rainfall. Sphagnum mosses, the dominant species of raised bogs, require uniformly damp, cool summers, and when these conditions prevail they can colonize the acid, infertile

**Figure 3.3**  Spruce-fir stand in the Canadian boreal forest.

Spodosols of the region. (3) *Soligenous* mires receive water from rainfall and that draining from surrounding slopes. Once a bog is drained peat mosses cease to grow and heath and dwarf shrubs and, later, birch, pine, or spruce occupy the site. However, the organic soils are too deficient in minerals to support good tree growth.

Wind-throw mounds are numerous on some Spodosols of the boreal zone, occupying up to 30 percent of the forest floor area in parts of the Great Lakes region of North America (Buol et al., 1973). They are formed by uprooting of trees in storms and they typically have relief of as much as 1 m and a diameter of 3 m.

Humans have had little impact on the boreal forests; the soil and climate are not well suited for agriculture and much of the area remains unsettled. However, forest fires, often caused by people, are an important factor in the boreal forest.

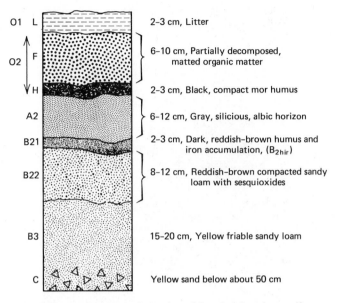

| | | |
|---|---|---|
| O1 | L | 2–3 cm, Litter |
| O2 | F | 6–10 cm, Partially decomposed, matted organic matter |
| | H | 2–3 cm, Black, compact mor humus |
| A2 | | 6–12 cm, Gray, silicious, albic horizon |
| B21 | | 2–3 cm, Dark, reddish-brown humus and iron accumulation, ($B_{2hir}$) |
| B22 | | 8–12 cm, Reddish-brown compacted sandy loam with sesquioxides |
| B3 | | 15–20 cm, Yellow friable sandy loam |
| C | | Yellow sand below about 50 cm |

**Figure 3.4** Profile of a well-developed Spodosol under coniferous cover in the boreal forest.

Most human inhabitants are dependent in part, or in whole, upon the forests for their livelihood. Many migratory birds and animals abound in the summer, but only the few permanent residents, such as moose, caribou, lynx, wolverine, snowshoe hare, and grouse remain to face the winter food scarcity.

## OTHER CONIFEROUS FORESTS

Although the boreal is by far the largest coniferous forest community, there are several other significant coniferous formation types in the Northern Hemisphere. These include the (1) transition or subalpine forests in North America and Europe, and the (2) lowland coniferous forests of the United States.

### Transitional Coniferous Forests (Subalpine Forests)

These forests occupy zones in the mountains lying between the southern extension of the boreal forests and the warmer communities of hardwood, grassland or chaparral found at lower elevations. Because of their generally mountainous locations, they are often called subalpine formations and the altitude at which they are found increases as the latitude decreases. Some of the species in the transition formations are identical to those of the boreal forests, but most species are quite distinct; though they have closely related counterparts in the boreal forest. For example, pine *(Pinus sylvestris)* and spruce *(Picea excelsa)*, as well

**Figure 3.5** Organic soil (Histosol) in a *Sphagnum* bog in the boreal zone.

as species of larch and fir, compose the subalpine forests of western and central Europe. Spruce and fir are found between 1000 and 2000 m in the Alps and 1500 to 2500 m in the Pyrenees. Similar associations are found at subalpine heights on the mountains of central and eastern Asia, extending southward to the Himalayas.

Subalpine forests of Europe extend southwest into the Mediterranean peninsulas and even as far as the Atlas Mountains in north Africa and the mountains of Turkey and Spain. Pines (*Pinus sylvestris* and *P. nigra*) and firs (*Abies* spp.) are generally the dominant species with cedars (*Cedrus* spp.) of local importance.

In North America there is an almost continuous belt of subalpine forest extending the whole length of the Sierra Mountains from Canada to California in the west, and along the entire length of the Rocky Mountains as far south as New Mexico in the interior. In the Appalachians, they are at lower elevations and are separated from the boreal in the north by a low area dominated by hardwoods

(Figure 3.6). The height of the subalpine forests varies from between 600 and 1200 m in southern Alaska, 1000 and 2000 m in British Columbia, and 2700 and 4000 m in New Mexico. Engelmann spruce, subalpine fir, and lodgepole and limber pines are widely distributed in these western formations.

## Lowland Coniferous Forests

These forests are well represented in the United States where they make up several distinct formations, all of which are largely the result of human activity. Extending across the lake states into southern Ontario and northern New England, a belt of white spruce, white pine, red pine, and hemlock flourished until the middle of the nineteenth century. After this mid-continent forest was cut over for its valuable timber, much of the land was diverted to agricultural use or was revegetated by hardier deciduous species. Pine may be found in pure stands on dry sands, but it is usually mixed with maple, aspen, and other hardwoods.

**Figure 3.6** A natural stand of white pine in the transitional coniferous forest in New York.

There is, consequently, some question as to whether the conifers of this area are true climax species.

The intermountain, or montane, forests of the Pacific Northwest occur on the middle slopes of mountains from the Rockies westward, usually interposed between the subalpine forest and the temperate rain forest along the coast. In the southwest they give way at lower elevations to scrub communities. The dominant species are ponderosa and lodgepole pines, Douglas-fir, and Engelman spruce.

The pine forest of the coastal plain and piedmont of the southeastern United States is comprised of several species collectively called "southern pines." The dominant species are loblolly, shortleaf, longleaf, and slash pines. These species often occur on low-lying, sandy sites, and it is generally believed that they are not true climax species, even though they occupy roughly one-half of the 90 million hectares of commercial forests of that region. They are probably a kind of disclimax maintained by frequent burning of the forest. In recent years, southern pines have been widely used in intensively managed pulpwood plantations.

The climate and tree species vary markedly among the subalpine and lowland coniferous formations and, as a consequence, a great variety of soils are found beneath them. Site requirements differ among species but conifers are generally less demanding of the soil than most hardwoods. Nutrient cycling in the midlatitude coniferous forests is relatively slow, although not so slow as that in the boreal. The dominant pedogenic processes of most of these areas is podzolization. There is a positive water balance for much of the year, and leaching of bases has generally resulted in soils of slight to strong acidity. Soils vary from weakly leached Cryandepts, Eutrochrepts, Fragiorthods, and Haplorthods (Brown Podzolic and Brown Forest) soils in the northeastern and the Great Lake regions to Fragiudalfs, Hapludalfs, Hapludults (Gray-Brown Podzolic) and mountain soil complexes in the western states. While the high base status Alfisols more commonly support hardwoods, they may also be found under mixed coniferous-hardwoods and some pure coniferous forests of the transition zones.

Pine forests of southeastern United States are on highly weathered low base status soils, mostly Ultisols (Red and Yellow Podzolics) and the Aquod group of Spodosols (Groundwater Podzols). Aquods have developed on sandy soils in humid areas under a wide range of temperature conditions, from temperate to tropical zones (Figure 3.7).

## MIXED FORESTS

The high altitude coniferous forests of the Northern Hemisphere generally grade into broad-leaved deciduous forests at lower elevations. This change is usually achieved through a transition zone where the two forest formation types exist side by side. These zones with competing species may vary from a few miles to

**Figure 3.7** An Aquod (Ground-Water Podzol) in the North Island of New Zealand.

several hundred miles in width, and are found in Europe and Asia as well as North America. Species of birch, beech, maple, and oak are commonly found growing with the conifers. Mollisols, Inceptisols, and Alfisols (Brown Forest and Gray-Brown Podzolic) soils are generally found under these mixed forests. Since they are desirable agricultural soils, many of these forests have been destroyed by mankind.

In several lowland areas of the Southern Hemisphere there are mixed forests of broad-leaved evergreens and conifers. They occur in several climatic regions in Chile, southern Brazil, Tasmania, northern New Zealand, and South Africa, and may be true climax formations. The species in this formation vary rather widely among the continents, as might be expected, but they are quite similar in life form and general appearance. Such genera as *Araucaria, Podocarpus,* and *Agathis* are found in many of these mixed evergreen forests.

## DECIDUOUS FORESTS

The broad-leaved summer deciduous forest of eastern and northcentral North America, western and central Europe, and north China generally lies south of the mixed coniferous-deciduous forests. The areas are heavily populated and intensively farmed and, consequently, much of the climax forest has disappeared. The forests that remain today have been so altered by human activity that the original distribution of component associations cannot be accurately determined.

In France and the British lowlands, *Quercus robur, Fagus sylvatica,* and *Fraxinus excelsior* are the dominant species, with other species of oak, birch, and elm becoming important in local areas. The tree associations gradually change when moving to southern Europe so that *Quercus lusitanica, Q. pubescens, Acer platenoids, Castanea sativa,* and *Fraxinus ornus* are more plentiful. The loess soils of the northern Ukraine across to the southern Urals once carried deciduous forest dominated by various oak species.

A similar but more diverse formation of hardwoods is found in the United States, extending from the Appalachian Mountains to beyond the Mississippi River, from the mixed forests of New England and the lake states southward to the southern coniferous and mixed forests of the coastal plain. It reaches westward into the prairies along the valleys of the several rivers. Although noted for its diversity, the American formation is dominated by species of oaks and hickory with basswood, maple, beech, and ash species found in the eastern regions. Along the lower slopes of the southern Appalachians several other species of oaks, sweetgum, and yellow-poplar make their appearance, often with an understory of wild cherry, dogwood, and alder. The east Asian formation of deciduous hardwoods has essentially the same association of genera as those of Europe and America, except that the hickory is notably absent. On the other hand, Chile has a formation that is morphologically more closely related to those of the nearby evergreen forest than to the deciduous genera so common in the Northern Hemisphere formations.

Precipitation is relatively heavy and well distributed throughout the year in the deciduous forest formation. Summers are generally warm and humid and winters are cool to cold, with heavy snowfall in the northern parts of the region in the Northern Hemisphere. A difference in nutrient demands of the various species, plus a diversity of parent material on which they grow, have resulted in a wide range in soil properties of the deciduous formations. Generally speaking, mild acids formed from the decomposition of temperate-zone deciduous forest litter are carried into the soil by the abundant rainfall, which results in rather extensive leaching. However, the results of the leaching are less severe in deciduous forests than in coniferous forests, because the decomposition products are not as acid. The mineral content of deciduous leaves is generally higher than that of coniferous foliage. Spodosols (Podzols) have developed in the acid sands of the oak-birch forests in France and parts of northern Europe. On the other hand, Mollisols (Brown Forest) and associated soils developed on calcic clays of central and southern Europe and they are nearly neutral in reaction. Alfisol (Noncalcic Brown and Gray Forest) soils are sometimes found in a narrow zone along the margins of the wooded steppes in the central European basins and in the Ukraine. They appear to reflect the earlier influence of a grassland vegetation on their development (Eyre, 1963).

One-quarter to one-third of Scotland, which originally supported deciduous forests has long been covered by *Calluna* heath, which is regularly burned. The Spodosol (iron Podzol) soils often have a cemented hardpan (Ortstein) B horizon. They apparently supported a vigorous forest vegetation as long as they remained untouched. As soon as the forests were cut, a large part of the mineralized nutrients were lost and an impoverished soil resulted (Walter, 1973).

Most Alfisol (Gray-Brown Podzolic) soils of the northern part of the United States (Figure 3.8), as well as the Ultisol (Red and Yellow Podzolic) soils in the south, were initially fertile and easily cultivated. They were exploited early in the settlement of the country, often abandoned after depletion of native fertility, and many of them have reverted to inferior hardwood or mixed pine-hardwood forests during the past century.

## TEMPERATE ZONE BROAD-LEAVED EVERGREEN FORESTS

These evergreen broad-leaved forests are locally important formations north of the tropics in Japan, south China, parts of Australia, and southeastern United States. In the latter formation, live oak, other evergreen oaks, magnolia, bay, and holly are limited to isolated stands, locally referred to as *hammocks*. Soils supporting these stands are generally reputed to be more fertile than the surrounding areas on which mixed deciduous-pine stands are found.

Evergreen oaks make up the preponderance of the species of Asian broad-leaved formations. Vast forests of broad-leaved evergreen trees once covered southwest and southeast Australia before that continent was settled. They were composed almost entirely of *Eucalyptus* species and the cut-over forests are still widespread, although the plant communities are greatly altered. The eucalypts are quite different from the broad-leaved evergreen of the northern mid-latitudes in both appearance and habitat. Many soils of Australia are very old and they are generally much drier than those of the northern formations (Stephens, 1961). Moreover, the widely varying properties of these soils are due to some distinctive geomorphic characteristics of the country. Australia has suffered only limited uplift, dissection has been modest, and basement rocks have been subjected to little juvenile weathering. Complementary to these conditions is a widespread survival of late Tertiary and Pleistocene land surfaces, on which relic soils exhibiting senescent features are preserved to a varying degree. Thus there is a wide range in parent materials, climatic environment, and age, but there is an unusually high proportion of soils of great antiquity. Furthermore, Pleistocene

**Figure 3.8**  A Hapludalf (Gray-Brown Podzolic) soil of mixed northern hardwood forest in Massachusetts.

glaciations and periglaciation conditions exposed only very limited new surfaces to postglacial weathering. Soils generally range from Udults (Reddish-Brown Lateritic) and Ustalfs (Reddish Brown soils) to various members of the order of Vertisols (Soil Survey Staff, 1975).

## SHRUB AND WOODLAND FORMATIONS
Developed in areas of dry or uncertain rainfall, these forests have little timber value but may be, nonetheless, managed as forest and range lands. Formations of woody vegetation vary from small trees forming a more or less open canopy, or woodland, to dense, low-growing shrubs. They may be conveniently divided into two rather distinct formations on the basis of climate and morphological adaptations: (1) the sclerophyllous communities of the area with Mediterranean-type climate of mild, rainy winters and summer droughts, and (2) the xerophytic shrubland where drought may persist throughout the year and winters may be quite severe (Dasmann, 1968).

### Mediterranean Sclerophyllous
This vegetation type is found in the lower flanks of the coastal range in southern California, in southeast to southwest Australia, the coastal fringe and central

valley of Chile, and southwest South Africa, as well as in the Mediterranean region. The common denominators in all of these areas are the mild winters of moderate rainfall with warm and dry summers, the medium height evergreen sclerophyllous forest formations, and a terrarossa soil profile. Although the species composition varies widely among formations, they generally have small, hard, cutinized leaves that are rolled at the margins and possess other adaptations to reduce transpiration rates.

In Europe, the original zonal vegetation was dominated by *Quercus ilex* and remnants of this association still exist. The formation is termed *maquis* and its location was the setting for much of the early development of western civilization, following its spread from desert river valleys. It is probable that the area once supported dense forests, but today the maquis is made up of several species of wild olive, arbutus, heather, gorse, and other shrub of similar life form. Species of evergreen oak are found throughout the Mediterranean area, and isolated stands of pines, such as *Pinus pinaster* and *P. halepensis* are scattered in the maquis. Some of the communities may be true climax vegetation, but others are apparently maintained by human activity, animals, and fires. This formation borders both sides of the Mediterranean and extends eastward to Afghanistan.

Vegetation similar to the maquis is found in California, where it is termed "coastal chaparral." Among the rather varied communities are to be found several bush oaks (most of which are evergreen), many species of *Ceanothus,* and members of the heather (Ericaceae) family. The chaparral has survived repeated burnings and is the apparent climax vegetation in areas of 25 to 50 cm annual rainfall. The roots of sclerophyllous species may penetrate from 4 to 8 m of soil and fissures in the bedrock to obtain sufficient moisture to exist during the hot dry summers. The sclerophyllous formation does not extend north beyond about latitude 36°N (a fog belt resulting from the cool ocean currents borders the coast further north).

The wide zone of sclerophyllous forest that extends across southern Australia is composed largely of shrub members of the genus *Eucalyptus*. In southwestern Australia the "jarrah" forest, which is completely dominated by *E. marginata,* has a Mediterranean-type climate, with 65 to 125 cm rainfall and a summer drought. However, these trees can reach a height of 15 to 20 m and an age of up to 200 years (Walter, 1973). In southern Australia areas with rainfall of 35 to 65 cm support low-growing woodland species collectively termed "mallee," which include such *Eucalyptus* species as *oleosa, dumosa,* and *uncinata*; species found in no other place in the world. The Australian bush was once almost inpenetrable, but large areas of it are now rather open scrubland.

The soils of the Australian sclerophyllous winter-rain region are acid and poor in nutrients; many contain silica and iron concentrations which are the oxic crusts of an earlier geological age when the climate was tropical. The parent rocks are among the oldest geological formations on earth (Stephens, 1961). Soils of the sclerophyllous zone belong principally to the Mollisol, Alfisol, and Aridisol orders.

Soils of the sclerophyllous formation in the Cape Province of South Africa are similar to those of the Australian formation. The vegetation of the two formations, although of different genera, are also alike in appearance.

In Chile, the sclerophyllous region occupies the central part of the country, adjoining the arid regions on the north and a temperate rain forest on the south. The cold Humbolt current flowing along the coast modifies the summer drought so that the temperatures are lower than those of the chaparral in California. The formation extends up to about 1400 m in the Andes and consists of remnants of xerophytic families forming woodlands of Rosaceae, Monimiaceae, or Anacardiaceae.

## Semidesert Woodlands

These formations are found in rather extensive areas of interior southwestern United States, between the Caspian and the Great Khingan Mountains in Eurasia, in the southern Sahara, in central Australia, and on the western side of the Andes in central and southern South America. These xerophyllous woodlands generally occur between montane forests or prairies on the one hand and deserts on the other.

In North America the formation extends upward to about 2100 m on the several mountain ranges of southwestern United States. The vegetation is composed of coniferous and broad-leaved evergreens of which several species of live oaks, junipers, and piñyon and ponderosa pines are prominent. The formation is mostly open woodland today, although it may have originally formed a fairly complete canopy over the soil. Nevertheless, sparseness of plant cover is characteristic of semidesert woodlands. Walter (1973) pointed out that the amount of water available to plants is less dependent on total rainfall than on the transpiring surface per square meter of soil surface. Consequently, the water supply per unit of transpiring surface is more or less the same in arid as in humid regions. While a greater percentage of the total plant biomass is below ground in arid regions, this does not mean that roots penetrate deeper than in wet areas, unless the plants are dependent on deep ground water. Where the rainfall is so sparse that it penetrates only to surface horizons, root systems are flat and extensive.

In semidesert regions part of the rainfall runs off and a further part evaporates. How much of the water remains in the soil and thus available to plants is determined by the texture of the soil. In humid regions, sandy soils are dry because they retain only small amounts of rainfall, whereas clay soils are wetter. But in flat arid regions, rainfall does not penetrate to the water table and only upper layers are affected. Water penetrates to a greater depth in sandy soils and because of low capillary movement, they do not dry as fast as clay soils.

Contrary to the situation in humid regions, clay soils form the driest habitats, whereas sandy soils offer better water supplies (Walter, 1973).

Aridisol and Mollisol (Terra Rossa, Red-Yellow Mediterranean, and Brown Forest) soils are generally associated with the semidesert forest formations. In these areas where the evapotranspiration losses are greater than the amount of rainfall, there tends to be an accumulation of calcium in the profile. The amount of calcium accumulated largely depends on the parent rock from which the soils were derived. The Durixeralf-Palexeralf (Noncalcic Brown) soils were derived from calcium-poor material, while many soils of south Australia were aeolian calcareous and sandy deposits of the Pleistocene period and they may have thick pans of calcium carbonate at shallow depths.

The red color of many soils of this vegetation formation is due to the presence of oxides of iron exposed by erosion of the surface soil. The freely drained soils of the interior semidesert generally contain a fair amount of humus and they are likely to be light brown to grey in color with a well developed crumb structure. They are mostly classed as Aridisols, Mollisols, Inceptisols, or forested Alfisols (Soil Survey Staff, 1975).

Soils developed in depressions or under impeded drainage in arid regions have a high accumulation of salts in the upper horizons and are called halomorphic, or alkali soils. When there is an accumulation of soluble salts of sodium, calcium, magnesium, and potassium, they are termed saline soils, or Salorthids (Solonchaks). The associated anions are mostly chloride and sulfate with some carbonate and bicarbonate, and since calcium and magnesium are the dominant cation, the soils are only slightly alkaline and the soil particles are not dispersed. On the other hand, where high concentrations of sodium are associated with the colloid, the matric groups of Aridisol or Alfisol (Solonetz) soils are formed. Under this condition, both the organic and mineral colloids become dispersed, giving rise to a black surface crust. Few halomorphic soils support any significant forestry.

## TEMPERATE RAIN FORESTS

Associated with the sclerophyllous winter-rain regions are forests without pronounced summer drought. For example, on the northwest coast of North America is an area of relatively high rainfall, well distributed throughout the year, and with mild temperatures. This marine west coast climate favors the development of unusually tall, dense, and luxuriant conifers. The dominant species range from Sitka spruce *(Picea sitchensis)* in northern British Columbia to western cedar *(Thuja plicata)*, western hemlock *(Tsuga heterophylla)*, and Douglas-fir *(Pseudotsuga menziesii)* in Washington and Oregon. Without the help of fire and cuttings, however, the Douglas-fir would probably not be a dominant species in this formation and would be confined to drier sites. From Oregon to northern California, the giant coastal redwoods *(Sequoia sempervirens)* are a common forest dominant (Figure 3.9). They apparently survive the summer drought in the southern part of their range because of the frequent fog that envelops the area. Western species of hemlock, pine, spruce, firs, and cypresses are also found in this coastal region.

Other temperate rain forest formations are found along the western side of the South Island of New Zealand, the Valdivian forests of southern Chile, and in southeastern Australia. These Southern Hemisphere formations are largely broadleaf evergreens. Apart from this fact, the beech *(Notofagus)* forests of New Zealand and the *Eucalyptus regnans* forests of Australia resemble in size and general appearance the temperate rain forest of other regions.

Soils developed under these forests are rather well supplied with organic matter, have a moderate abundance of layer lattice clay, and often possess argillic horizons. They include members of the Udalf, Ustalf, Umbrept, and Humult suborders.

## TROPICAL RAIN FORESTS

In undisturbed areas of the tropics, where the year-round rainfall and temperatures are favorable to a high level of biological activity, tropical rain forests have developed. The mean daily temperature and day length vary only slightly in the equatorial zone and the seasons are distinguished on the basis of the distribution of rainfall. Theoretically, two rainy and two dry periods should exist, but as a result of the influence of tradewinds, the monsoons, and the presence of leeward

**Figure 3.9** Parklike stand of mature redwood in temperate rainforest of northern California.

and windward slopes of mountains, enormous variations in amount and distribution of annual rainfall occur.

There are three distinct formations of tropical rain forests; (1) the American, (2) African, and (3) Indo-Malaysian. The term "tropics" is applied broadly here to the regions between the tropics of Cancer and Capricorn. Because of the great variations in elevation and rainfall of this zone, there is a wide range of climax vegetation from permanently humid evergreen rain forests through various types of semideciduous and deciduous forests to savanna, semidesert, and desert. However, where there are like environments, the three tropical rain forest formations are remarkably similar in structure and general appearance, although they each have their own distinct assemblage of plant families. Almost all of the thousands of species of woody plants found in the tropical rain forests are evergreen. Hence, they have a distinct advantage over deciduous trees in an environment that permits year-round growth.

Many trees have leathery, dark-green, laurel-type leaves, and the dominant species often develop large plank buttresses on the lower trunk. The trees in turn support a variety of plants that can survive without soil contact (epiphytes), such as orchids, bromeliads, lianas, ferns, mosses, and lichens. A dense climax rain forest has a compact, multistoried canopy that allows little light to penetrate to the ground. Therefore the forest floor is often relatively free of undergrowth. Rain forests that have been opened up, either by natural or human causes, quickly grow into dense, second-growth, successional forests. Similar "dense, impenetrable jungles" occur naturally along edges of streams and other natural clearings.

The high temperatures and humidity of the rain forest insure that the never-ending supply of litter reaching the forest floor is rapidly decomposed. The nutrient cycle is thus a rich and rapid one. The impoverished, acid soils support a luxuriant vegetation, but almost the entire nutrient reserves required by the forest are contained in aboveground biomass (Table 3.1). Bases leaching from the decaying litter keep the base status and pH of the surface soil at low but reasonable levels. Parent material weathering in tropical rain forests is deeper than in any other groups of soils. Weathering and intense leaching gradually remove a large part of the silica from the silicate minerals in the thick subsurface *oxic* horizon, leaving a high proportion of hydrous oxides of iron and aluminum. The products of laterization are generally found only among older soils. The clay content of these latter soils is very high, but the clays are of a nonsticky type. While Oxisol (Reddish-Brown Latosol) soils, high in sesquioxides, are typically associated with rain forests (Figure 3.10), it should be noted that laterization apparently does not occur in all free-draining soils of these regions. Where the parent material is base-poor sandstone and percolation is rapid, the soil may be quite acid, similar to the Ultisols of the southeastern United States. On less well-drained sandy soils, Spodosols (Aquods) may develop.

Areas of impeded drainage are not unusual in tropical rain forests, but deep

**Table 3.1**   Content of Nutrient Elements in the
Surface (0–30 cm) Soil and the Percent of
the Total Inventory Stored in the
Vegetation of a Tropical Moist Forest at
Rio Lara, Panama (Golley et al., 1975)[a]

| Element | Soil Content | Amt. of Total in Vegetation |
|---------|--------------|------------------------------|
|         | kg/ha        | %                            |
| P       | 33           | 88                           |
| K       | 508          | 90                           |
| Ca      | 18582        | 20                           |
| Mg      | 2830         | 13                           |
| Cu      | 9            | 18                           |
| Fe      | 26           | 43                           |
| Mn      | 35           | 55                           |
| Zn      | 225          | 4                            |

[a]Used with permission by the University of Ga. Press.

peat accumulations appear to be rather rare. On the other hand, the organic matter content of mineral soils is generally higher than might be expected in an area of such high biological activity. Nitrogen contents of 2500 to 5000 kg per hectare were found in the surface 20 cm of French Guiana soils. It appears likely that the decomposition rate of soil organic matter is reduced by the presence of aluminum compounds or the low level of available phosphorus, thus slowing the activity of decomposing organisms. Nevertheless, Oxisols are not extensively used for agriculture, because the organic matter is rapidly oxidized and the soil becomes nutrient impoverished once the forest canopy is removed and the soil manipulated. As a consequence, they have commonly been subjected to a form of "swidden" or shifting agriculture. While relatively little is known concerning the management of Oxisols, either for agriculture or forestry, it appears that *if properly managed* their net productivity could be much higher than presently realized. For forestry, proper management probably means some form of shelterwood or seed-tree harvesting, to prevent excessive soil perturbation or long periods of exposure, as happens following clearcutting. Where plantations are to be established on these fragile soils, care should be taken to conserve as much of the litter layer-root mats as possible to reduce erosion and loss of nutrients.

## TROPICAL MONSOON FORESTS

In tropical areas where significant seasonal droughts are experienced, the rain forest gives way to "rain green" or monsoon deciduous forests. The formation is not as distinct as the evergreen rain forest and, in fact, consists of a transitional

**Figure 3.10** Deep profile of Oxisol in tropical rain forest in French Guiana.

zone of semievergreen seasonal forest, as well as deciduous seasonal forest. The transitional forest is similar in structure and appearance to the rain forest, except that the trees are not so tall and the forest may have only two tiers. Monsoon forests are found in areas with relatively mild drought periods, usually no more than five months of less than 9 cm of rain each. As one passes on to areas with more distinctive drought periods, the forest becomes predominantly seasonal deciduous. Deciduous trees, such as teak *(Tectona grandis)*, with a seasonal opening of the canopy and the subsequent show of flowers in the lower tier, are common to both the transitional semievergreen and the seasonal deciduous communities of this formation type. Lianas and epiphytes also become scarcer as one passes into areas with distinctive dry periods.

Formations of monsoon forests have been described in Central and South America, Indo-Malaysia, and Africa (Eyre, 1963). The American formation is typically found in a belt skirting both the northern and southern fringe of the Amazon basin, along the east coast of Central America, and some West Indies Islands. The Indo-Malaysian formation stretches from northeast India and Indo-china, through Indonesia to northern Australia. In the wetter portions of the areas, teak and a deciduous legume, *Xylia xylocarpa,* are dominant, while in drier sections *Dipterocarpus tuberculatus* (eng tree) and bamboo, and *Acacia* spp., *Butea,* and other leguminous species are abundant. The African formation is rather indistinct. Except for a discontinuous belt along the northern edge of the west African tropical rainforest, the other communities of this African formation merge rapidly into savanna woodlands, perhaps kept in check by frequent burning.

Soils of these formations may be less highly weathered than those of the tropical rain forests. They generally include various groups of Ultisols and Oxisols (Red Podzolics, Reddish-Brown Lateritic soils, and Latosols).

## OTHER TROPICAL FORMATIONS

The deciduous seasonal forests of Asia and America often grade into drier regions with no sharp change in vegetation. As one approaches drier areas, many dominants disappear and those that persist have a lower, spreading canopy. Many of these low-growing bushy trees and shrubs possess thorns and, where the woody plants form a closed canopy, the formation is generally called a thorn forest. These thorn forests are found in northern Venezuela and Colombia, much of northeast Brazil, the West Indies, and Mexico. Elsewhere they show up in India, Burma, and Thailand in Asia, and in vast areas of Africa and Australia. Both deciduous and evergreen species are found in this formation, and leaves of both types are generally small and cutinized and possess other xeromorphic adaptations.

Soils developed under these thorn forests are often light in color, low in organic matter, and not severely leached. They may even have an accumulation of calcium salts in the subsurface horizon. These soils are mostly classed as Aridisols (Reddish-Brown or Reddish Desert soils).

Passing on from the tropical thorn forests into yet drier areas, one often encounters semidesert shrub and savanna woodlands before encountering desert areas. Soils developed under these latter formations are not forest soils and will not be considered here. Other major ecosystems in which trees play little or no part in soil formation are the tundra and grass prairies.

## SOIL ORDERS AND MAJOR ECOSYSTEMS OF THE WORLD

When examining major vegetation associations of the world it becomes apparent that climate, topography, drainage, parent material, and other factors that influence vegetation development have an equally profound influence on soil devel-

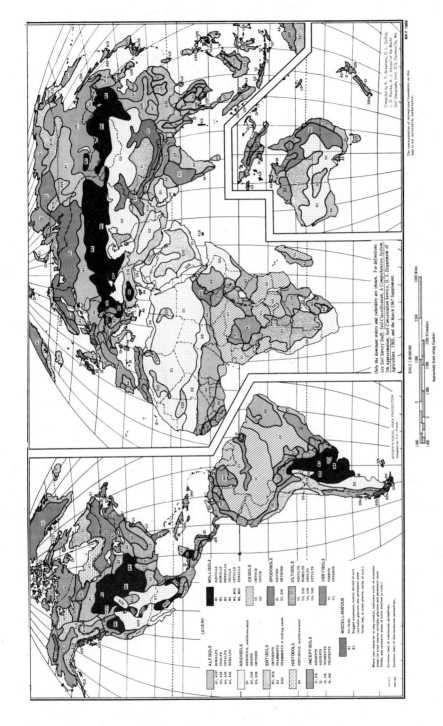

**Figure 3.11**  Probable occurrence of soil orders and suborders in the world (Soil Survey Staff, 1975).

**Table 3.2**   Some Soil Orders Commonly Associated with Major Ecosystems

| Major Ecosystems | Soil Orders |
| --- | --- |
| Boreal forests | Spodosols, Histosols |
| Other coniferous forests | Spodosols, Inceptisols, Ultisols, Alfisols |
| Mixed forests | Mollisols, Inceptisols, Ultisols |
| Deciduous forests | Alfisols, Ultisols, Mollisols |
| Scrub and woodlands | Aridisols, Mollisols, Alfisols |
| Temperate rain forests | Inceptisols, Ultisols, Alfisols |
| Tropical rain forests | Oxisols, Ultisols, Inceptisols |
| Tropical monsoon forests | Ultisols, Oxisols |

opment and taxonomy. It should not be surprising, therefore, that soils possessing certain properties are generally found in conjunction with certain
ecosystems, and if one ignores local site-to-site variations, soil distribution
patterns are quite similar to those of vegetation distribution. When soils are
grouped into 10 major orders, as in Figure 3.11, the world distribution is
strikingly similar to that of the distribution of major ecosystems in Figure 3.1. A
summary of the principal orders associated with the major ecosystems is given in
Table 3.2.

# 4

---

# THE FOREST FLOOR

The forest floor is undoubtedly the most distinctive feature of forest soils and a feature that contributes greatly to their unique properties. The term *forest floor* is generally used to designate all organic matter, including litter and decomposing organic layers, resting on the mineral soil surface. These organic matter layers and their characteristic microflora and fauna are the real dynamic phase of the forest environment and the most important criterion distinguishing forest soils from agricultural (cultivated) soils.

The forest floor is the zone in which vast quantities of plant and animal remains and forest litter disintegrate above the surface of the mineral soil (Figure 4.1). Much of these plant remains, plus animal tissue and excretory products, gradually become mixed in the mineral soil and, together with the subterranean portions of the plant, form the soil organic matter fraction. During the decomposition process, the soil organic matter including cells of dead microorganisms, serves as a source of carbon for following generations of organisms to use. The forest floor is not only a source of food and habitat for many microflora and fauna, but the continuing additions of litter to the floor represent a revolving fund of nutrients, particularly nitrogen, phosphorus, and sulfur, for higher plants. The removal of forest litter as bedding for farm animals in Germany prompted concern over site degradation and led to Ebermayer's (1876) classical work on litter production properties. Furthermore, forest litter layers physically insulate surfaces from extremes in temperature and moisture and offer mechanical protection from raindrop impact and erosional forces and improve water infiltration rate (Wooldridge, 1970). Foresters, as well as soil microbiologists, generally use the term forest humus rather broadly to mean any organic portion of the soil profile, but they have long known that there are differences in humus beneath the various forest cover types and, perhaps, beneath the same forest types growing on different kinds of soils (Wollum, 1973).

## CLASSIFICATION SCHEMES

There is little unanimity among soil scientists on systems of classification of the layers of organic materials in forest soils. The term *forest floor* generally refers to all organic materials resting on but not mixed with the mineral soil surface, while "humus layer" is defined as the surface soil below, but not including, the undecomposed litter. Specifically, the humus layer includes the decomposing organic remains beneath the litter layer as well as the A1 soil horizon, but not other deeper horizons containing a mixture of organic and mineral matter, such as the $B_2h$. At the risk of confusing the terminology even further, both litter layer and humus layers are discussed herein under the broad heading of "forest floor."

### Stratification in the Forest Floor

Three horizon layers, or strata, of the forest floor are customarily designated by forest soil scientists, although they do not all appear on all soils (Hesselman, 1926). They are:

**L** or litter layer consisting of unaltered dead remains of plants and animals. It must be recognized that while the litter is essentially unaltered, it is in some stage of decomposition from the moment it hits the floor. Although an early definition of humus (Waksman, 1936) as "all the plant and animal residues brought upon or into the soil and undergoing decomposition" would include litter as part of soil humus, the litter layer is considered part of the forest floor, but generally not part of the "humus layers."

**F** layer is a zone immediately below the litter consisting of fragmented, partly decomposed organic materials that are sufficiently well preserved to permit identification as to origin.

**H** layer consists largely of well-decomposed, amorphous organic matter. The H layer is largely coprogenic, whereas the F layer has not yet passed through the bodies of soil fauna. The humified H layer is often not recognized as such in mull humus types. Because the lower layer of mull humus often has a friable crumb structure and contains considerable mineral materials, it may be designated as the A1 horizon of the mineral soil and not a part of the forest floor as such.

The United States Soil Conservation Service (Soil Survey Staff, 1975) compromised on a relatively simple procedure for designating surface organic horizons of mineral soils. The two subdivisions are described below:

**01** Organic horizons in which essentially the original form of most vegetative matter is visible to the naked eye. Identifiable remains of soil fauna and considerable fungal hyphae may be present. The vegetative matter may be essentially unaltered, as freshly fallen leaves, but may be leached of its most soluble constituents and discolored. The 01 corresponds to the L and some

**Figure 4.1** Forest floor under old growth fir-birch stand in Quebec. Note large amount of fallen logs and debris in this boreal forest.

F layers mentioned in forest soils literature, and was formerly called Aoo by the Soil Conservation Service.

02　Organic horizons in which the original form of most plant and animal matter cannot be recognized with the naked eye. Excrement of soil fauna is commonly a large part of the material present. The 02 corresponds to the H layer and some F layers of forest soil literature, and it was previously designated as Ao.

## Types of Humus Layers

The difficulty in classifying these organic layers into humus type largely stems from differences in nomenclature used in describing the layers by scientists in various parts of the world, and to the difficulties in characterizing them under many different conditions within the same country.

**Historical.**   Heiberg and Chandler (1941) and Remezov and Pogrebnyak (1969) gave historical accounts of the evolution of terminology and classification of humus layer types and pointed out some of the variations in nomenclature that have been used in various parts of the world. For example, the terms raw humus, peat, acid humus, duff, and mor have been used to describe one type of humus layer; while mild humus, leaf mold, and mull have all been used to designate another type.

Müller (1879) proposed the first generally acceptable classification of forest floors based on field experiences in Denmark. Müller's classification scheme was largely based on morphological characteristics, but he recognized that variations in properties were largely due to differences in biological activities in the humus layer. He divided humus types into two broad groups and described their relation to forest growth and their effects on soil development. He termed the superficial deposit of organic remains as *mor* humus, and the intricate mixture of amorphous humus and mineral soil as *mull* humus. Müller described mor humus as compacted and sharply delineated from the mineral soil below, while mull humus possessed a diffuse lower boundary and a crumblike or granular structure. Furthermore, mulls are generally less acid than mors and, consequently, bacteria are more abundant in mulls, while fungi are the most important microorganisms in the mor humus type. Mulls generally support nitrification rather well but mors do not. Large earthworm populations are typically associated with mull humus and, in fact, earthworms and arthropods are believed to be essential in its formation. As a consequence, mull humus often contains significant amounts of mineral materials, while mor humus is essentially all organic matter.

Russian scientists have also described humus formation on the basis of the action of associated organisms. On sites where earthworms ingest plant residues and pass them through their digestive system, the materials undergo complex transformations that assist the actinomycetes, bacteria, and fungi to decompose these materials more rapidly and completely. This more complete humification of plant residues gives rise to soft, mull-type humus. On the other hand, decomposition through fungal action is predominant in soils poor in bases and nutrients, which then gives rise to plant residues that are similarly poor in bases. Typically, there are insufficient bases for the neutralization of the humic acids that arise, resulting in a strong acid reaction and less complete decompositon of organic residues associated with raw (mor) humus. They recognized a series of gradual transitions between the two extremes. These were collectively called moder in the fashion of an earlier system proposed by Kubiena (1953).

Some general conclusions can be drawn concerning the origin of the distinctive features of mor and mull humus types, but probably none of them hold for all soil conditions and plant communities. For example, mull humus types are generally formed under hardwood forests and under forest growing on soils well supplied with bases, while mor types are most often found under coniferous forests and heath plants often growing on spodic soils, but they are by no means found exclusively under these forests types nor on those soils (Romell, 1935). In fact, other factors in addition to the type of litter also influence the development of the humus type. Climate has an influence apart from its effect on soil and vegetation development. Furthermore, the fragmentation, decomposition, and mixing brought about by organisms associated with the litter layer have a powerful effect on humus development. For example, on soils where mor humus develops, acid-tolerant fungi, protozoa, collembola, and mites are largely responsible for litter decomposition and, as a consequence, decomposition is often slow, with a resulting stratification of the layer of organic matter and very little mixing of the organic materials with the mineral soil.

**Classification in North America.**   It is evident that all humus layers do not fit easily into one or the other of the two broad groups, mor and mull. The heterogenous nature of soils in the United States, as elsewhere, has been recognized and various subdivisions of the two humus layer types have been proposed. For example, Romell and Heiberg (1931), as revised by Heiberg and Chandler (1941), described humus layer types for northeastern United States upland soils. They defined mull as a humus layer consisting of mixed organic and mineral matter in which the transition to a lower horizon is not sharp. They divided mull humus into (1) coarse, (2) medium, (3) fine, (4) firm, and (5) twin types. They described mor humus as a "layer of unincorporated organic material, usually matted or compacted or both, distinctly delimited from the mineral soil unless the latter has been blackened by the washing in of organic matter" They divided mor humus into (1) matted, (2) laminated, (3) granular, (4) greasy, and (5) fibrous types on the basis of morphological properties. (An example of mor humus is shown in Figure 4.2). While this classification was adequate for the northeastern United States, some humus types in other regions were not clearly identified under this system.

In a report for the Soil Science Society of America, Hoover and Lunt (1952) proposed a more comprehensive system based on the presence or absence of a humified layer, the degree of incorporation of organic matter into the upper mineral soil layer, and the structure, thickness, and organic matter content of the humus layer and/or A1 horizon. They proposed three major categories—mull, duff mull, and mor—with the mull humus divided into (1) firm, (2) sand, (3) coarse, (4) medium, (5) fine, and (6) twin mulls; and mor humus grouped into (1) thick, (2) thin, (3) greasy, (4) felty , and (5) imperfect mors. They described duff mull as having a humified layer with an underlying A1 horizon essentially similar to a true mull and possessing characteristics of both mulls and mors. They divided duff mulls into thick and thin types. While these and other detailed systems of forest humus classification (Romell and Heiberg, 1931; Mader, 1953;

**Figure 4.2** Thick mor humus in spruce stand in central Norway. Note gray leached A2 layer over spodic horizon.

Wilde, 1958 and 1966) have merit for technical reporting and some specialized conditions, they are probably too complex for general field use.

In practice, humus found in transition zones of overlapping environmental conditions that could produce either mull or mor types has been termed "duff mull" by Hoover and Lunt (1952) and others, or moder (morlike mull) by Edwards, Reichle, and Crossley (1970) as described earlier by Kubiena (1953). This humus of transition zones has characteristics of both mor and mull, such as stratification into layers of varying degrees of decomposition, but with some mixing of the lower humus layer with the mineral soil. It is often intermediate in acidity, base saturation, and nitrogen content between the two extreme mor and mull types. Most of the plant remains below the litter have been converted into animal feces or other residues, although some plant tissues and cellular structure may still be discerned. Mineral matter, animal feces, and organic residues form a fairly loose mixture and the production of humic substances has not proceeded very far (Edwards et al., 1970). Lyford (1963) considered that moder humus results from continued mixing of mineral soil into the lower part of the forest floor by ants or rodents, without destroying the organic horizons.

The terms duff mull and moder are apparently used interchangeably for describing humus in most transition zones, but regional preferences for one term or the other have also been expressed. Moder may be more often used for humus having a preponderance of characteristics of mor humus, while duff mull may be more like mull humus. For field purposes, grouping of forest floor humus into three broad types—(1) mor, (2) moder or duff mull, and (3) mull—is probably adequate. In North America, it appears that duff mull is more widely used than moder for the intermediate types.

A mor humus type is generally found under the spruce-fir forests of the boreal region of northern and eastern Canada and under much of the coniferous forests of Scandinavia and Siberia. Duff mull (moder) humus exists under most natural hardwood and mixed pine-hardwood stands such as those of the New England transition and northern hardwood regions, while mull humus is often found in the central hardwood region of the United States and central Europe. However, it should be emphasized that local site conditions may alter the type of humus layer developed from that normally expected. The major morphological types may be further divided on the basis of quantitative physical and chemical properties, if there is need for such subdivision. For example, Wooldridge (1970) reported that forest floors under mixed conifers and ponderosa pines of central Washington were generally typed as felty mors with well-developed H layers and abundant fungi, but that they included granular mors on the drier sites of mixed conifers and imperfect mors on the driest sites and in semi-open pine stand. However, he more meaningfully characterized them on the basis of their physical and chemical properties.

**The Forest Floor Under Southern Pines.**   Heywood and Barnette (1936) studied 42 well- to moderately well-drained longleaf and slash pine sites in the United States lower coastal plain. The stands had been protected from fire for 10 to 50 years and the humus layer possessed characteristics intermediate between those of mull and mor types. The 2 to 4 cm litter (L) layers generally contained needles shed during the previous 2.5 to 3.0 years, after which time they began to break up and form part of the F layer. The F layers were well expressed, and often rested directly on the mineral soil, as the H layers were indistinct and discontinuous. The F layers were 2 to 3 cm thick, rather loosely arranged, and permeated with fungus mycelia. Immediately overlying the mineral soil was a thin layer of excrement from soil fauna, but there was little mixing of this material into the mineral soil. Although they found rather uniform humus layers over a fairly wide range of pine sites in the southeast United States, they avoided classifying them. Because these sites appear to have some characteristics of both mor and mull types, they would probably be classed as duff mulls according to Hoover and Lunt (1952) or as moder types by Kubiena (1953). The forest floors under four pine species in the Virginia piedmont were found to have similar characteristic as those described for longleaf and slash pines (Metz, Wells, and Komanik, 1970).

## DECOMPOSITION AND ACCUMULATION

The amount and character of the forest floor depends to a large extent on the decomposition rate of the organic debris. The rate of breakdown of the floor material is influenced by the physical and chemical nature of the fresh tissue; the aeration, temperature, and moisture conditions of the floor; and the kinds and numbers of microflora and fauna present. Because the decomposition processes are largely biological, these rates are influenced by the same factors that govern the activity of the organisms. Phosphorus and base concentrations and the ratio of carbon to nitrogen in the debris affect the activity of microorganisms. Because the influence of environment will be discussed in some detail in the following chapter, only the general pattern of organic matter breakdown will be outlined here.

### Forest Floor Decomposition

The rate of breakdown of fresh litter can be quite rapid, with a turnover rate varying from one to three years in temperate and cool climates to a few months in the tropics. The percent loss of dry leaf weight of four tree species in Tennessee during one year was mulberry, 90 percent; redbud, 70 percent; white oak, 58 percent; and pine, 40 percent (Edwards et al., 1970). In a similar study the authors reported that beech leaves lost 64 percent of their weight, oak lost 80 percent, and elm, birch, and ash leaves broke down completely after one year.

As noted earlier, the process of decomposition often begins even before the plant debris is added to the forest floor. Leaf exudates promote the invasion of pathogens while the leaves are still on the trees, and further invasion by fungi occurs during the first few weeks of weathering after the litter reaches the forest floor. The leaves darken, and much of the water-soluble sugars, organic acids, and polyphenols are leached out during that period. As the water-soluble poly-phenols are removed by weathering, the litter becomes more palatable to arthropods and earthworms *(Lumbricus terrestris),* according to Edwards et al. (1970). In these early stages of litter breakdown, there may be a large microbial population present, but it is largely inactive. Apparently, without preliminary fragmentation by soil animals, the ubiquitous microbes cannot decompose the constituents of many leaf species. In temperate regions, earthworms, rotifers, and arthropods, including acarinas and crustaceans, are responsible for most fragmentation, and if for any reason this fragmentation is retarded, the entire process of decomposition is slowed.

Estimates of annual amounts of deciduous leaf litter converted to animal feces range from 20 to 100 percent, according to a review of the subject by Edwards et al. (1970). A succession of species of microflora were identified on fecal pellets, but only slight chemical changes occurred after passage of litter through the gut of soil arthropods. The increase in surface area resulting from ingestion of the litter by microarthropods is considered the most important contribution to its ultimate breakdown. While some soil fauna are capable of breaking down cellulose with the aid of enzymes in their alimentary tract, most evidence indicates that the chemical processes of humification are caused more by microbes than by soil fauna.

In some regions, termites are particularly important in the reduction and decomposition of the large amounts of wood that reach the forest floor. Fungal mycelia ramify through cracks in the wood and soften the tissue, and then many insects and larvae invade the moist tissue. The feces of these animals are a rich substrate for microorganisms. The spectrum of organisms involved depends to a large extent on the nature of the organic debris, but they all work toward the gradual transformation of complex compounds into such simple materials as carbon dioxide, gas, water, nitrogen, mineral elements, and complex amorphous humus material. Materials accumulate on the forest floor and, even under aerobic conditions, complete oxidation seldom takes place. While the carbohydrates, proteins, and pectins disappear rather rapidly, waxes, resins, and lignins persist for years. As a result, considerable cell substance is synthesized, and together with the modified lignins, they constitute the bulk of the soil humus. The latter is a dark, structureless material composed of complex polymers of phenolic substances. Humus is more or less stable, and, in soils rich in fauna, it becomes throughly mixed in the upper horizon of mineral soil.

## Forest Floor Accumulation

The accumulation of organic materials on the forest floor is largely the function of the annual amount of litter fall minus the annual rate of decomposition. Although many environmental factors affect the rate of litter decomposition, the rate of litter fall is remarkably uniform among tree species growing under similar soil and climatic conditions.

Periodic litter fall is determined by collecting all debris falling to the forest floor in a litter trap and, after drying and weighing, converting to an area basis. A trap generally consists of a square frame with low sides and a mesh bottom to allow for drainage. Total accumulated forest floor material is determined by collecting all organic materials above a specified area of soil and separating in component layers.

An annual return of 2.0 to 6.0 tons per hectare appears to hold for most conifers and hardwoods in cool temperate regions; however, up to 12 tons per hectare per year are produced in tropical rain forests (Bray and Gorham, 1964). The relationship of litter production to latitude is shown in Figure 4.3. The annual litter fall in a mature northern hardwood stand in New Hampshire averaged 5.7 tons per hectare, with leaves, branches, stems and bark contributing 49.1, 22.2, 14.1, and 1.7 percent, respectively (Gosz, Lichens, and Bormann, 1972). Other structures (bud scales, fruit, and flowers), insect frass, and miscellaneous tissue contributed 10.9 percent. Overstory contributed 98 percent of the litter fall. Bray

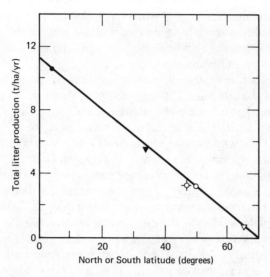

**Figure 4.3**  Annual production of litter in relation to
latitude. Line fitted to means for climatic zones: ●—
equatorial; ▼—warm temperate;✧—cool temperate
(North America); ○—cool temperate (European); ▽—
artic-alpine (Bray and Gorham, 1964). Used with permission. Copyright in Academic Press, Inc. (London)
Ltd. Advances in Ecological Research, Vol. 2.

and Gorham (1964) noted that nonleaf litter made up an average of 30 percent of
the total litter in forests worldwide. They also reported an average of 9 percent of
total litter to derive from understory vegetation, but they pointed out that the
amount of this litter varied with the density of the forest canopy. Heyward and
Barnette (1936) found an annual leaf fall of 2.7 to 3.9 tons per hectare in mature
stands of longleaf and slash pine stands on well-drained soils, with total accumulations of material on the forest floor averaging about 61 tons. Lutz and Chandler
(1946) reviewed reports on annual litter fall from a variety of species and
locations and found the amounts to vary from 2.2 to 3.4 tons of dry material.

   Forest floor accumulation is influenced not only by the annual rate of
decomposition but also by the age of the floor or the elapsed time since the last
fire or other disturbance. The increase in organic accumulations is rather rapid in
the early stages of stand development and in the first decade or so following
burning, but eventually a condition of near equilibrium is reached in which the
rate of decomposition is about equal to the annual input of organic materials
(Bray and Gorham, 1964; Wells, 1971). The balance is a dynamic one, and is
most nearly reached in climax communities. Hayward and Barnette (1936)
thought that a balance in input and output from the forest floor was probably

reached in about 10 years in a mature stand of southern pine protected from fire. The progress toward an equilibrium is much more rapid under favorable growing conditions than under extremes of temperature or moisture. For example, McFee and Stone (1965) reported that weight of the organic matter in forest floors under mature yellow birch-red spruce stands growing on well-drained outwash sands in the Adirondack Mountains of New York increased from a mean of 131 tons at 90 years of age to 265 tons at 325 years or more. Available soil moisture influences the accumulation of forest floor material as a result of its effect on tree growth. Wollum (1973) reported that dry weights of forest floors increased from 9.4 to 80.8 tons per hectare along a moisture gradient from the dry (piñon-juniper) site to the wet end of the gradient (white fir) in New Mexico. To be sure, the accumulation of forest floor organic matter, as well as the development of peat soils, may result from reduced microbial activity due to site wetness—especially where there is a plentiful supply of organic litter and yet a limited period with temperatures favorable for decay. Some B horizon organic materials in Spodosols (podzols) of Scandinavia have been estimated to be more than a thousand years old by carbon-14 dating, and the humus in cold-climate forest floors may be more than 100 years old. In fact, the slow rate of decomposition of organic debris is a major reason for the chronic nitrogen deficiencies noted in some mature stands on these soils and similar soils in the Canadian boreal forest (Weetman and Webber, 1971).

Olson (1963) pointed out that, in contrast to the high levels of carbon and energy accumulation in the relatively unproductive cool temperate forests, the highly productive tropical forests have low storages of carbon due to the high rate at which organic matter is broken down and incorporated into the mineral soil. While the annual litter fall under a tropical rain forest may be several times greater than that under a temperate zone forest, the rate of turnover is very rapid and the accumulation of litter on the forest floor is never large, except on very wet sites. Nevertheless, some tropical forest soils contain greater amounts of organic matter in the A1 horizon than might be expected under the prevailing favorable temperature and moisture conditions for decomposition. This situation apparently results from a complexing of soil aluminum with the organic matter which, in turn, reduces the rate of microbial activity (Mutatkar and Pritchett, 1967). A deficiency of available phosphorus was also found to limit mineralization of organic matter in Andepts of Colombia (Munevar and Wollum, 1977).

Romezov and Pogrebnyak (1969) reported that the amount of dry material in the forest floor (unincorporated organic debris) varied from 22 to 35 tons under conifers, 27 to 77 tons under mixed conifer-hardwoods, and 35 to 95 tons per hectare under hardwood forests in Russia. However, most investigators have reported a greater accumulation of mor humus than of mull humus. This apparently results from the slow rate of litter decomposition under mor conditions. For example, both mor and duff mull forest floors were found under old growth conifer stands in the Cascades (Gessel and Balci, 1965), with the former possessing a total weight of 158 tons and the latter 103 tons per hectare. These investigators also found that the weights of forest floors under immature Douglas-fir in eastern and western Washington averaged 28 and 14 tons per hectare, respectively. Similar values have been found for unincorporated organic debris in forest soils in Minnesota. In mature stands of hardwood in Minnesota dry matter averaged about 45 tons per hectare, while in conifer stands, the dry weight of forest floors averaged about 112 tons. The forest floor under a fertilized stand of 15-year-old slash pine on a wet savanna soil in Florida averaged 46 tons per hectare, with approximately $2/5$ in the L layer and $3/5$ in the F layer, as shown in Table 4.1. Lyford (personal communication) found the dry weight of organic matter of the forest floor under hardwood, conifer, and mixed hardwood-conifer forests in the northeastern United States to vary from 40 to 109 tons for duff mull humus and from 53 to 82 tons per hectare for mor humus (Table 4.1). Bray and Gorham (1964) compiled a comprehensive review of litter accumulation in forests of the world which is summarized in Figure 4.3 by climatic zones.

The amount of unincorporated organic matter accumulated on the surface of mineral soils is influenced by the soil texture, as well as the nature of the layer itself. Fine-textured mineral soils normally accumulate greater quantities of organic matter than coarse textured soils. Although it is generally reputed that grassland soils contain a greater amount of organic matter than forest soils, this is probably true only for the mineral horizons. The differences in organic matter content between those two groups of soils are due more to the character and vertical distribution in the soil than to actual amounts present (Lutz and Chandler, 1946).

## PROPERTIES OF THE FOREST FLOOR

Many factors that influence tree growth and the rate of organic matter decomposition, such as soil characteristics, climate, and tree species, also affect the physical and chemical properties of the forest floor humus.

### Physical Properties

Among other things, stage of litter decomposition and compaction affect the physical properties of the forest floor. Gessel and Balci (1965) reported that the mean bulk density of mor humus was 0.12 g per cc while that of duff mull was 0.14 in old growth conifer stands. McFee and Stone (1965) found that the mean bulk density of the forest floor under birch-spruce in New York varied from 0.12 to 0.16 and was not influenced by the age of the stand. In the Gessel and Balci (1965) study, mean values for field capacity and saturation capacity were greater for mor than for duff mull humus. They pointed out that the mor forest floor can hold 4.26 cm of water as compared to 2.53 cm for duff mull. Wooldridge (1970) reported the maximum water-holding capacity of forest floors to range from 1.9 cm under pine forests to 3.2 cm under mixed conifer forests in central Washington. These values are in line with those found by Remezov and Pogrebnyak (1969). Their values fell within a range of 1.0 to 5.0 cm with the water-holding capacity increasing with compactness of the litter.

## Chemical Properties

Fresh litter is composed of a large number of complex organic compounds, the relative percentages of which vary in different plant parts and in different species, age of material, and the soils on which they were produced. Some general statements can be made about the composition of the raw material for the forest floor. The relative ash content of tree leaves is generally greater than that of the bark, which in turn is often greater than that of branches, with the lowest content in the bole wood. For example, the phosphorus concentrations in components of slash pine averaged 0.07, 0.016, 0.018, and 0.006 percent in foliage, bark, branches, and bolewood, respectively. The ash content of bole wood usually falls within the range of 0.2 to 1.0 percent, and sapwood normally has a higher content than heartwood.

Bray and Gorham (1964) reported that the percentage of ash of a majority of gymnosperm litter was in the range of 2 to 6 percent while that of angiosperm litter ranged from 4 to 14 percent. The leaves of hardwood species generally contain higher concentrations of nitrogen, phosphorus, potassium, calcium, and magnesium than do leaves of conifers, although not under all conditions. For example, leaves of oak or beech growing on soils low in bases may contain less calcium than needles of spruce or fir growing on fertile soils. Ash contents of litter of species that usually pioneer in forest development and that often occur on the more infertile soils are generally lower than ash content of species in more developed (climax) communities and on the more mesic and fertile sites (Bray and Gorham, 1964). The age of the leaves at the time they reach the forest floor also influences their composition. In most species, the percentage of nitrogen, phosphorus, and potassium decreases as the growing season progresses, due largely to translocation to more actively growing tissue. However, a decrease in concentration does not always mean a decrease in absolute amounts, because the dry weight of individual leaves often increases throughout the season (Table 4.2).

The chemical composition of the forest floor has a significant effect on the rate of litter decomposition and nutrient release, and on the soil population and tree growth. The pH, carbon:nitrogen ratio, and concentration of mineral constituents in the various organic layers are influenced by both the kind of soil and the type of vegetation from which the layers were developed. These properties influence soil development and stand composition. As a general rule, mor humus is more acid than mull humus, but because these two types of humus often develop under wide ranges of tree species and soil types, the reaction (pH) of

**Table 4.1** The Weights and Properties of Forest Floor Layers Under Several Forest Types

| Forest Type | Layer | Oven-dry weight | N | Ca | Mg | P | K | S | Al |
|---|---|---|---|---|---|---|---|---|---|
| | | ton/ha | percent (oven-dry basis) | | | | | | |
| Spruce stand (Class 1) | L | 2.9 | 1.17 | 1.23 | 0.31 | 0.14 | 0.24 | 0.13 | 0.32 |
| Old stand, Russia | F | 8.2 | 1.45 | 1.18 | 0.30 | 0.11 | 0.13 | 0.08 | 0.58 |
| (Remezov and Pogrebnyak, 1969) | H | 10.1 | 1.32 | 0.76 | 0.30 | 0.09 | 0.09 | 0.08 | 1.09 |
| Birch stand (Class 11) | L | 1.1 | 1.37 | 1.36 | 0.32 | 0.21 | 0.31 | 0.13 | 0.43 |
| Old stand, Russia | F | 6.1 | 1.66 | 1.38 | 0.32 | 0.21 | 0.18 | 0.21 | 0.48 |
| (Remezov and Pogrebnyak, 1969) | H | 10.9 | 1.18 | 1.25 | 0.27 | 0.17 | 0.05 | 0.42 | 1.26 |
| Longleaf–slash pine—moder | L | 10.2 | 0.52 | 0.44 | 0.12 | 0.05 | 0.06 | — | — |
| Ave. 19 stands (10–50 yr. old) | F | 22.8 | 0.54 | 0.42 | 0.09 | 0.06 | 0.04 | — | — |
| (Heyward and Barnette, 1936) | | | | | | | | | |
| Slash pine (15 yr. old)—moder | L | 15.0 | 0.45 | 0.66 | 0.07 | 0.03 | 0.04 | — | 0.05 |
| (Pritchett and Smith, 1974) | F | 24.5 | 0.56 | 1.15 | 0.05 | 0.02 | 0.04 | — | 0.43 |
| Old growth conifers—mor | L | 14.4 | 1.10 | — | — | 0.10 | 0.10 | — | — |
| (Gessel and Balci, 1965) | F | 22.4 | 1.35 | — | — | 0.11 | 0.11 | — | — |
| | H | 121.0 | 1.22 | — | — | 0.10 | 0.10 | — | — |

| | | | | | | | |
|---|---|---|---|---|---|---|---|
| Old growth conifers—duff mull (Gessel and Balci, 1965) | | | | | | | |
| L | 13.6 | 1.27 | — | — | 0.11 | 0.11 | — |
| F | 18.0 | 1.46 | — | — | 0.12 | 0.12 | — |
| H | 71.7 | 1.30 | — | — | 0.11 | 0.12 | — |
| Red pine—duff mull 45 yr. old—New England (Lyford, personal comm) | | | | | | | |
| L | 5.3 | 0.63 | 0.55 | 0.06 | 0.12 | 0.13 | — |
| F | 31.7 | 1.43 | 0.27 | 0.04 | 0.08 | 0.08 | — |
| H | 31.2 | 1.13 | 0.14 | 0.07 | 0.08 | 0.08 | — |
| Hemlock-maple—mor Old growth—New England (Lyford, personal comm) | | | | | | | |
| L | 5.8 | 0.68 | 0.38 | 0.02 | 0.06 | 0.07 | — |
| F | 44.9 | 1.49 | 0.21 | 0.03 | 0.10 | 0.10 | — |
| H | 31.4 | 1.17 | 0.10 | 0.04 | 0.09 | 0.60 | — |
| Loblolly pine—mull No. Carolina (26 yr. old) (Lyford, personal comm) | | | | | | | |
| L | — | 0.59 | 0.26 | 0.06 | 0.06 | 0.06 | — |
| $F_1$ | — | 0.67 | 0.29 | 0.06 | 0.07 | 0.05 | — |
| $F_2$ | — | 0.87 | 0.23 | 0.05 | 0.08 | 0.05 | — |
| Ponderosa pine (mature) White fir (80 yr. old) (Wollum, 1973) | | | | | | | |
| All | 25.1 | 0.75 | 1.71 | 0.33 | 0.06 | 0.32 | 0.15 |
| All | 80.8 | 1.09 | 3.32 | 0.42 | 0.10 | 0.29 | 0.21 |

**Table 4.2** Elemental Content of Some Coniferous Forest Floors

| Forest Types | Layers | N | P | K | Ca | Mg |
|---|---|---|---|---|---|---|
| | | kilograms per hectare | | | | |
| Old growth mixed | L | 162 | 15 | 14 | — | — |
| conifers—mor humus | F | 313 | 24 | 24 | — | — |
| (Gessel and Balci, 1965) | H | 1565 | 102 | 89 | — | — |
| Old growth mixed | L | 171 | 15 | 15 | — | — |
| conifers—duff humus | F | 266 | 22 | 21 | — | — |
| (Gessel and Balci, 1965) | H | 956 | 77 | 67 | — | — |
| Immature (E. Wash.) | FF[a] | 327 | 29 | 42 | — | — |
| Douglas-fir (W. Wash.) | FF[a] | 193 | 16 | 15 | — | — |
| (Gessel and Balci, 1965) | | | | | | |
| Black spruce (65 yr) | FF[a] | 1214 | 213 | 382 | 102 | 430 |
| Red spruce-fir (mixed) | FF[a] | 1465 | 100 | 1052 | 253 | 154 |
| (Weetman and Webber, 1972) | | | | | | |
| Birch-spruce (300 yr) | FF[a] | 3187 | 152 | 91 | 617 | — |
| (McFee and Stone, 1965) | | | | | | |
| Piñon pine-juniper (mature) | FF[a] | 80 | 12 | 21 | 216 | 41 |
| Ponderosa pine (mature) | FF[a] | 191 | 15 | 80 | 432 | 83 |
| White fir (80 yr) | FF[a] | 883 | 85 | 235 | 2674 | 339 |
| (Wollum, 1973) | | | | | | |
| Southern pine (10–50 yr) | L | 53 | 5 | 6 | 45 | 11 |
| (Heyward and Barnette, 1936) | F | 123 | 13 | 8 | 96 | 20 |
| Slash pine (14 yr) | FF[a] | 183 | 8 | 14 | 341 | 20 |
| (Pritchett and Smith, 1974) | | | | | | |
| Virginia pine (17 yr)[b] | L | 18 | 1.4 | 3 | 11 | 2 |
| (Metz et al., 1970) | F | 217 | 13 | 10 | 77 | 9 |
| Shortleaf pine (17 yr)[b] | L | 18 | 1.6 | 3 | 12 | 2 |
| (Metz et al., 1970) | F | 159 | 12 | 11 | 73 | 10 |
| Loblolly pine (17 yr)[b] | L | 28 | 2.5 | 4 | 17 | 4 |
| (Metz et al., 1970) | F | 178 | 13 | 10 | 65 | 9 |
| | H | 59 | 4 | 4 | 20 | 3 |
| Eastern white pine (17 yr)[b] | L | 23 | 2.1 | 3 | 28 | 3 |
| (Metz et al., 1970) | F | 124 | 10 | 7 | 82 | 8 |

[a]All forest floor humus.
[b]Average of 19 stands in southeastern United States.

their humus layers also varies within wide limits. Wilde (1958) suggested that the acidity of mull humus could vary from pH 3.0 to 8.0. Heyward and Barnette (1936) found the acidity of the F layer under longleaf pine to be in the range of pH 3.4 to 5.0. These values were from 0.25 to 1.0 pH units lower than those of the A1 horizon of the corresponding soil. Lutz and Chandler (1946) reported average values of pH 4.3, 4.5, and 4.9 for the L, F, and H layers of mor humus under jack pine and 4.5, 5.9, and 6.5 for the same layers in mull humus under a maple-basswood stand. These values are slightly higher than found under old growth conifers in the Cascades (Gessel and Balci, 1965). The reaction of the L, F, and H layers under spruce averaged pH 5.1, 4.9, and 4.7 while the corresponding layers under birch averaged 5.9, 5.7, and 5.7, respectively, in Russian forests. Nitrogen concentration in the humus layers appears to be correlated with the pH of the material. The nitrogen concentrations in forest floor materials often vary from 1.5 to 2.0 percent in hardwoods and some northern conifers and 0.5 to 1.5 percent under southern pines, as shown by Voigt (1965b) in Table 4.3.

The carbon:nitrogen ratio gives an indication of availability of nitrogen in floor material and of its rate of decay. The ratios for longleaf and slash pine forest floors (Heywood and Barnette, 1936) ranged from 70 to 142 in the litter layers, 38 to 64 in the F layers, and 20 to 46 in the A1 horizons of the sandy soil. The averages for 19 sites were 101, 47, and 33 in the L, F, and A1 layers, respectively. Lutz and Chandler (1946) reported that the carbon:nitrogen ratio in the forest floor materials of Douglas-fir stands averaged about 57; and that of the F layer of

**Table 4.3**  Nitrogen Concentrations in Leaf Litter and Weight Loss, Nitrogen Deficit, and pH Following Three Months of Decomposition in Laboratory (Voigt, 1965b).[a]

| Species | N | After Incubation | | |
|---------|---|----------------|---|---|
| | | Weight Loss | N Deficit | pH |
| | % | % of original | | |
| Hemlock | 1.65 | 36.8 | 12.9 | 6.3 |
| Juniper | 1.40 | 31.5 | 13.1 | 7.3 |
| Red pine | 1.50 | 18.2 | 12.7 | 5.7 |
| Alder | 2.18 | 34.2 | 24.3 | 5.5 |
| Dogwood | 1.63 | 44.6 | 17.8 | 7.2 |
| Tulip poplar | 1.87 | 48.8 | 36.8 | 7.1 |
| | | Average Values | | |
| Conifers | 1.52 | 28.8 | 12.9 | 5.9 |
| Hardwoods | 1.89 | 42.5 | 26.3 | 6.5 |

[a]Reproduced from *Soil Science Society of America Proceedings*, Volume 29, page 757, 1965 by permission of the Soil Science Society of America.

mor humus to be 29, while the corresponding layer in mull humus was 33. The carbon:nitrogen ratio of the L, F, and H layers of mor humus under old growth conifers averaged 45, 36, and 38 while those of duff mull averaged 38, 31, and 29 (Gessel and Balci, 1965). In general, the ratios of carbon to nitrogen are wide in forest floors and decrease as decomposition proceeds and the ratio is greater in mor humus than in mull humus. However, the carbon:nitrogen ratio of the forest floor seldom narrows to that of humus of agricultural soils. Ratios of the latter humus materials may approach 12, a value at which nitrogen mineralization proceeds at a rapid rate.

The chemical compositions of the principal layers of the forest floors of some conifer and hardwood stands are given in Table 4.1. All of these stands had been protected from fire for extended periods. It can be noted that the humus from birch stands was generally higher in all nutrients than that from spruce stands in Russia. The nutrient content of humus from stands of northern conifer-ous species was generally higher than humus from three southern pine stands. The low concentration of nutrients in the southern pine forest floor probably reflects the low level of fertility of the soils of the lower coastal plain, and may explain the relatively large accumulation of litter in the latter stands. There is a slower rate of decomposition than might be expected from the favorable temper-ature and moisture conditions in this area, apparently due to the poor base status of the litter.

The concentration of potassium, calcium, and magnesium generally decreased from the surface litter to that of the lower humus layers under some stands, while aluminum concentrations increased with depth. This indicates that the bases are eluviated to a greater extent than some other elements. On the other hand, the increases in aluminum concentration, as well as that of iron and manganese, in the more decomposed layers reflect a concentration of these elements and, perhaps, some contamination from the mineral soil.

The organic constituents of the plant residues, such as hemicellulose and cellulose, decrease with depth (increase in decomposition); while most com-pounds synthesized in the course of humification, such as humic and fulvic acids, increase with depth within the floor. Values for humic acid of 2.9, 6.4, and 7.2 percent have been found in the L, F, and H layers of a mor humus under Russian spruce stands. The corresponding values for fulvic acid were 4.3, 13.7, and 15.2 percent (Remezov and Pogrebnyak, 1969).

Relatively large quantities of nutrients are stored in the forest floor. The total content of nutrients is dictated by the amount and composition of the floor, which is influenced by the forest vegetation, climate, mineral soil, and the accumulation period following a major disturbance of the floor. In some forest soils, such as glacial outwash sands or sandy soils of coastal regions, the forest floor represents the major reserve of nutrients for tree growth. Although the total nutrient content of forest floors in warm regions may be only a fraction of that in forest floors of cooler climates (Table 4.2), the rate of decomposition and nutrient turnover is much more rapid in the warm temperate forests than in northern forests. It is reported that organic matter accumulation due to the slow rate of decomposition in soils of the boreal forest and other cool climate areas sometimes results in nitrogen deficiencies (Weetman and Webber, 1972). Regardless of whether the forest floor is developed under cool or warm climates, it is the home of most soil organisms, the reservoir of most nutrients involved in the cycling process, and the very life of the soil itself. It should be regarded as vital to the continued productivity of any forest ecosystem and managed so as to produce an adequate and long-term supply of nutrients to all forest components.

## ALTERING THE FOREST FLOOR

The usual explanation of factors influencing decomposition and accumulation of organic debris comprising the forest floor is, understandably, oversimplified. Both natural and man-made forces act independently and in concert to disrupt an otherwise orderly process. Wildfires are perhaps the most dramatic of nature's weapons of change. A hot fire can reduce the surface organic layers of the floor to a thin coating of ash in only a few minutes. The insulating blanket of the floor is not only destroyed by such an uncontrolled fire, but the resulting ash may significantly affect the nutrient status of the underlying mineral soil. In contrast to wildfires, prescribed fires are normally controlled in such a fashion that little more than the litter layer is disturbed and no permanent damage is rendered the mineral soil (Wells, 1971).

### Natural Forces

Lyford (1973) considered the uprooting of trees by windthrow and other natural forces a feature of all forested areas and a common disturbance of the organic layers. The mound and pit microrelief resulting from uprooted trees may persist for several decades, particularly in northern climates. Thin or discontinuous mineral soil horizons occur on the mounds and deep layers of organic matter form in the pits. Mass movement of soil by gravity may also occur on unstable areas of steep terrain.

Among nature's other agents operating to alter the forest floor are fossorial mammals, such as gophers, moles, and shrews (Troedsson and Lyford, 1973), as seen in Figure 4.4. These animals often pile soil around burrow entrances and move and mix soil when tunnels are made. Abandoned runways collapse or become filled with soil material from above. Transport of mineral soil into the organic horizons may also occur as a result of activity of ants, termites, earthworms, rodents and other small animals. Where the mineral soil materials are deposited by these fauna, there is likely to be sudden and sharp contrast in soil properties. The reverse process of transport of organic matter into the mineral soil also occurs as a result of animal activity.

**Figure 4.4**  Mineral soil transported into forest floor material by a crustacean in the coastal plain.

## Human Activities

Human influence on the forest floor has often been more dramatic than that of nature. In their efforts to increase site productivity by intensive management, humans have altered the equilibrium that is normally established under mature forests by harvesting, slash burning, site preparation, prescribed fire, fertilization, and other practices. Burning can be considered as an acceleration of the natural process of oxidation that the floor continuously undergoes. It is a rather drastic operation, but when properly controlled, burning does little long-term damage to the floor, and may result in some benefits by reducing understory competition and improving conditions for both symbiotic and non-symbiotic dinitrogen fixation. Burning has also been used to reduce excessive buildup of mor humus and to increase mineralization in these acid organic layers. The residual ash results in an increase in pH and base content of the surface layer (Viro, 1974).

Alteration in organic layers can sometimes be made by increasing the amount of sunlight and precipitation reaching the floor. Thinning or removal of the forest stand by harvesting normally results in higher soil temperature and moisture and increased decomposition and mineralization of the organic layers. Presumably the increase in decomposition rate results from increased microbial activity in general, with bacterial activity assuming a more important role in the latter stages of decomposition.

Site preparation for intensively managed forests largely destroys the forest floor by mixing the organic layers with the mineral soil by discing or ridging. Such manipulation may concentrate the humus and increase aeration and oxidation of organic matter in these soils. Schultz and Wilhite (1974) found that the organic matter content of the 0 to 15 cm layer of a flatwood soil in north Florida was not significantly affected by shallow discing, but was increased by 33 percent in low beds after four years. Bedding of these soils also significantly increased the levels of available nitrogen, potassium, calcium, and magnesium in the tree rooting zone during the first few years after planting (Haines and Pritchett, 1965).

Attempts have been made to improve the nutrient status and rate of cycling from the forest floor by altering the stand composition. It is generally believed that nutrients in humus under mixed stands are more readily cycled than from humus under pure stands. European research has indicated that soil productivity can be improved by introducing some hardwood species into coniferous stands. The resultant debris was higher in bases and pH and more rapidly decomposed. However, interplantings of some hardwood species, such as beech, were not particularly beneficial to the productivity of spruce stands. It is doubtful that the interplanting of hardwood species on sandy soils, or on other sites not favorable for the hardwood, will substantially improve site conditions.

Leguminous plants have been used in Germany to improve the soil nitrogen status and increase the rate of mineral cycling in forests impoverished earlier by litter removals (Rehfuess and Schmidt, 1971). However, root damage resulting from forest floor tillage in connection with the legume establishment apparently prevented significant increases in tree volume growth. Alder and other nonleguminous plants capable of symbiotic fixation of nitrogen, have also been planted in forests to improve the forest floor and site productivity. Plants of this type may play a significant role in maintaining adequate levels of nitrogen in some forest ecosystems, and dinitrogen fixing species should be given consideration in devising any forest management scheme (Schultz, 1971).

Applications of chemical fertilizers may be necessary in the establishment and maintenance of dinitrogen fixing plants. Apart from this possible beneficial effect of fertilizers, they may have a more direct effect on the forest floor through increased litter fall. For example, dry weight of the forest floor under a 15-year-old slash pine stand on a wet coastal soil fertilized with 45 kg nitrogen per hectare and 267 kg each of phosphorus and potassium was 39.6 tons per hectare as compared to 8.2 tons per hectare in unfertilized plots. Nutrients in the forest floor were also increased by fertilization. For example, nitrogen was increased from 53 to 205 kg per hectare, even though only 45 kg were added as fertilizer (Pritchett and Smith, 1974).

It appears that most manipulations of the forest floor increase biological activity and rate of organic matter oxidation. This undoubtedly increases cycling and nutrient availability. However, it is not known whether all such increases are beneficial to long-term tree growth and site productivity.

# 5

## FOREST SOIL BIOLOGY

The kinds of organisms found in forest soils may not differ significantly from those found in other soils. The variety, numbers, and activity, however, are generally much greater in forest soils than in agricultural soils. The favorable environment of the forest floor encourages the proliferation of a myriad of microorganisms that perform many complex tasks relating to soil formation, slash and litter disposal, nutrient availability and recycling, and tree metabolism and growth.

Most soil organisms are ubiquitous, but the inoculation, or introduction, of organisms to a new environment is largely a haphazard occurrence. How well the organisms thrive in new surroundings depends on characteristics of the organisms and on such soil factors as moisture, temperature, aeration, acidity, and nutrient and energy supplies. The same factors greatly influence the spatial distribution of organisms in the soil. Conditions favorable for most organisms can be found at some level within the litter layer and underlying soil horizons.

Most soil animals make their home in the surface litter or humus layers, where space and light conditions fit their particular needs. They are rather mobile, but because their distribution is dependent on an organic food base, they generally move only within the organic horizons. Microflora, on the other hand, are found throughout the soil profile, with each type claiming its own niche. The photosynthetic organisms, such as blue-green algae, are found only on surface materials where light is not limiting for photosynthesis. Other organisms get their carbon from $CO_2$ and their energy from the oxidation of inorganic substances and therefore they are less restricted in their spatial distribution. The vast majority of organisms get their carbon from complex organic materials and their abundance is directly dependent on the presence of a proper organic substrate. Some of the smallest of these microorganisms are not free in the soil, but are held to clay or organic colloids by cation exchange forces. Organic exudates stimulate the growth of other microorganisms at the root surface. They are often 10 to 100 times more numerous in the rhizosphere, that portion of the soil in the immediate vicinity of roots, than in other nearby areas of mineral soil. Generally, the numbers of organisms are greatest in the forest floor and the rhizosphere with a decreasing gradient with soil depth.

## KINDS OF ORGANISMS AND THEIR FUNCTIONS IN SOILS

Forest soils contain a multitude of organic and mineral substances, available as carbon and energy sources, and the physical environment suitable for a vast array of plant and animal populations ranging in size from microscopic bacteria to fairly large animals. Perhaps the only feature that many of those organisms have in common is that they spend all or a major part of their life in the soil. Because of the diverse and complex nature of these populations, a number of classification schemes have been devised. The most commonly used system divides the organisms into two broad groups: the plant and animal kingdoms. However, they may be further arbitrarily grouped on a taxonomic or morphological basis or, in the case of microorganisms, on some physiological differentiation using a variety of metabolic characteristics or oxygen requirements. Grouping according to the functions that organisms perform in the environment is of particular interest in forest soils because of its relevance to soil formation, properties, and management.

### Soil Microflora

Soils contain four major groups of microflora: bacteria, actinomycetes, fungi, and algae. A fifth group consisting of very small (less than 1 $\mu$ in diameter) protobacteria is sometimes included. However, because their function in soil is not well understood, this group will be omitted from further discussion.

Soil microorganisms are conveniently divided into two broad classes with respect to their energy and carbon sources: *heterotrophic* forms, which require preformed organic nutrients to serve as sources of energy and carbon, and *autotrophic* forms, which obtain their energy from sunlight or by the oxidation of inorganic compounds and their carbon by the assimilation of $CO_2$ (Alexander, 1977). Most types of bacteria and all fungi, protozoa, and larger animals are heterotrophs. Only algae and a few types of bacteria share with higher plants the capacity to use sunlight as a source of energy.

**Bacteria.**   Although bacteria are small, rarely more than several micrometers in length, they are especially prominent in soils because of their great numbers. Bacteria and fungi dominate in well-aerated soils, but bacteria alone account for most biological and chemical changes in anaerobic environments. In fact, the ability to grow in the absence of $O_2$ is the basis for further grouping of bacteria into three distinct categories: *aerobes*, those that live only in the presence of $O_2$;

*anaerobes*, forms that grow only in the absence of the gas; and *facultative anaerobes*, those organisms that can develop either in the presence or absence of $O_2$. However, the more meaningful grouping for understanding soil processes is that based on energy and carbon source requirements.

*Autotrophic Bacteria.* Most soil organisms require complex organic compounds as sources of energy and carbon and are classed as heterotrophs, while only a small group of microorganisms use $CO_2$ as their sole source of carbon. These autotrophs are of two general types: *photoautotrophs* whose energy is derived from sunlight, and *chemoautotrophs* whose energy comes from the oxidation of inorganic materials. The chemoautotrophs are limited to a relatively few bacterial species, but their importance to soils greatly exceeds their numbers. The chemoautotrophic bacteria are grouped on the basis of the element whose oxidation provides energy for their growth.

**1.** Bacteria using *nitrogen* compounds as energy sources include those that oxidize ammonium to nitrite *(Nitrosomonas* and *Nitrosococcus)* and bacteria that oxidize nitrite to nitrate *(Nitrobacter)*. The energy-yielding reactions involving these organisms are the following:

$$2NH_4^+ + 3O_2 \rightarrow 2NO_2^- + 4H^+ + 2H_2O \quad (Nitrosomonas)$$
$$2NO_2^- + O_2 \rightarrow 2NO_3^- \qquad (Nitrobacter)$$

The numbers of these organisms in forest soils are often quite low and the mineralization of nitrogen proceeds at a very slow pace under most acid conditions. The population of these bacteria rapidly expands when acidity is corrected and a source of nitrogen is available. Apart from the influence of soil acidity, the composition of organic debris affects the rate of nitrification. Because litter layers generally have wide carbon-to-nitrogen ratios, a large part of the nitrogen compounds mineralized during decomposition are used by the nitrifying organisms in maintaining their own population. This disappearance of inorganic nitrogen following additions of nitrogen-poor litter is termed nitrogen immobilization. Only after the carbon content of certain materials has been reduced to below a ratio of about 20:1 does any given significant amount of the nitrogen become available to higher plants (Waksman, 1952). As materials with such ratios are uncommon in acid forest soils, organic compounds and ammonium are apparently principal sources of soil nitrogen available to trees (McFee and Stone, 1968). A low level of nitrates in a forest soil at any given time does not necessarily mean an absence of nitrification in that soil, because nitrates can be lost by leaching, plant uptake, immobilization, and denitrification.

Nitrification has long been assumed to be solely a chemoautotrophic process, but Alexander (1977) has provided evidence of heterotrophic nitrification, at least in vitro. A large number of heterotrophic bacteria and actinomycetes are able to generate traces of nitrite from ammonium salts, and several fungi are capable of oxidizing small amounts of nitrite to nitrate. The inefficiency of the nitrifying heterotrophs may be compensated by their large numbers, and they may exert significant influence on the rate of nitrate synthesis under some conditions. This may be particularly true in acid forest soils that often contain few autotrophs.

2. Of the *mineral oxidizers*, bacteria involved in the oxidation of inorganic *sulfur* are perhaps the most important in forest soils. Sulfur exists as sulfide in several primary minerals, and it is added to forest soils as plant and animal residues and in rainwater. Elemental sulfur is sometimes added to nursery soils to increase the acidity for the control of certain plant pathogens. The major part of the sulfur in soils is in organic combinations and, like nitrogen, it must be mineralized to be useful to trees. Sulfur is largely absorbed by tree roots as sulfate. While the initial decomposition of these organic materials and conversion to inorganic sulfur compounds is accomplished by heterotrophic organisms, the oxidation of sulfides and elemental sulfur to sulfates can be by both heterotrophs and chemoautotrophs. Bacteria of the genus *Thiobacillus* are the principal autotrophic oxidizers in well aerated soils. This genus contains eight species with widely varying habitat requirements, but the acid-loving aerobe, *T. thiooxidans* is probably the most prevalent in forest soils. The reaction involved in sulfur oxidation by this organism is generally expressed as follows (Alexander, 1977):

$$2S + 3O_2 + 2H_2O \rightarrow 2H_2SO_4 \ (T. \ thiooxidans)$$

The oxidation of elemental sulfur can result in the mobilization of some slowly soluble soil minerals as the result of the sulfuric acid formed. The solubility of phosphorus, potassium, calcium, and several micronutrients may be increased as a result of the acidification resulting from this reaction.

Oxidation of inorganic sulfur compounds can also be accomplished by heterotrophic bacteria, actinomycetes, and fungi under some conditions, and, in the case of *Thiobacillus denitrificans*, oxidation can take place anaerobically. This chemoautotroph can use nitrate as a terminal electron acceptor and convert nitrate to gaseous nitrogen while oxidizing sulfur compounds. However, under some anaerobic conditions in water-saturated soils, inorganic sulfur compounds are reduced to sulfides instead of being oxidized. The anaerobic *Desulfovibrio desulfuricans* is largely responsible for the reduction of sulfates, but it has a narrow acidity range. Because it is usually limited to a pH of 6.0 and above, the formation of sulfides in acid forest soils is not common, even in wet areas. However, hydrogen sulfide may accumulate to toxic levels for tree roots in stagnant marshes where soil pH is near neutral. *Desulfovibrio* bacteria use sulfates as electron acceptors, but since they use carbohydrates as electron donors (energy sources) they are not considered autotrophs.

3. A number of types of bacteria are involved in the transformations of *iron* and *manganese* and other heavy metal compounds. The most important chemo-

autotroph for iron oxidation is *Thiobacillus ferrooxidans*. As the genus name indicates, the bacterium may also derive energy by oxidizing inorganic sulfur, when such compounds are present. Because iron is readily available to higher plants only in the reduced state, the action of this organism can result in iron deficiency in some well-aerated soils. It is in the ferrous state that iron is leached in soils, and when oxidized or complexed with organic molecules in a lower horizon a kind of iron pan may form in soils. (The reduction of ferric iron by such aerobes and facultative anaerobes as *Bacillus, Clostridum,* and *Pseudomonas* is heterotrophic, but it is conventionally mentioned with other transformations of metals).

*Heterotrophic Organisms.*   These microorganisms, which require preformed organic compounds as sources of energy and carbon, comprise the largest group of soil bacteria. The diverse group of heterotrophic bacteria includes free-living and symbiotic nitrogen-fixers and bacteria that decompose fats, proteins, cellulose, and other carbohydrates. They include both aerobic and anaerobic forms.

   **1.** *Biological nitrogen ($N_2$) fixation* is accomplished by free-living bacteria (or blue-green algae) and by symbiotic associations composed of a microorganism and a higher plant. It is primarily through the action of these organisms that part of the huge reservoir of atmospheric nitrogen is rendered available to higher plants.

   The free-living bacteria capable of utilizing $N_2$ are primarily the aerobic species of *Azomonas, Azotobacter, Beijerinckia, Spirillum* and anaerobic species of *Clostridium* and *Desulfovibrio*, although strains of several other genera are capable of the transformation under some conditions. Most are apparently not obligate, since they can obtain nitrogen from organic and inorganic nitrogenous compounds, as well as from the atmosphere. In other words, they can maintain themselves over a wide range of soil conditions, but the conditions under which they fix atmospheric nitrogen are limited. For example, *Azotobacter* is an obligate aerobe, prefers temperatures around 30°C, and generally fails to fix $N_2$ below pH 6.0. *Beijerinckia* spp. are also aerobic; they grow well in acid conditions (perhaps as low as pH 3.0), but they are apparently confined largely to tropical soils. The dominant anaerobes are of the genus *Clostridium*. They are like the blue-green algae in that they are most numerous in wet and flooded soils and they are active over a pH range of 5 to 9. The wide carbon-to-nitrogen ratios of forest floor materials would appear to favor $N_2$ fixation by nonsymbiotic organisms, but this type of fixation is not believed to be of much practical significance in forest soils. The amount of nitrogen fixed annually probably averages less than 1 or 2 kg per hectare.

   *Symbiotic* $N_2$ fixation can be of considerable importance in many forest soils. The classical example of such a symbiosis is that between leguminous plants and bacteria of the genus *Rhizobium* in the nodules on legume roots. This is a true symbiosis but *Rhizobium* may also be free-living $N_2$ fixers. There are more than 10,000 species of legumes, many of which grow wild in forests. Furthermore, legumes are sometimes planted in young forests as a source of nitrogen under special silvicultural conditions (Rehfuess and Schmidt, 1971). Lupines have been planted in conjunction with grasses and pine to stabilize sand dunes in New Zealand (Gadgil, 1972). Planted legumes may have a place in intensively managed young plantations on infertile soils such as glacial outwash and coastal sands of many parts of the world. Some reseeding legumes may be maintained for 5 to 7 years and fix 50 to 200 kg per hectare of nitrogen in plantations where soil phosphorus and potassium are not deficient, or in fertilized areas. Since forest soils are often very acid, tolerant species must be selected for planting in unamended sites. The expense usually involved in establishing an effective stand of legumes in forests has probably been the greatest deterrent to their use.

Several genera of nonleguminous plants (such as *Alnus, Myrica, Hippophae, Elaeagnus, Shepherdia, Casuarina, Coriaria,* and *Ceanothus*) possess root nodules and are capable of $N_2$ fixation. In addition to certain angiosperms, a few gymnosperms such as *Podocarpus* and *Cycas* possess nodulelike structures, but proof of $N_2$ fixation is confirmed only for the latter. *Cycas* can apparently affect $N_2$ fixation in association with blue-green algae under some conditions. These plants, mostly trees and shrubs, are more abundant than legumes in many forests and may have considerable ecological significance on some sites. Annual fixation of $N_2$ varies widely among these eight genera—from a few kilograms by *Myrica gale* to more than 100 kg per hectare by pure stands of alder (Youngberg and Wollum, 1970).

**2.** *Organc matter decomposing* bacteria play a major role in the degradation of vast amounts of forest litter, plant roots, animal tissue, excretory products, and cells of other microorganisms. These materials are both physically and chemically heterogeneous, and include such constituents as cellulose, hemicellulose, lignin, starch, waxes, fats, oils, resins, and proteins. With such a diversity of organic materials, it is not surprising that a complex population of heterotrophic bacteria, as well as fungi and actinomycetes, are involved. They include both aerobic and anaerobic forms, and the mechanisms of decomposition vary depending on environmental conditions and participating organisms. In all cases, the organic matter provides the microflora with energy for growth and carbon for cell formation. The end products are carbon dioxide, methane, organic acids, and alcohols, in addition to bacterial cells and resistant materials.

The aerobic cellulose-decomposing bacteria are intolerant of poor aeration and soil acidity. Since their activity ceases below about pH 5.5, they are found chiefly in hardwood mull humus and that developed under mixed pine-hardwood stands (Waksman, 1952). The anaerobic bacteria tolerate strongly acid and poorly drained soil, but organic matter breakdown by these organisms is consistently slower than under aerobic conditions. Consequently, organic matter accumulates in many poorly drained soils. Although the subsidence of organic soils after drainage may be due primarily to shrinkage in cold climates, subsidence of drained peat lands results mostly from biological oxidation in warm climates.

The rate of decomposition of plant materials depends largely on the nitrogen content, with protein-rich substrates metabolized most readily. Aside from the high acidity of many forest soils, forest litter tends to be slowly decomposed because of its wide carbon-to-nitrogen ratio, resulting in litter accumulations on the forest floor. Because extra nitrogen (and other elements) is needed for an expanding microbial population whenever carbonaceous materials are added to the soil, the addition of materials with a wide C:N ratio may result in temporary nitrogen deficiency. For example, raw organic materials such as sawdust, applied to forest nurseries, may result in a deficiency of nitrogen for the seedlings unless extra nitrogen fertilizer is applied to replace that immobilized in the tissue of cellulose-decomposing bacteria.

3. In some wet soils, particularly where the pH is near neutral, the activity of *denitrifying* organisms may exceed that of the nitrifying bacteria. Under anaerobic conditions certain bacteria derive their oxygen supply from the oxides of nitrogen (anaerobic respiration), reducing nitrates to nitrite and then to nitrous oxide or elemental nitrogen. True denitrification is largely limited to the genera *Pseudomonas, Achromobacter, Bacillus,* and *Micrococcus.* Because these organisms are facultative anaerobes they can survive under a wide range of soil conditions and the presence of a large population gives most soils a large denitrifying potential. However, conditions must be favorable for the organisms to change from aerobic respiration to a denitrifying type of metabolism. This would normally occur in the presence of nitrates and a source of readily available carbohydrates when the demand for $O_2$ by the microflora exceeds the supply. Anaerobic conditions that favor denitrification are found in flooded soils, and even in microsites of drained soils, but very little volatile loss of nitrogen would normally be expected in acid forest soils because of the scarcity of nitrates and a source of readily available carbohydrates when the demand for $O_2$ by the microflora exceeds the supply. Anaerobic conditions that favor denitrification are found in flooded soils, and even in microsites of drained soils, but very little volatile loss of nitrogen would normally be expected in acid forest soils because of the scarcity of nitrates in such soils, unless they had been fertilized.

**Actinomycetes.** This group of heterotrophic organisms is morphologically transitional between the simple bacteria and the filamentous fungi. They are unicellular microorganisms that produce a slender, branched mycelium that may undergo fragmentation or may subdivide to form asexual spores. Numerically the actinomycetes are second only to bacteria in most soils. They are not only taxonomically similar to bacteria, but many of their environmental requirements are also similar. They are typically aerobic organisms found less commonly in wet than in dry areas. Peats, waterlogged areas, and soils whose pH is less than about 5.0 are unfavorable habitats. They are more numerous in warm climates than in cooler areas. Because of these environmental influences, actinomycetes are not believed to be as important in cellulose decomposition in forest soils as in prairie or pasture soils.

1. The activities of actinomycetes in soil transformations are not well understood, but in many respects these activities are similar to those of fungi. They are active in the *decomposition* of cellulose and a range of other organic materials. They are not good competitors and appear to play their role in decomposing resistant components of plant and animal tissue. Species of the genus *Streptomyces* are also capable of chitin hydrolysis, while *Nocardia* species metabolize paraffins, phenols, steroids, and pyrimidines. The end products of their activity are complex molecules assumed to be important in the humus fraction of mineral soils (Alexander, 1977). Their absence in strongly acid soils may account for the development of layers which are highly resistant to decomposition and high in lignin content in forest humus (Wilde, 1958).

2. It has been reported that certain actinomycetes are involved in symbiotic $N_2$ fixation. Torrey (1978) speculated that the actinomycete-induced nodulation of alders, bogplants like sweet gale *(Myrica gale),* and roadside and disturbed area invaders such as sweet fern *(Comptonia),* bayberry *(M. pensylvanicum),* and various species of *Ceanothus* comprises one of the largest sources of biological fixation of atmospheric nitrogen, at least comparable to the legume symbioses. The endophyte associated with nodules on some nonleguminous angiosperms, such as *Alnus* and *Ceanothus* species are believed to be *Streptomycetes* (Youngberg and Wollum, 1970).

3. Certain actinomycetes are capable of synthesizing *antibiotics*, but the significance of such compounds under field conditions is not clear. The ability to excrete antibiotics or their capacity to produce enzymes that are responsible for lysis (killing) of bacteria and fungi may play an important role in microbial antagonism and regulating the composition of the soil community. Actinomycetes also cause certain soil-borne diseases of plants, such as potato scab.

**Fungi.** The microbial biomass with the decomposing litter of forest soils is predominantly fungal, and fungi are probably the major agent of decay in all acid environments. The large variety and numbers of mushrooms (fruiting bodies) seen in the forest during wet periods attests to the wide distribution of fungi. Fungi possess a filamentous network of hyphal strands in the soil. The mycelium may be subdivided into individual cells by cross walls, or septa, but many species are nonseptate. Fungal mycelia permeate the F and H layers of the forest floor, and are readily seen in mor and moder humus types. Taxonomically, most soil fungi are placed in one of two broad classes: Hyphomycetes and Zygomycetes. Species of Hyphomycetes produce spores only asexually, the mycelium is septate, and the conidial type of asexual spores is borne on special structures known as conidiosphores. Zygomycetes and other fungi, on the contrary, produce spores both sexually and asexually. Alexander (1977) lists six other classes of fungi found less frequently in soils. However, fungi are so diverse it is difficult to classify them on the basis of morphology or on their souce of carbon, since the dominant soil genera can utilize a variety of carbonaceous substrates. They can be more rationally divided into functional groups important in forest soils:

1. Decomposition of cellulose and related compounds is one of the most important activities of fungi. They are active in the early stages of aerobic decomposition of wood and other organic debris on the forest floor. These materials include hemicelluloses, pectins, starch, fats, and the lignin compounds particularly resistant to bacterial attack. By the degradation of plant and animal remains the fungi participate in the formation of humus from raw residues and aid in nutrient cycling and aggregate stabilization.

2. Proteinaceous materials are utilized for both nitrogen and carbon by fungi; as a consequence of proteolysis they are sources of ammonium and simple nitrogen compounds in the soil. Under some conditions, fungi compete with higher plants for nitrate and ammonium, which leads to a reduction in available nitrogen for the higher plants.

3. Some fungi are predators on such soil fauna as protozoa, nematodes, and certain rhizopods and may, thereby, contribute to the microbiological balance in soil.

4. Pathogenicity is another attribute of some soil fungi. There are both obligate and facultative parasites among this group. Members of the genera *Rhizoctonia, Pythium,* and *Phytophthora* cause damping-off disease among nursery seedlings. *Fusarium* species may cause root rot in the nursery and in older plants. Roots of older trees may also be invaded by representatives of the genera *Fomes, Armillaria, Verticillium, Phymatotrichum,* and *Endoconidiophora,* among others.

5. Fungi form symbiotic associations called mycorrhizae or "fungus root" with roots of higher plants. The adaptation to root tissue may be associated with the complex nutrient demand of the microorganism, and many of them have been cultivated in artificial media. The mycorrhizal fungi are very important to the nutrition and growth of trees, and because of their critical relationship with higher plants they will be discussed in a separate chapter.

Mycelia of fungi may, in some way, be responsible for the development of a hydrophobic property in some forest soils. These soils, mostly sands, are slow to wet once they become air-dry and as a consequence their capacity for water retention is greatly impaired. The mechanism of this water repellancy is not well understood.

**Algae.** Soil algae are commonly unicellular but may occur also in short filaments or colonies. Taxonomically they are divided into the green, blue-green, yellow-green, and rod-shaped diatoms. Typically they possess chlorophyll, which enables them to use light as an energy source for fixation of carbon dioxide (photosynthesis), thus giving them a photoautotrophic nutrition. Algae are most commonly found in fertile soils, well supplied with bases, available nitrogen and phosphorus, and they tend to be sparse in infertile, acid sands. In temperate regions, green algae and diatoms may be most common, while blue-green algae are most common in tropical areas.

Algae aid in solubilization of soil minerals and thus hasten the process of weathering. They generate organic matter from inorganic substances and increase the humus content in soils. Strains of blue-green algae can assimilate atmospheric nitrogen, thus adding to the nitrogen supply of some soils. These organisms are particularly active in wet and flooded soils and in surface soils whose alkalinity has been increased following burning. Since they do not depend on organic matter as an energy source, they are early colonizers in barren or sandy areas, preparing the way for later invasion by higher plants.

Lichens are the result of a symbiotic association between fungi and algae. They usually form crustlike colonies and are often pioneer life forms on freshly exposed mineral soil, providing organic matter for succeeding higher forms of plant life. They also colonize trees, as illustrated by the encrusted branches of spruce in Norway (Figure 5-1). Green algae are apparently the most common symbiont in lichen formation, but blue-green algae may be predominant in some temperate climate forests. Denison (1973) concluded that the lichen, *Lobaria oregona*, contributed from 2 to 10 kg nitrogen per hectare annually in Douglas-fir forests of Oregon.

## Soil Animals

Essentially all fauna that inhabit the forest environment influence soil properties in some way which eventually affects tree growth. They range in size from wild beasts (and sometimes domestic animals) to simple one-celled protozoa. Their importance to the soil, however, is essentially inversely proportional to their size. Except on overgrazed forest rangeland, where a reduction in groundcover vegetation or alteration in species composition, plus soil compaction and reduced water infiltration, result in soil erosion, the effects of large animals on forest soils are minimal.

**Vertebrates.** Vertebrates, largely consisting of four-legged animals, influence the soil through fertilization, trampling, scarification, and a form of cultivation.

**Figure 5.1**   Lichen enshrouded spruce limb in cool, damp region of Norway.

Many such animals burrow in the soil and aid in the breakdown of its organic matter. They assist in the mixing of this organic material with the inorganic surface soil and with the transport of the former into the soil profile. Decomposition of organic debris is hastened as a result of its use as a food by animals and the mixing with soil mineral materials. Woodchucks, moles, gophers, mice, shrews, and ground squirrels are particularly important in soil development. Moles are especially active in European forests supporting a mull humus layer, and it has been suggested (Lutz and Chandler, 1946) that the names *mole* and *mull* are of common origin. While large animals may trample and compact the soil surface, penetration of water and air into the soil is greatly facilitated by the actions of the smaller animals. These small quadrupeds are chiefly carnivorous, devouring worms, larvae, and insects. However, a few rodents are destructive of seedlings and young trees, and their beneficial actions as soil cultivators may be overshadowed by this destructive behavior.

Mice, shrews, and other small animals are abundant and have a pronounced influence on microrelief and other properties of forest soils in the northeastern United States and other temperate forests. Their labyrinth of interconnecting tunnels allow for ready penetration of air and water; while the nests, dung, stored food, and dead animals all enhance organic matter content and fertility of the soil (Troedsson and Lyford, 1973). It was suggested that the absence of an A2 horizon in some Eutrochrept (Brown Podzolic) soils of that area may be largely due to the mixing by fauna of materials in the surface horizons (Lyford, 1963). Burrowing rodents may perform functions in dry areas similar to those of the earthworm in humid forests. Rodent tunnels are also found in the longleaf pine forests of the southern United States that have been protected from fire. For example, pocket gophers (*Geomys* sp.) accomplish a large amount of soil cultivation in the excessively drained stands of that region.

**Arthropods.** Arthropods comprise a broad group of soil animals with articulated bodies and limbs that populate the forest floor. These beetles, ants, centipedes, millipedes, springtails, sow bugs, spiders, ticks, and mites are particularly important in forest litter decomposition. The primary consumers chew and move plant parts on the surface and into the soil. Some, such as sow bugs, are active feeders on dead leaves and wood and may be important in the disintegration of freshly fallen leaves. Another Crustacean, the crayfish, may aid in mixing, aeration, and drainage of some wet soils of the coastal flatlands.

Saprophagous mites constitute one of the most important groups of *Arachnida*, and by virtue of their great numbers, often exceeding several thousands per square foot, they play a major role in producing a crumblike structure of some surface organic layers. They feed on decaying leaves, wood, fungal hyphae, and feces of other animals.

Centipedes are predatory animals feeding on other members of the soil fauna, and play only a minor part in soil formation. Millipedes, on the other hand, are largely saprophagous, feeding on dead organic matter. They are generally confined to soils with mull humus layers, especially those supporting stands of deciduous trees. Lyford (1943) noted that leaves of certain trees, particularly leaves of high calcium content, were favored by millipedes. They are considered important in the formation of mull humus layers, though possibly not as important as earthworms.

The springtails and bristletails are prevalent small, wingless saprophagous insects that feed on decaying materials of the forest floor. Adults and larvae of many beetles and flies contribute to the breakdown of organic debris and improve the structure of surface soil. Ants and termites are important because of their tunneling and transporting activities. They are generally considered to be more numerous in tropical and warm climates than in temperate or cool climates. However, Lyford (1963) reported that the return of subsoil material to the forest floor in small litter-concealed mounds by ants is an important process in the development of some Spodosols (Brown Podzolic soils) of New England, and possibly in other temperate zone soils. He considered that the mineral material of the entire A horizon of some, if not most, of the forest soils of that region consists of material returned by ants to the surface from the B horizon over a period of many years. Fine-textured materials returned to the surface of coarse textured soils provides increased cation exchange capacity and available moisture, and may be important in burial of seeds, roots, and charcoal.

Fires generally reduce the numbers of arthropods in the forest floor, but the reduction is largely temporary in nature and not all genera are reduced equally by such burns. Carabids were the most numerous arthropods in protected forests of northern Idaho, but *Acarina, Chilopoda, Thysanoptera, Protura,* and *Thysanura* were most numerous in recent prescribed burns (Fellin and Kennedy, 1972).

**Worms.**   Worms that inhabit soil can be grouped as (1) segmented earthworms or as (2) nonsegmented round worms, or nematodes. The former group is by far the most important in soil formation. In fact, the ordinary earthworm is probably the most important component of soil macrofauna. Although there are a number of species of earthworms, all large and middle-sized worms are often lumped into the Lumbricidae, while the smaller, light colored "potworms" are considered Enchytraeidae (Wilde, 1958).

The large, reddish *Lumbricus terrestris* occurs widely in North America, but it is probably a species introduced from Europe. A high population of earthworms is generally associated with a mull humus formation, and this is particularly true of *L. terrestris*, which makes up as much as 80 percent of the total soil fauna weight. They feed on fallen leaves and organic debris and pass it, together with fine mineral particles, through their bodies. Each year, earthworms may pass as much as 30 tons per hectare of soil material through their bodies where it is subjected to digestive enzymes and to a grinding action within the animals. The casts are higher in total and nitrate nitrogen, available phosphorus, potassium, calcium, and magnesium, pH, and cation exchange capacity than is the soil proper. Earthworms serve to mix bits of organic materials into the mineral soils, and they promote good soil structure and aeration through their burrowing action. As a result of this transporting and mixing action, the upper layer of certain forest floors takes on a crumbly structure of "earthworm mull." The concentration of earthworms in forest soils has been estimated to be from one-half million to more than two and a half million per hectare—the actual numbers depending on several climatic and soil factors. For example, highly acid soils support fewer earthworms than less acid soils. The optimum range appears to be from about pH 6.0 to 8.0. Sandy soils and soils that dry excessively are not favorable habitats for earthworms. *Lumbricus rubellus* and *L. festivus* appear to be more acid tolerant than *L. terrestris* and may be more common than the latter in coniferous forests and in mor humus.

About the same size but somewhat lighter in color are the *Allobophora* spp. They are active in forest soils of Europe and North America and are particularly important in the development of hardwood mull humus.

The smaller worms, such as *Octolaseum* spp. and *Dendrobaena* spp., also devour organic debris and thus improve the physical and chemical properties of the surface soil. Because of their smaller size, generally no more than a few centimeters long, they are not usually considered as important as the other worms in forest soil formation. However, since potworms have less stringent environmental requirements than their larger relatives, they may be about as numerous in mor as in mull humus layers.

Nematodes, the near microscopic-sized nonsegmented roundworms, are fairly widely distributed in forest soils. They include both saprophytic and parasitic groups. The saprophytes are free living and generally beneficial to the process of humification. Some of the predatory species may also be beneficial by virtue of their appetite for certain bacteria, protozoa, fungi, small arthropods, and other nematodes. However, some species attack the roots of trees and may inflict considerable damage. In a survey of three-year-old slash pine plantations in the lower coastal plains of the United States, plant parasitic nematodes were found in all 34 sampled sites. Spiral *(Helicotylenchus)* and ring *(Criconemoides)* nematodes were found most frequently, although 11 other plant parasitic genera were also identified.

The actual damage to tree vigor and growth is not known, but several of the genera found in these plantations have been reported to cause damage to pine. In addition to the two genera mentioned above, others included sheath *(Hemicycliophora)*, lance *(Hoplolaimus)*, stunt *(Tylenchorhynchus)*, and dagger *(Xiphinema)* nematodes. Rather spectacular increases in growth have been reported for seedlings planted in some fumigated soils. This response is presumed to have resulted from a reduction in the population of nematodes that attack pine roots, and tree responses have been particularly notable on dry sandy sites. However, neither the actual causes of the increased growth nor the longevity of the effect from fumigation are well understood. The total concentration of nematodes in the soil may return to normal in a year or two following fumigation, but the spectrum of genera may be altered for longer periods. Although the benefits derived from fumigation of nursery soils are obvious, it does not appear to be an economic practice for forest plantations at the present time.

**Protozoa.**    Protozoa are the most abundant of soil fauna, and Waksman (1952) reported that their number varied from 1500 to 10,000 per gram of forest soil. These one-celled organisms may exist in either an active or a cyst stage, but they are generally aerobic and occur in the upper horizons. Their diet consists largely of decomposing organic materials and bacteria. Soil conditions that favor their development are similar to those that favor bacteria. They are found in soils supporting both hardwood and coniferous forests. In a strict sense, they are the only major group of soil fauna classed as microorganisms.

## Plants—the Rhizosphere as a Modifier of Microbial Activity
The roots of higher plants exert a profound influence on the development and activity of soil microorganisms. Roots grow and die in the soil and supply soil fauna and microflora with food and energy. More importantly, live roots create a

unique niche for soil microorganisms, resulting in a population distinctly different from the characteristic soil community (Alexander, 1977). This difference is due primarily to the liberation by the root of organic and inorganic substances that are readily consumed by the organisms in its vicinity. The rhizosphere effect has been demonstrated with a wide variety of forest trees and other plants (Katznelson, Rouatt, and Peterson, 1962). It is influenced by many factors, such as the kind and stage of development of the trees, the physical and chemical properties and moisture content of the soil, and such other environmental conditions as light and temperature. These factors may act directly on the soil microflora or indirectly by influencing plant growth.

The most important influence of the growing plant on the rhizosphere flora results from the root excretion products and sloughed-off tissue that serve as sources of energy, carbon, nitrogen, or growth factors. The plant root absorbs inorganic nutrients from the rhizosphere, thereby lowering the concentration available for both plant and microbial development. On the other hand, root respiration may increase rhizosphere acidity and hasten the solubilization of less soluble inorganic compounds. By this means, the amount of available phosphorus, potassium, magnesium, and calcium may increase. Because the microflora are strong competitors for these nutrients, there can be a temporary deficiency due to immobilization in spite of increased solubility.

There is a high level of ammonifying bacteria in the rhizosphere of many plants, apparently stimulated by the presence of organic nitrogen compounds. Although they contribute to increased mineralization in the rhizosphere, net mineralization (and plant available nitrogen) may be low due to the large amount of immobilization by microorganisms themselves. Fisher and Stone (1969) suggested that, in secondary plant succession, conifer rhizospheres mineralized or otherwise extracted some fraction of the soil nitrogen and phosphorus that had been resistant to microbial action under the previous vegetation (Figure 5.2).

There is good evidence that $N_2$ fixation by some free-living bacteria is greater in the rhizosphere of nonleguminous plant roots than in adjacent soil (Richards and Voigt, 1965). The identity of the excretions that stimulate strains of *Azotobacter, Beijerinckia,* and *Sprillum (Azospirillum)* to grow in some rhizospheres, but not in others, is not resolved. Coinciding with the increase in $N_2$ in the rhizosphere is a greater density of denitrifying organisms, but since the population of autotrophic nitrifiers is not stimulated by root exudates, the level of nitrates in the rhizosphere may not be sufficiently large to result in significant denitrification.

Root excretions have an effect on the germination of the resting structures of several fungi, perhaps as a result of the energy sources in the rhizosphere. The stimulation can be deleterious to the host plant if the fungus is pathogenic. There are allelopathic relationships between several tree species. In some instances this apparently results from root exudates of one species inhibiting the root development and growth of another species in the same soil mass. Allelopathy may result

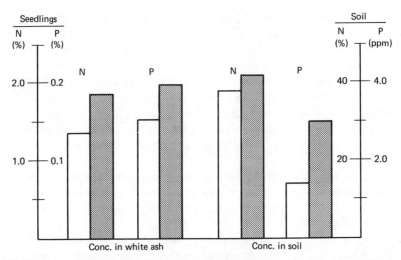

**Figure 5.2** Concentrations of nitrogen and phosphorus in white ash seedlings planted in open □ and at edge of conifer canopy ■; and of HF-extractable organic nitrogen (as percent of total nitrogen) and Morgan-extractable phosphorus in soil samples taken from above locations (average of four locations). From Fisher and Stone (1969). Adapted from *Soil Science Soc. of America Proceedings* 33:955–961 (1969) by permission of the Soil Science Society of America.

from rhizosphere microorganisms altering exudates to form materials that are toxin to some species (Tubbs, 1973). On the other hand, there is evidence that the rhizosphere flora provides protection to the root from some soil-borne pathogens. Some antibiotic-producing microorganisms are found in abundance in the rhizosphere, but it is not known what effect these organisms have on pathogens. It appears certain that exudates from some rhizosphere organisms form a kind of buffer zone protecting roots against the attack of soil pathogens. An example is protection afforded pine roots against *Phytophthora cinnamomi* by mycorrhizal fungi (Marx, 1977). Rhizosphere microflora also produce considerable amounts of growth substances, such as indoleacetic acid, gibberellins and cytokinins, that may influence growth of the host plant. Bowen and Rovira (1976) prepared an excellent review on microbial colonization of plant roots and the influence of root exudates on microbial activity.

Although much remains to be learned about the root-rhizosphere relationship, it is obvious that this relationship is a very close and significant one that is mutually beneficial.

## CONDITIONS INFLUENCING BIOLOGICAL ACTIVITY IN SOILS

Many factors have an effect on the density and composition of soil organisms—both animals and microflora. Among the most important factors are the supplies of oxygen and moisture, soil temperature, levels of inorganic nutrients, and the amount and nature of soil organic matter.

Soil animals are less affected by soil conditions than are the microflora. Animals often occupy the transition zone between the forest floor and the mineral soil and some of the larger animals spend a significant part of their life outside the confines of the soil. As a consequence, they may be more affected by such environmental factors as weather conditions, flooding, fire, and site disturbances, than by soil conditions. Adverse temperature and site perturbations may destroy many animals or drive the more mobile animals from their normal habitats.

Since soil animals include both primary consumers and predators, they obtain their energy by consuming plant parts or preying on other organisms. Soil acidity and level of inorganic nutrients indirectly affect animal numbers and activity as a consequence of the animal food supply. For example, soils rich in calcium and other bases produce plants with higher concentrations of these elements, which render them more palatable to such animals as the common earthworm. Animal populations in general are large in fertile soils capable of producing abundant food supplies, but soil chemical properties normally have less direct influence on animals than do physical properties.

Some fine textured and compacted soils discourage burrowing animals. High water tables and impervious layers may completely restrict their activity. Soil temperature and other physical factors significantly influence the animal populations, but not to the extent that they influence the relatively immobile microflora.

In addition to the primary environmental variables of moisture, aeration, temperature, organic matter, acidity, and inorganic nutrient supply, activity of soil microorganisms are influenced by season of the year and soil depth.

Moisture affects microbial activity because, as a component of protoplasm, it must be available for vegetative development. However, an overabundance of water restricts gaseous exchange, lowers the available oxygen supply, and creates anaerobic conditions. Aerobes, anaerobes, and facultative anaerobes may all function in the soil at the same time. Because microbial populations are sensitive to soil moisture conditions, the community size in a given soil varies with fluctuations in moisture; aerobes predominate in well aerated soils, but this changes to a largely anaerobic population under waterlogged condition. The maximum density of microorganisms, however, is usually found at 50 to 75 percent of the water-holding capacity of the soil (Alexander, 1977).

Temperature affects the activity of all soil organisms, but not all to the same extent. Each microorganism has an optimum temperature for growth and a range outside of which development ceases. Most soil organisms grow best in the 25 to 35°C range, but they can survive and develop at both higher and lower temperatures. Some organisms grow readily at temperatures up to 65°C and have temperature optima at 35 to 45°C. Temperatures affect population size and the rate of biochemical processes carried out by the microflora, up to the optimum

**Table 5.1**  Summer and Winter Microbial Populations in Two Swamp Soils in South Carolina (Priester and Harms, 1971)

| Microorganism | Summer | | Winter | |
|---|---|---|---|---|
| | Mucky Clay | Loam | Mucky Clay | Loam |
| | Millions per gram of oven-dry soil | | | |
| Aerobic bacteria | 77.8 | 67.5 | 26.1 | 20.8 |
| Anaerobic bacteria | 3.4 | 2.8 | 2.6 | 1.8 |
| Actinomycetes | 6.2 | 4.4 | 1.6 | 1.1 |
| Fungi | 0.6 | 0.6 | 0.6 | 0.5 |
| Total[a] | 88.0 | 75.3 | 30.9 | 24.2 |

[a]The differences among soils and between seasons are significant at the 1 percent level.

temperature for the transformation. Ordinary soil temperatures seldom kill bacteria.

The addition of carbonaceous materials directly affects the numbers and activities of all heterotrophic organisms and indirectly affects autotrophic organisms. The application of sawdust or turning under of a green manure crop in a nursery, or the incorporation of forest floor material during site preparation stimulates population numbers and activities, sometimes resulting in nutrient immobilization.

Highly acid conditions inhibit activities of many common bacteria, algae, and actinomycetes, but most fungi are able to function over a wider pH range. Consequently, the microbial population of acid forest soils is commonly dominated by fungi. This is not necessarily because fungi prefer acid conditions, but rather it is a consequence of the lack of microbiological competition for the available food supply.

Inorganic nutrients are required by soil microorganisms, but the addition of fertilizers may affect the activity of these organisms only as it stimulates plant growth. The exceptions are those instances when the supply of nutrients in the soil does not meet microbiological demands.

Season of year influences microbial activity as a secondary effect on temperature and moisture and on food supplies. Activities are usually greatest in spring and fall, with declines in numbers during hot, dry summers and cold winters. Numbers of organisms fluctuate closely with seasonal changes in temperature and moisture. The average numbers for four types of microorganisms in two swamp soils in South Carolina are shown in Table 5.1 for winter and summer.

Another secondary ecological variable that influences soil microorganisms is soil depth. The greatest concentration is in the top few centimeters of forest soils, with a rapid decline in numbers of most organisms with depth. The decline with depth is probably due to the decrease in organic matter and oxygen.

# 6
# CHEMICAL PROPERTIES OF FOREST SOILS

Chemical properties of forest soils have held only passing interest for forest managers until the last two or three decades. They have generally been considered to have less influence on tree growth than soil physical properties. This belief apparently resulted from the widely held, but erroneous, opinions that forest trees have a much lower annual nutrient requirement than most agricultural crops and that these requirements could be met by even the most impoverished forest soil without seriously impeding tree growth. Toumey and Korstian (1947) stated "the amounts of essential constituents in the soil and the amounts used by forest vegetation are such that even soils low in them contain an excess." These authors, not unlike other scientists of that period (Lutz and Chandler, 1946; Coile, 1952), subscribed to the widely held belief that soil physical properties completely overshadowed the rather minor contributions of soil chemical properties to tree growth. These assumptions were largely based on the observations that most forest trees possess deep, efficient root systems capable of exploiting large volumes of soil and that the nutrients obtained from soil depths are effectively cycled to promote long-term growth with a minimum drain on nutrient reserves.

Many of the early soil-site concepts were well founded. However, most ignored the highly important interrelationships of the physical, chemical, and biological site factors that influence tree growth. An example of this interrelationship is the poor tree growth on wet (cold) mineral and organic soils that results from induced nutrient deficiencies caused by restricted rooting or by slow nutrient cycling, rather than actual deficiencies in the soil. On the other hand, slow tree growth in highly weathered soils, quartz sands, and some alkaline soils often results from strictly chemical factors.

The importance of soil chemical properties to tree growth has gained considerable attention in recent years, largely because of greater demands placed on forest soils as a result of the increasing popularity of short rotations and intensively managed forests, seed orchards, and forest nurseries in many areas of the world. The development of more reliable soil and tissue tests as diagnostic tools in forest fertilization programs and a general proliferation of knowledge of soil chemistry have also contributed to the increased interest in the subject.

This chapter is not intended to be an exhaustive treatment of the complex and voluminous subject of soil chemistry; this information can best be obtained from basic courses in soil science. Instead, this material reviews some of the basic principles of soil chemistry, particularly as they pertain to the chemistry of forest soils, and demonstrates how these principles can be applied to problems in forest soil management.

## SOIL ACIDITY

### Soil pH Measurements

The most commonly used method of expressing soil acidity is pH. It was originally conceived as the logarithm of the reciprocal of the hydrogen (H) ion concentration (and later as its activity):

$$pH = \log \frac{1}{A_{H^+}}$$

where $A_{H^+}$ is the hydrogen ion activity in moles per liter. Therefore the greater the hydrogen ion activity (the more acid the solution), the smaller the value on the pH scale. A solution containing 0.0001 g $H^+$ ions per liter is pH 4.0, while that of pure water, or a neutral solution containing 0.0000001 g $H^+$ ions per liter, is pH 7.0. The range of most forest soils is from about pH 3.5 to pH 6.5. Values below pH 3.5 generally exist only when free acids are present.

The most accurate method of measuring soil acidity is with a pH meter. In this electronic method the hydrogen concentration of the soil solution is balanced against a standard hydrogen electrode.

Soil acidity is usually measured in a slurry of soil and water. In routine procedures, one part of soil is mixed with two parts of water and the electrodes are immersed in the stirred suspension. Only active acidity is measured in such a system, but it should be pointed out that a reserve of exchangeable (or potential) acidity also exists in soils. The sum of the active and potential acidity is the total acidity of a soil. An estimate of the potential acidity can be obtained by making the pH measurement using a salt solution instead of water. The pH reading is generally lower when made in a salt solution and it goes down as the salt concentration increases because of the greater displacement of hydrogen and aluminum ions from the exchange site. In extremely weathered soils, such as

Oxisols where the pH reading is below the zero point of charge, the addition of salt to a solution will increase the reading. Exchangeable acidity may be of little consequence in well-limed and fertilized agricultural soils, but most forest soils have a substantial reserve of potential acidity. The measurement of soil acidity in a solution with a neutral salt, such as potassium chloride, is particularly useful for estimating the effects of fertilizer use on soil reaction.

Coleman and Thomas (1967) pointed out that soluble acids that result from biological activity may produce significant fluctuations in soil acidity. They mention three circumstances that could lead to acid formation and possible leaching of calcium and magnesium and subsequent development of soil acidity.

1. The first situation, which has been of little concern to foresters in the past, is the accumulation of high concentrations of nitric and sulfuric acids after heavy applications of ammonium fertilizers to slightly buffered soils. Furthermore, monocalcium phosphate added as a fertilizer hydrolyzes in the soil solution to produce dicalcium phosphate and phosphoric acid. The latter diffuses away and produces a zone around the fertilizer particle where the acidity can be as low as pH 1.5.

2. When soils containing ferrous sulfide are drained and exposed to oxidizing conditions, high concentrations of sulfuric acids may be produced. Examples are the coastal flats and marine floodplains of the temperate and tropical zones that often develop "cat clays" when these sulfide sediments are drained. Most cat clays are formed in organic soils containing sufficient amounts of mineral material to yield large quantities of soluble aluminum, manganese, and iron upon dissolving in the sulfuric acid. While draining of wet soils usually increases soil acidity, resubmergence of the soils lowers the acidity (raises the pH) due to the precipitation of aluminum hydroxide, the reduction of ferric iron, and the absorption of ferrous iron by the clay.

3. Organic acids, produced from the decomposition of forest litter, are important weathering agents and producers of soil acidity. Some decomposition products of organic matter may act as chelating agents that facilitate the hydrolysis of aluminum:

$$[Al \cdot 6H_2O]^{+3} + H_2O \rightleftharpoons [Al(OH) \cdot 5H_2O]^{+2} + H_3O^+$$

The acidity of most forest soils changes only slightly with seasonal changes. Some seasonal changes have been observed with the highest pH generally found in mineral soils during the winter period and the lowest in the summer (Nehring, 1934). The degree of variation is influenced by the magnitude of the seasonal changes, but changes are seldom more than 1.0 pH unit. The pH of the forest floor may be highest in the fall, particularly in deciduous forests, because of the release of bases from the freshly fallen leaves.

Considerable differences in acidity are often found among horizons of the same soil. The H layer of the forest floor and Al horizon (where it exists) of

Spodosols are often very acid with some decrease in acidity with depth. Soils derived from basic materials are often more acid in the surface layers because these layers are subjected to more leaching than those at greater depth. In areas of moderate rainfall a concentration of bases may occur at the surface of some acid soils as a result of nutrient cycling, primarily through root absorption of bases from lower horizons and their return to the soil surface through litter fall. Forest fires result in the surface deposition of nonvolatile remains of the destroyed organic matter. The ash deposit decreases the acidity of the soil surface, but this effect may be short lived in sandy soils due to the leaching of bases from the ash into the soil profile. Some variations in soil acidity and other chemical properties by sampling date are shown for two central Washington soils in Table 6-1.

## Vegetation Effects

As previously mentioned, most forest soils are moderately to extremely acid as a result of the release of organic acids during the decomposition of the litter layer and the subsequent leaching of bases from the surface mineral soil. As a consequence, the types of vegetation growing on a soil are likely to have a marked influence on soil acidity because of inherent differences in base content of their litter. Soils supporting conifers tend to be more acid than those support-ing hardwood species, partly because conifer leaves and litter have a lower base content. This association is not always evident nor is the cause and effect relationship always clear. Because of species differences in tolerances to soil acidity, soil conditions may influence the makeup of plant communities more than the communities influence soil reaction. For example, many hardwoods such as *Platanus occidentials, Liriodendron tulipifera,* and some *Quercus* spe-cies appear to have an optimum range near neutrality, while other hardwoods grow best under moderately acid conditions. On the other hand, most *Tsuga, Picea, Abies,* and *Pinus* species grow best in quite acid soils and, as a conse-quence, their litter is acid.

With few exceptions, forest species are well adapted to acid soil conditions and, in fact, grow best in fairly acid media. Soil reaction, nevertheless, may dictate the distributions of more acid sensitive plants. Species such as *Sabal palmetto* and *Juniperus silicicola* are sometimes used as indicator plants, because they are often found on soils containing limestone at shallow depths. However, the distribution of such indicator plants is not always a reliable guide to soil acidity because of the modifying effects of climate and soil nutrient or moisture supplies on tree growth. Furthermore, some of the more acid-tolerant tree genera, such as *Tsuga* and *Abies,* are sometimes found growing successfully on calcareous soils, just as "acidophilous" understory vegetation *(Rumex, Ledum,* and *Chamaedaphne)* may thrive at relatively high pH levels if competi-tion from more tolerant species is not severe.

**Table 6.1**  Variations in Some Physical and Chemical Properties by Sampling Dates of Two Central Washington Forest Soils Derived from Different Parent Materials (Anderson and Tiedeman, 1970).

| Sampling Date | Porosity | Bulk Density | Organic Matter | Cation Exchange Capacity | Exchangeable | | | | Ext. P | pH |
|---|---|---|---|---|---|---|---|---|---|---|
| | | | | | Na | K | Ca | Mg | | |
| | % | g/cm³ | % | | meq/100 g | | | | ppm | |
| *Basaltic Parent Material* | | | | | | | | | | |
| June 10 | 77.6 | 0.61 | 10.6 | 29.7 | 0.18 | 0.44 | 1.4 | 0.17 | 14.2 | 4.8 |
| July 8 | 75.1 | 0.61 | 14.7 | 37.2 | 0.18 | 0.48 | 2.5 | 0.25 | 15.4 | 4.8 |
| Aug. 20 | 72.8 | 0.65 | 11.0 | 29.2 | 0.13 | 0.34 | 1.3 | 0.14 | 14.5 | 5.1 |
| *Sandstone Parent Material* | | | | | | | | | | |
| May 27 | 57.5 | 1.06 | 3.70 | 15.7 | 0.19 | 0.67 | 8.3 | 1.00 | 71.7 | 5.9 |
| June 10 | 60.7 | 1.03 | 3.80 | 14.8 | 0.19 | 0.53 | 8.5 | 0.92 | 64.6 | 6.1 |
| July 8 | 61.8 | 1.01 | 3.76 | 14.1 | 0.23 | 0.72 | 8.0 | 0.99 | 75.0 | 5.8 |
| Aug. 20 | 59.2 | 1.04 | 2.89 | 12.8 | 0.12 | 0.82 | 6.8 | 1.00 | 76.3 | 5.7 |

## Indirect Effects of Acidity

Many of the apparent direct effects of soil acidity on tree growth may, in fact, result from its indirect effects on such soil conditions as microbial activity and nutrient availability. The relative availability of most plant nutrients as a function of soil acidity is shown graphically in Figure 6.1.

The availability of such micronutrients as boron, copper, manganese, and iron is generally increased when soil acidity is increased (pH lowered). In soils where the reserves of these nutrients are inherently low, a significant reduction in soil acidity may result in a deficiency of one or more of these essential elements. For example, instances of iron and boron deficiencies have been noted in local areas of young slash pine plantations of the southeastern United States coastal plain. Wind and water movement of fine particles from limestone-topped forest roads was found to decrease acidity of sandy soils sufficiently to cause micronutrient deficiencies in trees up to 10 to 20 m from the road. Deficiency symptoms have generally disappeared within a few years as root systems developed into acid soil.

The acidity of nursery soils irrigated with water containing a moderate level of dissolved bases may increase sufficiently to cause iron deficiencies, particularly in nurseries located on poorly buffered sandy soils. Furthermore, "damping off" organisms and other pathogens often become problems in coniferous nursery soils if the reaction is allowed to increase appreciably above about pH 5.5. For these reasons, soils of coniferous nurseries are usually maintained in the range of pH 5.2 to 5.6, while soils of deciduous nurseries are generally kept in the range pH 5.6 to 6.0. The acidity of nursery soils may be maintained or lowered by applying acid-forming sources of nitrogen, such as ammonium sulfate. In extreme cases, applications of elemental sulfur or aluminum sulfates may need to be made to lower the soil reactions to the desirable range. In rare instances, nursery soils are so acid that they require adjustment by limestone applications. Fungi are active over a wide range of acidity conditions, but they do not compete as well as other soil organisms at acidity about pH 6. Bacteria are less tolerant of acid conditions than fungi. Their activity in relation to acidity is similar to that for nitrogen availability in Figure 6.1.

Correcting soil acidity is seldom recommended in forest plantations because of the expenses involved and the relatively small growth responses resulting from additions of limestone needed to change soil reaction. Most tree species can be grown in soils within the range between pH 4.5 and 6.5, but if both conifers and hardwoods are to be planted, soils with values below pH 5.0 should be reserved for the more acid tolerant species, which includes most conifers. Limestone has sometimes been applied in older coniferous plantings in central Europe. The value of such treatments presumably resulted from acceleration in the decomposition rate of the litter layer and in the subsequent mineralization of nitrogen and phosphorus. Although the rate of nitrification is rather slow in soils under

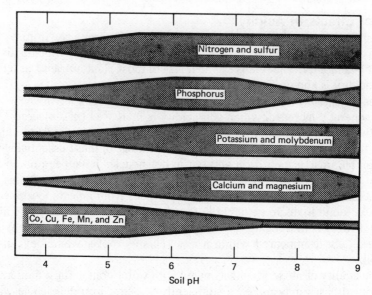

Nitrogen and sulfur

Phosphorus

Potassium and molybdenum

Calcium and magnesium

Co, Cu, Fe, Mn, and Zn

4        5        6        7        8        9

Soil pH

**Figure 6.1**  Relative effects of soil acidity on the availability of plant nutrients.

coniferous cover, no particular advantage in tree growth has been observed from liming soils in most temperate zone forests.

## ION EXCHANGE

The capacity of soils to retain nutrient ions in a form available for plant use is of particular importance to forests growing on sands and other soils with extremely low nutrient reserves. This phenomenon, called ion exchange, is a reversible process by which both cations and anions are exchanged between liquid and soil phases. It is primarily associated with the soil colloidal organic and mineral materials. Generally, soils with low content of clay-sized particles can be expected to have a low exchange capacity but that small capacity is extremely important to tree nutrition and the efficient cycling of nutrients in sandy soils.

### Cation Exchange

Cation exchange is considered to be of greater importance in soils than anion exchange, because most of the essential minerals are absorbed by plants as cations. Such cations as calcium, magnesium, potassium, sodium ammonium, aluminum, iron, and hydrogen are positively charged and are attracted to the negatively charged surfaces of the colloidal mineral and organic soil particles. The negative charge on the organic particles arises primarily from the -COOH and -OH groups, while charges on the inorganic clay fraction mostly derive from isomorphous substitution, or from ionization of hydroxyl groups attached to the silicon atoms along the edges of tetrahedral planes.

The negative charge that develops on organic and mineral colloids may be neutralized by cations attracted to the surfaces of these colloids. The amount of cations thus attracted, expressed as milliequivalents per 100 g of oven-dry soil, is termed the *cation exchange capacity* of the soil. The cations thus held do not all exchange with the same degree of facility. For example, most divalent and trivalent cations are held more tightly than monovalent cations (except for H+ ions) and the greater the degree of hydration, the less tightly an ion is held.

The cation exchange capacity of a soil is generally determined by mass displacement of the various cations with an aqueous salt solution (such as normal ammonium acetate). The ammonium ions, which then saturate the exchange sites, are extracted by a different salt, such as potassium chloride. The quantity of ammonium ions displaced into the suspension by the second salt is a measure of the cation exchange capacity of the soil in question. However, it should be pointed out that the cation exchange capacity of a soil is not a fixed quantity but is dependent on the pH of the extracting solution used. The total charge on soil colloids is made up of two components: (1) a permanent charge and (2) a pH-dependent charge. The former is considered to arise from isomorphous substitution in the clay lattice, while the latter is thought to result from the carboxyl and phenolic groups on soil organic matter and from ionization of hydrogen from OH groups along the broken edges of clay lattices (Coleman and Thomas, 1967). The cation exchange capacity of acid forest soils is largely composed of pH-dependent charges. This is particularly true of many forest soils whose colloidal fraction is composed of organic material and 1:1 clay minerals. As a consequence, the actual cation exchange capacity of those soils may be much lower than that measured in neutral normal ammonium acetate. Typical cation exchange capacities associated with various soil components are shown in Table 6.2.

The degree to which all cation exchange sites of a soil are occupied by bases, such as calcium, magnesium, potassium, and sodium, is termed the percent *base*

**Table 6.2**  Typical Cation Exchange Capacity of Various Soil Components

| Component | Cation Exchange Capacity |
|---|---|
| | meq/100 g |
| Humus | 200 |
| Montmorillonite | 100 |
| Vermiculite | 30 |
| Kaolinite | 8 |
| Hydrous oxides | 4 |

*saturation* of that soil. Most forest soils possess a very low degree of base saturation, but this depends greatly on the climate and materials from which the soils were formed, as well as the vegetation it supports. Soils in arid regions normally have a higher degree of base saturation than those in humid regions, and soils formed from limestone or basic igneous rock are more base saturated than soils formed from sandstone or acid igneous rock in the same climatic zone. As a general rule, the pH and fertility level of a given soil increases with an increase in the degree of base saturation. This results not only from the quantitative increase in bases in the soil but also from the ease with which the bases are released by the soil and absorbed by plants. The relationship between percent base saturation and cation availability is influenced by the type of soil colloids. Soils with a high percentage of organic and 1:1-type colloids can supply basic cations to plants at a lower level of base saturation than soils high in 2:1-type colloids—a fortunate situation for most forest trees.

## Anion Exchange

Anions, such as phosphates, sulfates, chlorides, and nitrates, are held on colloidal surfaces to varying degrees by anion exchange properties. Phosphates are held more tenaciously in most soils as precipitation products, largely as insoluble iron and aluminum phosphates in very acid soils and as calcium or magnesium phosphates in more neutral soils. The retention of phosphorus as a precipitation product is discussed in a later section of this chapter.

Nitrates and chlorides, and to a large extent sulfates, are readily leached from well drained soils. However, since anion exchange is believed to be largely a pH-dependent function in which the more acid the soil the greater the anion exchange capacity, forest soils have a greater capacity to retain anions than many agricultural soils. One must be aware that fertilizer nitrates readily leach from all soils, including acid forest soils, if not soon absorbed by plant roots or combined into microbial protoplasm.

## ESSENTIAL ELEMENTS

In the presence of light, all green plants are capable of using water, carbon dioxide, and several mineral elements as raw material in the manufacture of their food. Plants absorb many elements from the soil, but not all of them are essential to their existence and well being. The elements required by plants in order to complete their vegetative and reproductive cycles are called *essential* elements. Carbon, hydrogen, and oxygen are obtained from carbon dioxide and water. They are converted into simple carbohydrates by photosynthesis and, with nitrogen, phosphorus, and sulfur, they are elaborated into amino acids, proteins, and protoplasm. There are 14 other elements that are apparently essential to the growth of some plants, but not all of them are required by all plants. Boron, calcium, chlorine, cobalt, copper, iron, magnesium, manganese, molybdenum, potassium, silicon, sodium, vanadium, and zinc—along with nitrogen, phosphorus, and sulfur—are largely obtained from the soil and usually constitute the *plant ash,* except that nitrogen and sulfur may be lost by volatilization during burning (Tisdale and Nelson, 1966).

Three elements, nitrogen, phosphorus, and potassium, are often called *primary plant nutrients* because they are used in relatively large quantities by the plant and are the elements most often deficient in soils. Three others, calcium, magnesium, and sulfur, are also used in fairly large quantities but are not so often deficient in soils. They are frequently referred to as *secondary plant nutrients.* The remaining elements are termed *micronutrients* because they are needed in very small quantities by trees and are not as likely to be deficient in soils.

## Nitrogen

The element nitrogen makes up approximately 78 percent (by volume) of the atmosphere, but this gaseous nitrogen is largely unavailable to higher plants. Only through fixation by soil microorganisms and lightning discharges does a small part of this nitrogen reserve become available. The forms most commonly assimilated by plants are the nitrate and ammonium ions. Urea can be absorbed through the leaf epidermis when applied as a fertilizer, but it is unlikely that much urea nitrogen is absorbed by plants roots, due to the rapid hydrolysis it undergoes in the soil.

Nitrogen is accumulated in soils in the form of plant and animal residues, and a kind of equilibrium between the rate of accumulation and decomposition becomes established on each site over long periods of time. The amount of organic matter and nitrogen in the soil at any given time depends on many climatic and edaphic factors and natural and human disturbances that influence the ratio of plant and animal additions (input) to the rate of decomposition (output). Total nitrogen in forest soils is found largely in the humus layers of the forest floor and in the A1 horizon. The amount varies from no more than 1 ton per hectare in some excessively drained sands to as much as 30 tons in some boreal forests with a deep accumulation of humus (Weetman and Webber, 1972). Nitrogen accumulation in these latter soils is favored by cold weather and high soil moisture throughout much of the year.

Nitrogen in organic materials becomes available to higher plants only after the carbon:nitrogen ratio approaches 10:1. During the mineralization of carbonaceous materials, such as forest floor litter, the C:N ratio decreases with time. This results from the gaseous loss of carbon while the nitrogen remains bound in organic combination. The narrowing of the ratio in the decomposition of nitrogen-poor substrates is not linear, the curve approaching a ratio of 10:1 asymptotically (Alexander, 1977). This critical ratio is a reflection of the dynamic equilibrium that results from the presence of the micro-biological population because it is approximately that of the chemical composition of microbial cells. Table 6.3 illustrates C:N ratios in mineral soil under loblolly pine at 5, 11, and 15 years after they were planted in an abandoned agricultural field.

Mineralization rates in acid soils are extremely low in undisturbed forests. Soil disturbance such as occurs in seed bed preparation and harvesting operations generally increases rate of organic matter decomposition and nitrogen release. Increases in soil temperature following clearcutting also stimulate nitrification (Likens et al., 1970). Liming acid soils favors mineralization of nitrogen, but most tree species apparently grow well with the ammonium form as the predominant source of nitrogen and forest soils are seldom limed to promote nitrification.

Much of the nitrogen in the slash and litter layers is volatilized during burning, and the decreased acidity of the soil surface resulting from the ash deposits may stimulate mineralization of nitrogen in residual organic matter. Thus nitrogen availability to trees is often temporarily increased following a controlled burn of the forest floor. Nitrogen may also be lost by denitrification in wet, poorly aerated soils. While denitrification is probably of minor consequence in established forests, it could account for a significant loss of nitrogen in coastal flatwoods, and other potentially wet areas, following clear-cut harvesting and the associated rise in water table when transpiration is suddenly reduced.

In addition to its role in the formation of proteins, nitrogen is an integral part of the chlorophyll molecule. An adequate supply of the element is generally associated with vigorous vegetation growth and a deep green color. Added nitrogen is presumed to delay maturity of plants, to increase the percentage of early wood in relation to late wood in trees, and to increase their susceptibility to some diseases and insects. However, it is unlikely that nitrogen will adversely affect tree properties in the presence of adequate supplies of other essential nutrients. Nitrogen deficiencies are most often reported in coniferous forests of

**Table 6.3**  Carbon:Nitrogen Ratios of Organic matter in South Carolina Piedmont Mineral Soil 5, 11, and 15 Years After Planting Loblolly Pine in an Abandoned Agricultural Field (C. G. Wells, personal comm.)

| Soil Depth cm | Age | | |
|---|---|---|---|
| | 5 | 11 | 15 |
| 0–8 | 16.9 | 19.4 | 21.7 |
| 8–15 | 15.5 | 19.2 | 21.9 |
| 15–30 | 11.4 | — | 13.6 |
| 30–60 | 9.0 | — | 9.0 |

cold climates under conditions that favor accumulation of thick acid humus (Weetman, 1962). Incipient nitrogen deficiencies are also found in many sandy soils of warmer climates, including the flatwoods and sand hills of the United States coastal plain and the Douglas-fir region of the Pacific Northwest. Deficiencies in the latter areas and the tropics would doubtlessly be even more widespread except for the relatively rapid and efficient nutrient cycling and relatively large biological fixation of nitrogen in tropical areas. Soils of eroded and abandoned agricultural lands in many regions of the world are likely to have nitrogen deficiencies.

## Phosphorus

Phosphorus is an essential element in the energy transfer processes so vital to life and growth of all green plants. It is derived primarily from the calcium phosphates (apatites) and iron and aluminum phosphates in soils and is believed to be absorbed by plants mostly as the primary orthophosphate ion. It also occurs in soil organic matter, and certain soluble organic phosphates such as nucleic acid phosphates and phytin may be absorbed directly by plants. As a matter of fact, organic matter is the principal source of phosphorus for trees on many soils. The total phosphorus content of soils may vary from no more than 20 to 40 kg per hectare in the surface (A1) horizon of most sands to more than 2000 kg in some phosphate-rich soils.

The availability of inorganic phosphorus to trees depends largely on (1) soil acidity and its effects on the solubility of iron, aluminum, and manganese, which form insoluble precipitation products in very acid soil; (2) the availability of calcium, which may react with phosphorus to reduce its solubility in less acid soils; and (3) the activity of microorganisms that control the rate and amount of organic matter decomposition.

Some soluble iron, aluminum, and manganese are usually found in strongly acid mineral soils. Under such conditions reaction with phosphate ions would soon occur, rendering the phosphorus insoluble and unavailable for use by most plants. The chemical reactions that occur between soluble iron, aluminum, and manganese, and the phosphate ions result in the formation of hydroxyphosphates, as exemplified by the following equation for the aluminum ion:

$$Al^{3+} + H_2PO_4^- + 2H_2O \rightleftharpoons 2H^+ + Al(OH)2H_2PO_4$$

The surface horizon of some coastal acid sands and organic soils are particularly low in phosphorus because of their weak capacity for phosphorus retention. They contain very low concentrations of iron, aluminum, and manganese and, thus, most of the phosphorus in the surface layers has been leached to lower horizons, with only that fraction in organic combination remaining (Ballard and Pritchett, 1974). Much of the phosphorus leached from the surface of acid Spodosols may be found as iron and aluminum phosphates in the spodic horizon. The latter phosphates are not readily available to young trees and the proportion from this source that is available to older trees is not known.

Roots of forest trees possess mycorrhizal associations that increase the capacity of these trees to utilize less available forms of phosphates in soils. Soil tests based on the relationship of soil *total* phosphorus to tree response to fertilizers have generally not proven satisfactory as a means of delineating phosphorus deficient soils. The amount of phosphorus in the soil solution at any given time is very low, usually less than one part per million (ppm), and the amount available to plants is influenced by several soil factors. These include soil acidity and the presence of soluble iron and aluminum, the humus type and its rate of decomposition, as well as the total amounts and forms of mineral phosphorus in the soil. Less phosphorus is generally found in soils supporting coniferous stands than in those under hardwoods, probably reflecting the capacity of conifers to survive and compete in soils containing relatively low levels of phosphorus. Nevertheless, there are probably more reports of phosphorus deficiencies than for deficiencies of any other nutrient for conifers growing in plantations on poorly drained sands in the Northern Hemisphere and as exotics on a variety of soil conditions in the Southern Hemisphere.

Phosphorus is readily mobilized in plants and when a deficiency occurs, the element is rapidly transferred from older tissue to the active meristematic regions. The overall symptom of phosphorus deficiency is one of retarded growth rather than striking foliar symptoms. Because phosphate fertilizers apparently aid in seedling root development and often remain available in the soil for many years after fertilization, it is customary to apply these materials early in the life of plantations on deficient soils.

## Potassium

Potassium, unlike nitrogen, phosphorus, sulfur, and several other elements, apparently does not form an integral part of protoplasm, fats, and other plant components. It acts rather as a catalyst but is, nonetheless, essential to many physiological functions. These include carbohydrate metabolism, protein synthesis, activation of various enzymes, and growth of meristematic tissue. Potassium may be associated with the resistance of plants to certain diseases.

Potassium appears to be plentiful in most forest soils; the exceptions are some acid sands, such as glacial outwash sands of northeastern United States. The element is derived primarily from feldspars and micas and exists in soils in inorganic compounds. Deficiencies in agricultural crops are often due to the slow release by weathering of the unavailable forms to the available forms. However, the 20 to 100 ppm of potassium often found in the exchangeable form in forest soils is apparently adequate for good growth of most trees. For example, southern pines are able to make reasonable growth on some soils containing as little as 10 ppm of exchangeable potassium and, as a consequence, often fail to respond to applications of potassium fertilizers. It is suspected that trees are capable of absorbing potassium from unweathered feldspars and other potassium bearing minerals with the aid of mycorrhizal roots (Voigt, 1965). Furthermore, it appears that potassium is rapidly and efficiently cycled in established forest stands. Very little potassium appears to be leached below the surface root mat in undisturbed forests.

## Calcium, Magnesium, and Sulfur

These elements are used by trees in relatively large amounts but are normally found in soils in sufficient quantities for good tree growth. Few instances of actual deficiencies in forests have been reported in the literature, but this may result from the rather sparse information available on forest soil conditions, nutrient requirements, and deficiency symptoms.

**Calcium**　　The specific physiological functions performed by calcium in plants are not well understood. It is involved in the development of meristematic tissue and in root and shoot elongation and, perhaps, in protein formation. Calcium is considered to be an immobile element, but in at least one forest species (western white pine) previously deposited calcium moves from older tissue to developing tissue. Calcium exists in soils mostly in inorganic forms, and from 50 to 1000 ppm or more may be held in an exchangeable form in the surface soil. Soils developed in regions of relative low rainfall generally contain larger supplies of calcium than soils in humid regions, and lower horizons normally contain more calcium than surface horizons. Deep-rooted trees with a high calcium requirement, such as hardwoods, tap calcium reserves in the lower horizons and build up the concentration in the surface soil through annual leaf fall. Relatively high concentrations of calcium in the soil may favor certain species over less demanding types.

**Magnesium**　　Magnesium is the only mineral constituent of the chlorophyll molecule and is essential to photosynthesis. It is a mobile element that is transported from older to younger plant parts in the event of a deficiency. Therefore the symptoms of magnesium deficiency, like that of potassium, often appear first on the older leaves. Magnesium deficiencies have been noted in nurseries and in young seedlings growing on sandy soils but are not usually found in older trees. Most forest soils contain ample magnesium for good tree growth, and deficiencies, when encountered, are easily corrected. Dolomite, a double salt of calcium-magnesium carbonate, is an abundant and relatively inexpensive liming material often used for correcting magnesium deficiencies.

**Sulfur**　　Sulfur is primarily derived from pyrites and gypsum, but in sandy soils where these minerals are absent, the chief source may be the atmosphere. Sulfur from industrial wastes is washed from the air in amounts ranging from a few kilograms per hectare per year in isolated areas to more than 100 kg near large cities. In forest soils much of the sulfur is accumulated in the organic matter of the upper horizons. Although sulfur is used in approximately the same amounts as phosphorus, it is much more readily available in the soil. Sulfur-oxidizing bacteria can convert free sulfur and sulfur in organic compounds to sulfates and sulfuric acid. Sulfur can be readily absorbed as sulfate by plants or leached from the soil in the absence of plants. Under anaerobic conditions, such as in marshes, sulfates are readily reduced to sulfides. Sulfur has been found deficient in some areas of the ponderosa pine region, in coconut palms, and in some hardwoods.

## Micronutrients

Several other elements have been identified as essential to plant growth, but they are required in very small quantities by tree crops and are seldom deficient in forest soils. There are a number of reports, however, of isolated instances of deficiencies of one or more of these elements. The availability of the micronutrients, except for molybdenum, is reduced as soil acidity is decreased. Consequently, any practice that substantially reduces soil acidity may cause deficiencies of iron, manganese, copper, boron, or zinc in soils with low reserves of these elements. Iron is the element most frequently reported to be deficient, particularly in nurseries. Copper is sometimes deficient in organic soils and boron has been found deficient in New Zealand and Australia. Zinc deficiencies have been reported in exotic conifers in Australia.

Chlorine, cobalt, silicon, sodium, and vanadium have been identified as necessary in the life cycle of some plants, but no instances of deficiencies of these elements have been reported for forest soils.

## SILVICULTURAL IMPLICATIONS

The growth of a tree with a particular genetic makeup is a function of age and many site factors. Among the environmental components are temperature, moisture supply, radiant energy, composition of the atmosphere and soil air, biotic factors, and soil physical and chemical properties. Soil physical properties and soil water have been long considered of primary importance to site productivity and in more recent years the importance of chemical properties has become better understood and appreciated. Not only are soil acidity, cation exchange capacity, and nutrient availability of significance to tree growth, but they can be of overriding importance on some sites. For example, soil acidity appears to have little direct effect on tree growth, for most trees grow quite well over a wide range of reactions. However, acidity can have tremendous indirect influence on growth and development through its effect on nutrient availability, microbial activity, and the existence of toxic compounds. It can also give an indication of the base status of a soil.

In a similar fashion, it can be said that cation exchange capacity is of little direct importance to forest development. However, the capacity of a soil to retain nutrients in an available form against the leaching action of percolating waters can be of great significance to tree nutrition. In fact, nutrient availability has received little attention in forest soils because of the apparent ability of some forest trees to grow well on impoverished soils. It is well recognized now that the annual uptake of nutrients by forest trees is comparatively large, but that trees make very good use of the nutrients available through efficient recycling and deep soil exploitation, enabling them to survive on relatively infertile soil. Nevertheless as the intensity of forest management increases, the demands placed on the soil for nutrients will greatly increase.

# 7
## PHYSICAL PROPERTIES OF FOREST SOILS

Forest soil investigators have long recognized the profound influence of soil physical properties on the growth and distribution of trees. The view that physical properties were of primary importance to vegetation development, however, led early researches to minimize or even ignore the influence of soil chemical and biological properties on the site. This attitude has changed somewhat in recent years as the need for more intensive use of our forest resources has increased. A more balanced view has evolved of the importance of all soil properties to the forest environment. This position reflects an awareness of the close interrelationship of soil chemical, biological, and physical properties and that, while good physical properties may help compensate for poor chemical or biological properties, soil productivity can not be equated with physical characteristics alone.

Soil physical properties are less easily altered by the forest manager than are chemical properties. Soil structure and porosity, however, can be altered by manipulating the soil under certain conditions. By draining wet soils, the silviculturist can affect changes of such indirect properties as soil moisture, aeration, and temperature. Draining may also increase the effective depth of rooting. Bedding and deep plowing to fracture impervious layers are sometimes used to improve conditions for root development. But properties such as texture, color, and those related to profile characteristics are not easily changed.

### SOIL TEXTURE
The soil can be conveniently divided into three phases—solid, liquid, and gaseous. The solid phase makes up approximately 50 percent of the volume of

most surface soils and consists of a mixture of inorganic and organic particles varying greatly in size and shape. The proportionate distribution of different sizes of the mineral particles determines the *texture* of a given soil. The sizes of the mineral particles and the relative proportion of size groups vary greatly among soils, but they are not easily altered in a given soil. Thus, soil texture is considered a basic property of soil.

## Soil Particle Sizes

All mineral soils are made up of a mixture of soil separates, or grouping of particles of similar size. Schemes for classifying soil separates have been developed in a number of countries. The classification used by the United States Department of Agriculture, based on diameter limits in millimeters, is outlined in Table 7.1.

The determination of particle size distribution in soils is normally called a *mechanical analysis*. There are several techniques for determining the percentage distribution of particle sizes, but most of them involve the complete dispersion of soil particles in water (usually containing a detergent), separating them into size classes, and calculating the percentages of each class by weight. These methods are based on the principle that the particles suspended in water tend to settle in relation to their size. Sand fractions settle very rapidly and are separated into arbitrary groups by sieving. The hydrometer (Bouyoucos, 1927) and the pipette (Baver, 1956) methods are the two most widely used systems for determining silt and clay fractions. The hydrometer method is most useful in forest soils work because it is relatively rapid, requires a minimum of equipment, and is reasonably accurate.

## Textural Classes

Soils are composed of particles of widely varying sizes and shapes and the proportionate distribution of the different-sized mineral particles largely deter-

**Table 7.1.** The United States Department of Agriculture Classification of Soil Particle Sizes

| Name of Separate | Diameter Limits, mm |
|---|---|
| Sand | 0.05–2.0 |
|   very coarse | 1.0–2.0 |
|   coarse | 0.5–1.0 |
|   medium | 0.25–0.5 |
|   fine | 0.10–0.25 |
|   very fine | 0.05–0.10 |
| Silt | 0.002–0.05 |
| Clay | Less than 0.002 |

mine many of the basic properties of soils. Textural class names are used to identify groupings of soil with similar mixtures of mineral separates. Mineral soils can be generally grouped into three broad textural classes—sands, loams, and clays—and a combination of these class names are used to indicate intergrades. For example, *sands* must contain 70 percent or more of the sand separate, but those containing 15 to 30 percent silt and clay are called loamy sands. *Clays* contain more than 40 percent of the clay separate, but they may also contain up to 45 percent sand and up to 40 percent silt, in which event they are classed as sandy clays or silty clays. Soils containing sufficient colloidal material to be classed as clays are generally hard when dry and sticky and plastic when wet. *Loams* consist of diverse groupings of sand, silt, and clay separates ranging from sandy loams to silty clay loams. However, they have the appearance of containing about equal proportions of each.

The textural class of a soil can be estimated in the field with reasonable accuracy after some experience. The "feel" of moist soil rubbed between the thumb and fingers should be checked against that of known samples in the laboratory in order to gain skill. A more accurate method of determining textural class designations is by use of the textural triangle. This system is used in most parts of the world but its use depends first on the determination of particle-size distribution. The relationship between the class name of a soil and its particle-size distribution is shown diagrammatically in Figure 7.1.

The state of decomposition is used in the place of textural classes in describing organic soils (Histosols). The *Fibrist* suborder includes soils in which the undecomposed fibrous organic materials are easily identified. They are generally yellow and brown in color, have low bulk densities, and high water-holding capacities. On the other hand, the highly decomposed dark *Saprists* have relatively high bulk densities (often more than 0.2) and the lowest water capacities of organic soils. The *Hemist* suborder includes soils intermediate in decomposition and in their properties between those of the Fibrists and Saprists.

Cobbles and gravels are fragments larger than 2 mm in diameter. They are not included in the particle size designations of Table 7.1 because they normally play a minor role in agricultural soils. However, they may occupy a considerable percentage of the total volume of forest soils (up to 50 percent or more of some mountain soils)—space that would otherwise be occupied by soil, air, or moisture. A reasonable amount of rock in a fine textured soil may favor tree growth. The coarse fragments may increase penetration of air and water and increase soil temperatures during the growing season. Nevertheless, coarse skeleton dilutes the soil and can be detrimental to tree growth if it occupies a significant volume in sandy soils, as in the glacial till soil in Finland in Figure 2.2. It reduces the already low water-holding capacity and exchange capacity of these latter soils. The coarse textured materials, those particles settling out of a soil-water suspension after about one minute (cobbles, gravels, and sands larger than 0.05 mm in diameter), contribute little to plant nutrition.

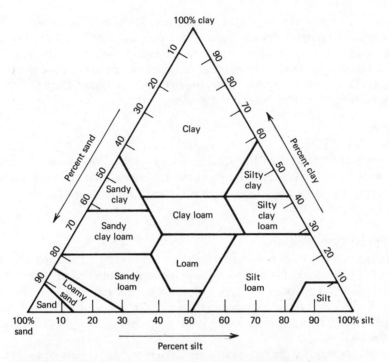

**Figure 7.1**  Soil textural triangle: a textural class may be found by projecting inward a line parallel to the clay apex and then the sand apex of the triangle and determining in which compartment the two lines intersect.

## Tree Growth as Influenced by Texture

Texture of a forest soil influences its productivity, but this influence may be more of an indirect than of a direct nature. For example, deep, coarse sandy soils often support relatively poor forest stands of pines, cedar, scrub oak, and other species that have low requirements for moisture and nutrients. The productivity of sandy soils, consequently, increases as the proportion of material smaller than 0.05 mm (silt and clay particles) increases to an optimal level. Because of this relationship loams and clay soils often support trees of high moisture and nutrient requirements, such as spruces, firs, hard maples, and basswoods in cool, moist climates, and a variety of hardwoods in more temperate climates.

While the indirect influence of texture on tree growth may be considerable in upland soils, its importance is often masked by more critical factors in other regions. Texture per se has little effect on tree growth as long as moisture, nutrients, and aeration are adequate. In coastal plains, changes in soil moisture conditions brought about by small differences in elevation may completely overshadow textural effects. Fertilization of moist, sandy soils can overcome the low capacity of these soils to retain nutrients. Furthermore, a forest stand tends to modify its environment to the extent that texture of the soil is of minor importance. Through species succession, soil conditions may be gradually changed so that they more nearly meet the requirements of the trees in the stands. In this way, pioneer trees create conditions for the establishment of more exacting climax species by increasing organic content, thus minimizing the effect of soil texture on establishment and growth.

## SOIL STRUCTURE

Soil structure is defined as the spatial arrangement of individual soil particles. Such soil characteristics as water movement, aeration, bulk density, and porosity are greatly influenced by the overall aggregation or arrangement of the primary soil separates.

### Structural Descriptions

Field descriptions of soil structure usually give the (1) *type* or shape, (2) *class* or size, and (3) *grade* or distinctness of soil materials contained in each horizon of the soil profile. The following types of structure are recognized by the United States Soil Conservation Service: (1) platy, (2) prismatic, (3) columnar, (4) angular blocky, (5) subangular blocky, (6) granular, (7) crumb, (8) single grain, and (9) massive.

Class, or size, of the aggregates are designated as (1) very fine, (2) fine, (3) medium, (4) coarse, and (5) very coarse. Structural classes are determined by comparing a representative group of peds with a set of standardized diagrams.

Grade is determined by the relative stability or durability of the aggregates and by the ease of separation of one from another. Grade varies with soil moisture content and is usually determined on nearly dry soil and designated by the following terms: (1) weak, (2) moderate, and (3) strong.

The complete description of soil structure, therefore, consists of a combination of the three variables, in reverse order of that given above, to form a type, such as "moderate fine crumb." A more complete description of structural properties can be found in most general soils texts (Brady, 1974).

The combining of individual particles into structural aggregates is influenced by a number of factors. Among these the nature and origin of the parent materials and the physical and biochemical processes of soil formation are most important. Of particular note are the presence of salts, growth and decay of roots, freezing and thawing, wetting and drying, and the activity of soil organisms. Soil texture has considerable influence on the development of aggregates. The absence of aggregation in sandy soils gives rise to "single grain" structure, but clay soils exhibit a wide variety of structural types. Fires may increase the formation of aggregates in fine-textured soil due to dehydration of the soil colloid. Soil animals, such as earthworms and millipedes, favor the formation of crumb structure in the surface soil by ingestion of mineral matter along with organic materials. A vigorous forest cover favors the maintenance of good soil structure, regardless of texture. Aggregate stability is influenced by the temporary binding action of microorganisms, particularly the mycelia of fungi common in forest soils. The intermediate products of microbial synthesis and decay are effective stabilizers and the cementing action of the more resistant humus components that form complexes with soil clays give the highest stability.

## Soil Disturbance

Mechanical disturbance often affects the physical condition of a surface soil as a result of altering its structure through compaction and puddling. The direct effects are on the soil air-water systems and soil strength properties affecting root penetration.

Compaction by heavy equipment or repeated passages of light equipment compresses the soil mass and breaks down surface aggregates, decreasing the macropore volume and increasing the volume proportion of solids and micropores. A reduction in air diffusion and water infiltration rates and an increase in soil strength are common in the compacted zone. Compaction is more evident on wet soils than on dry soils and more on clay soils than on sandy soils. However, the effects of compaction appear to be less permanent in fine-textured soils (especially those containing considerable montmorillonite clay) than in some coarse-textured soils, because of swelling and shrinking as a result of wetting and drying. Nonetheless, the effects of compaction persist for long periods on some soils as evidenced by stunted growth of trees on skid trails and log landings on these soils. Hatchell (1970) found that establishment and growth of loblolly pine seedlings were substantially less on compacted soil cores than noncompacted cores. Loosening compacted cores resulted in an establishment percentage almost as high as the percentage on cores of normal soil, but loosening compacted cores did not significantly improve growth.

Puddling results from the dispersal of soil particles in water and a differential rate of settling. This permits the orientation of clay particles so that they lie parallel to each other. The destruction of soil structure by this method may result in a dense crust that has the same affect on soil conditions as a thin compacted layer. It is most common on soil surfaces where the litter has been removed by burning or by mechanical means. Reduced germination and increased mortality rates of loblolly pine seedlings have been observed on soils compacted or puddled by logging equipment (Pomeroy, 1949).

Soil aggregates are generally more stable under forested conditions than under cultivated conditions. Continued cultivation tends to reduce aggregation in most soils through mechanical rupturing of aggregates and by a reduction in organic matter content and associated cementing action of microbial exudates and fungal hyphae (Figure 7.2).

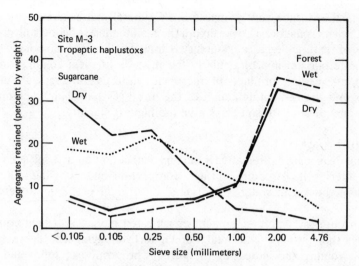

**Figure 7.2**   Stability of aggregate of an Oxisol used for forestry and for sugar cane in Hawaii (Wood, 1977). Reproduced from *Soil Science Society of America Journal,* Volume 41, No. 1, page 135, 1977 permission of the Soil Science Society of America.

## BULK DENSITY

*Bulk density* (or volume weight) is the ratio of dry weight of a given volume of undisturbed soil to the weight of an equal volume of water. Since it is a weight measurement by which the entire soil volume is taken into consideration, it is greatly influenced by soil structure. Unlike particle density, which is concerned with the solid particles only, bulk density measurements include air space, as well as soil volume, and these measurements are thus related to porosity. The particle density of most mineral soils varies between the narrow limits of 2.60 and 2.75 g per 100 cm³, but the bulk density of forest soils varies from 0.2 in some organic layers to almost 1.9 in coarse sands. Soils high in organic matter have lower bulk densities than soils low in this component. Soils that are loose and porous have low weights per unit volume (bulk densities) while those that are compacted have high values.

Excessive trampling by grazing animals, use of heavy logging machinery, intensive recreational use, or disturbance while soils are wet will increase bulk density, particularly of fine-textured soils. Compacted sands with bulk densities exceeding about 1.75, or clays with bulk densities exceeding 1.55, may prevent the penetration of tree roots. That is, soil density above which roots do not penetrate varies with soil texture, and a given bulk density in fine-textured soils limits root growth more than the same density in coarse soils. Furthermore, the roots of some species are able to grow in moderately high soil densities where the roots of others cannot grow. Smith and Woollard (1969) reported that roots of two-year-old seedlings of lodgepole pine, Douglas-fir, red alder, and Pacific silver fir penetrated soil columns compacted to densities of 1.32, 1.45, and 1.59 g per $cm^3$ while the roots of Sitka spruce, western hemlock, and western redcedar did not penetrate as well.

## PORE VOLUME
Pore volume is that part of the soil volume not occupied by solid particles. The soil pores normally contain both air and water, but the relative proportions of each are constantly changing. In a dry soil the pore spaces are largely occupied by air, but in a waterlogged soil they are filled with water. The coarse-textured soils have large pores but their total pore space is less than that of fine-textured soils (except that puddled clays may have even less porosity than sands). Because clay soils have greater total pore space than sands, they are normally lighter per unit volume (have a lower bulk density).

### Calculating Pore Volume
If bulk density is known, the pore space of mineral soil can be easily calculated by dividing bulk density by particle density and converting to percent:

$$\text{Pore space percentage} = 100 - \frac{\text{bulk density}}{\text{particle density}} \times 100$$

Using an average density of mineral particles of 2.65 g per 100 $cm^3$ and assuming a bulk density of 1.2, following calculations can be made:

$$\text{Pore space} = 100 - \frac{1.2}{2.65} \times 100 = 54.7\%$$

Pore volume is conveniently divided into capillary and noncapillary pores. Soils with a high proportion of capillary (small diameter) pores generally have high moisture-holding capacity, slow infiltration of water, and perhaps a tendency to waterlog. On the other hand, a soil with a large proportion of noncapillary (large diameter) pores generally has good aeration, rapid infiltration, and a low capacity to retain moisture.

### Factors Influencing Pore Volume
Sandy surface soils have a range in pore volume of approximately 35 to 50 percent, whereas medium- to fine-textured soils vary from 40 to 60 percent or higher. The amount and nature of soil organic matter and the activity of soil flora and fauna influence pore volume, as well as soil structure. Pore space is reduced by compaction and generally varies with depth. Some compact subsoils may have no more than 25 to 30 percent pore space.

Pore volume of forested soils is normally greater than similar soil used for agricultural purposes, because continuous cropping results in a reduction in organic matter and of micropore spaces (Figure 7.3). Porosity of most forest soils varies from 30 to 65 percent. Soils supporting mixed stands of trees may have higher pore volume than soils supporting pure stands. The short-term effects of clear-cut harvesting may be to increase pore volume because of the resultant decay of tree roots, and the mixing of surface organic matter into the mineral soil. However, the longer-term effects would probably be to decrease pore volume due to the destruction of the soil organic layers and compaction by machinery and rain action. Anderson and Tiedemann (1970) attributed a slight, but nonsignificant, decline in bulk density and porosity during June to August to a decline in soil moisture and shrinking in two central Washington soils, as shown previously in Table 6.1

## SOIL ATMOSPHERE

It has been pointed out that about 50 percent of a surface mineral soil is pore space and that the soil air varies inversely with soil water content. However, it is not sufficient that there are ample soil voids in order to have a well-aerated soil, for there must also be an opportunity for the ready movement of gases into and

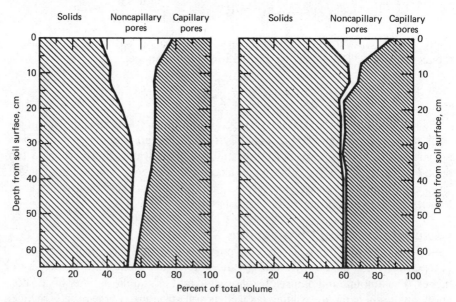

**Figure 7.3** Porosity as measured in the surface 60 cm of Vance soil (Typic Hapludult) in the South Carolina Piedmont: (left) in an undisturbed mixed hardwood forest, and (right) in abandoned farmland (Hoover, 1949).

out of these spaces. Soil air is important primarily as a source of oxygen to aerobic organisms, particularly for tree roots. Soil air composition, like air volume, is constantly changing in a well-aerated soil. Oxygen is used by plant roots and soil microorganisms, and carbon dioxide is liberated in root respiration and aerobic decomposition of organic matter.

## Soil Air Composition

Gaseous exchange between the soil and the atmosphere above it takes place by *mass flow* and *diffusion*. However, mass flow due to pressure differences is relatively unimportant, except as diurnal soil temperature changes may result in gaseous exchange in the upper surface soil. Diffusion accounts for the greatest portion of gaseous exchange in the soil. Because of diffusion exchange of gases between the soil air and atmospheric air, the oxygen content of well-drained surface soils seldom falls much below the 21 percent of atmospheric air. On the other hand, a deficit of oxygen may occur in compacted or waterlogged soils. Under these conditions, gas exchange is very slow due to the small and water-filled pore spaces. In very wet soils, carbon dioxide concentrations may rise to 5 or 6 percent and oxygen levels drop to 1 or 2 percent by volume (Romell, 1922). However, oxygen is not necessarily deficient in all wet soils in spite of the absence of voids. If the soil water is moving, it may have a reasonably high content of oxygen, but soils saturated with stagnant water are low in oxygen and they are very poor media for the growth of most higher plants.

Seasonal differences in the composition of soil air can be largely accounted for by soil moisture and soil temperature variations. Soils are generally drier during the summer months, providing a greater opportunity for gaseous exchange and a relatively higher content of oxygen. However, high summer temperatures also promote microbial activity, with a resultant increase in carbon dioxide release. Thus carbon dioxide contents of soil air may be higher in the summer than in winter months. Soil air usually is much higher in water vapor than is atmospheric air and it may also contain a higher concentration of such gases as methane and hydrogen sulfide, formed by organic matter decomposition.

## Importance of Soil Air

Good aeration not only favors water and oxygen absorption, but mineral nutrient absorption as well. For example, poor aeration of such soils as found in wet coastal areas may result in poor tree growth simply because root development is restricted to a small volume of impoverished surface soil. While there are notable differences in tolerance to oxygen deficiencies among tree species and within the same species at different stages of growth, a soil air content as low as 2 percent is generally not harmful to most trees for short periods. However, seedling root growth of many species is reduced when oxygen levels fall to 10 percent or lower. Some *Alnus, Taxodium, Nyssa,* and *Picea* species can thrive on lower levels of soil air. However, a restriction in gas exchange between the tree roots and the atmosphere will eventually result in the accumulation of carbon dioxide and other toxic materials.

Soil microorganism populations are greatly influenced by aeration. Aerobic organisms are unable to function properly in the absence of gaseous oxygen. The reduction in rate of organic matter decomposition and accumulation of plant residues in swampy areas is an excellent example of this fact. Anaerobic and facultative organisms can utilize combined oxygen and reduce such elements as iron and manganese, thus sometimes producing toxic conditions.

## SOIL COLOR

Color is perhaps the most obvious and one of the significant characteristics of soils. It is used in differentiating soil horizons and in some countries it is used extensively in soil classification. For example, the Red and Yellow Podzolics, Brown Earths, Brown Forest soils, and Chernozens (black soils) are great soils groups used in soil classification in many parts of the world.

### Factors Influencing Soil Color

Soil color depends on pedogenic processes and the parent material from which the soil was derived. Most soil minerals, such as quartz and feldspars, are light in color. Color is generally imparted by small amounts of colored materials, such as iron, manganese, and organic matter. Red colors due to ferric compounds are associated with well-aerated soils, while yellow colors signify intermediate aeration. Ferrous compounds of blue and green colors are often found under reduced conditions associated with poorly aerated soils. Mottling often indicates a zone of alternate good and poor aeration. Manganese compounds and organic matter produce dark colors in soils. In siliceous soils, color intensity is sometimes used as a means of estimating their organic matter content. However, this is not a foolproof system, even for the well trained, because the pigmentation of humus is less intense in humid zones than in arid regions. Brown colors predominate in slightly decomposed plant materials, but the more thoroughly decomposed amorphous material is nearly black.

### Significance to Tree Growth

Color in itself is of little importance to tree growth. However, it serves as an indicator of several important characteristics of soil, such as geologic origin and degree of weathering of the soil material, degree of oxidation and reduction, content of organic material, leaching or accumulation of such chemical compounds as iron, which may greatly influence site quality. Color is not an infallible indicator of soil conditions, due to other overriding factors. For example, the gray color of a gley horizon in poorly drained soils is due to ferrous compounds, but a similar color results from leaching of iron and organic matter from the A2 horizon of spodosols. Dark-colored surface soils absorb heat more readily than light-colored soils, but because of their generally higher content of organic matter, they often have higher moisture contents. Therefore dark soils warm less rapidly than well-drained light-colored soils. While it should be noted that soil color influences the temperature of bare soils, it has much less effect on the temperature of forested soils.

## SOIL TEMPERATURE

Soil temperature is a balance between heat gains and losses. Solar radiation is the principle source of heat, and losses are due to radiation, conduction, and convection. Surface soil temperature varies more or less according to the temperature of the air immediately above it. Therefore surface soils have greater temperature fluctuations than subsoils.

### Factors Influencing Soil Temperature

There is a general decrease in soil temperatures with increases in latitude and elevation. This, of course, accounts for the fact that the closer one gets to the equator, the greater the altitude at which trees survive in mountainous areas. However, soil temperatures do not decrease as rapidly or fluctuate as widely as air temperature, with increasing altitude. A point worth remembering is that as the angle of slope increases, the amount of radiant energy received per unit area decreases. Furthermore, aspect or direction of slope exposure influences soil temperature. In the Northern Hemisphere, soil temperatures are higher, on the average, on south and west exposures than on north and east slopes.

The tree canopy and forest floor moderate extremes in soil temperatures. They protect the soil from excessively high summer temperatures by interception of solar radiation, and they reduce the rate of loss of heat from the soil during the winter. However, the influence of stand canopy on soil temperature is greater in the summer than in the winter. In the summer, surface soil temperatures under a closed canopy may be 22° to 28°C cooler than in the open (Lutz and Chandler, 1946), and a dense canopy has a much greater influence on soil temperatures than an open canopy. Forest cover influences the development and persistence of frost in soils in cold climates. Freezing generally occurs earlier and penetrates deeper in bare soils than in soil under a forest cover.

Part of the ameliorating effect of the forest cover on soil temperature is due to the litter layer, as shown for a clay soil in Figure 7.4. The unincorporated organic matter not only lowers the maximum summer temperature and raises the minimum winter temperatures of the surface soil, but reduces the day-to-day and diurnal changes as well. The litter layer probably has less effect than the forest canopy on reducing maximum surface soil temperatures, but it has a much greater effect than the canopy in raising the minimum surface soil temperature

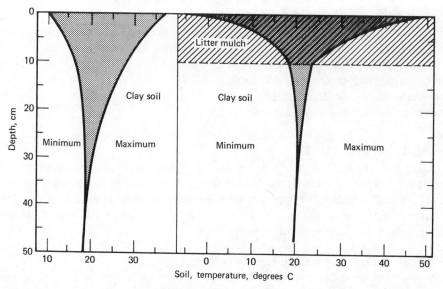

**Figure 7.4**  Daily temperature variations with depth for (left) an unmulched and (right) mulched clay soil where the litter mulch has a lower thermal conductivity (Cochran, 1969).

during cold weather. Litter also reduces the depth of frost penetration, in most instances. Frost penetration may be greater under coniferous cover than under hardwood cover due to greater snow depth under hardwoods. Differences in the forest floors and in soil horizon development between the two forest types may also influence frost penetration.

Snow also has a stabilizing effect on soil temperature because of the insulation effect resulting from its low thermal conductivity. A porous snow covering of about 46 cm was sufficient to prevent soil freezing in northern Sweden (Beskow, 1935), and a covering of 20 to 30 cm was sufficient to prevent freezing in southern Sweden. Winter soil temperatures may often be higher in the northern part of the United States, where the snow cover is thick, than at more southern locations where a snow cover is thin or lacking. Sartz (1973) reported that aspect influences soil freezing in Wisconsin because direction of slope affects the amount of solar radiation received at the ground or snow surface. Radiation at the snow surface affects soil freezing because it affects snow accumulation, and snow retards soil freezing by insulating the ground. Therefore depending on the conditions, frost may be deeper on southern slopes than on northern slopes, as shown in Table 7.2.

*Specific heat* and *conductivity of heat* are two inherent properties of soils that influence their daily temperature. Specific heat refers to the calories needed to raise a gram (or cubic centimeter) of a substance 1°C. It can be thought of as

**Table 7.2**   Snow and Frost Depths on North and South Slopes in Wisconsin, 1970–1973 (Sartz, 1973)

| Year and Date | Snow | | Frost | |
|---|---|---|---|---|
| | North | South | North | South |
| | Centimeters | | | |
| 1970 (Feb. 25) | 25 | 0 | 8 | 11 |
| 1971 (Feb. 25) | 55 | 25 | 0 | 0 |
| 1972 (Mar. 10) | 45 | 25 | 16 | 23 |
| 1973 (Mar. 1) | 5 | 0 | 21 | 12 |

the thermal capacity of a soil as compared to that of water. Conductance refers to the movement or penetration into the soil profile of thermal energy. Both specific heat and conductance are influenced by texture, as well as the moisture and organic matter content of a soil. That is, as the percentage of silt and clay, organic matter, or moisture increases the more heat is required to raise the temperature of a given weight of soil. However, calculated on a volume basis, specific heat decreases with increase in fine-textured materials or organic matter.

A dry soil has a lower specific heat than when it is wet (because the specific heat of water is four or five times that of a dry soil), but its thermal conductivity is also low, resulting in a rapid reduction in temperature with depth beneath a heated surface. Moist soils have relatively high thermal conductivity, but, because of their high specific heat, they do not get as hot as dry soils. This is why moist soils are generally considered to be *cold* soils. The high conductivity of moist soils results from the fact that water that fills pore spaces is a better conductor of heat than the air that fills pores of dry soils.

## Importance to Tree Growth

Favorable soil temperature is essential for the germination of seeds and the survival and growth of seedlings. It also influences microbial activity, nutrient availability, and metabolic activity and growth rate of trees. A dense canopy may keep soil temperature so low that it delays germination and slows up seedling development in some cold climates. On the other hand, the removal of the forest cover may result in increases in soil temperatures to lethal levels for germinating seedlings. These lethal soil temperatures may be no higher than 54°C for some species. However, this extreme will usually not be reached if there is a mulch or litter layer covering the soil.

Root growth is also affected by soil temperatures, as shown by Lyford and Wilson (1966). They found that day-to-day variations in growth rates of red maple root tips closely paralleled variations in surrounding soil temperatures.

Kaufman (1968) reported that in the coastal plains of the southeastern United States, root growth slows during the cool months but apparently does not completely cease. On the other hand, the rate of root elongation may significantly decrease during hot and dry midsummer periods. It is likely that the influences of extreme temperatures on root growth in the coastal plains are largely indirect through their effects on soil moisture content and on water uptake by the plant. In any event, it is obvious that the many physical properties of soils discussed here are closely interrelated, and they exert considerable influence on chemical and biological soil characteristics and on tree survival and growth.

## SILVICULTURAL IMPLICATIONS

Soil physical properties have long been considered to exert great influence on the distribution and growth of trees. Recently much attention has been directed toward the equally important influence of chemical and biological properties and the interrelation of all three groups of properties on each other and on site productivity. For example, such physical properties as texture, structure, and color have more indirect effects on plant growth than direct effects because it is through the influence of these physical properties on soil moisture, temperature, aeration, and nutrient availability that survival and growth of trees are possible.

Most physical properties are inherent basic properties that are not easily altered. The exceptions are those related to structure and porosity. There are many instances of problems resulting from mechanical soil disturbance such as compaction, puddling, displacement, and disruption of surface drainage. The direct effects of disturbances are on the soil air-water system and on soil strength properties affecting root penetration. The physical problems created by disturbance also affect the chemical and microbiological soil condition indirectly and, thus, they can be said to directly affect soil fertility and site productivity.

The fact that soil physical properties, including soil depth, are relatively stable and have decisive silvicultural importance makes them valuable criteria on which to base soil classification schemes. Many of these physical features can be readily determined in the field. Soil color, texture, and thickness of the A1 horizon can be accurately and rapidly gauged after moderate training. However, such features as effective rooting depth (depth to bedrock, waterlogged horizons, impervious pans, or layers containing toxic concentrations of salts) can usually be determined only by extensive probing. The accurate determination of such properties as particle-size distribution, pore volume, and water-holding capacity can only be made by laboratory analyses on appropriate samples.

It is usually not practical to use more than three or four soil physical properties in an operational classification and mapping program for site productivity because of the excessive time and labor involved. The goal of the forest soil scientist is to be able to select those few properties that have the greatest influence on tree survival, growth, and development in that area as the basis for delineating soils. In large mapping operations it is important that most or all of these properties be easily measured in the field. If the soil physical properties that meet these criteria can be identified for a given area, a very useful soil classification and mapping program can be developed for many forest management uses.

# 8

# SOIL WATER: MEASUREMENTS AND MOVEMENTS

The role of soil water in forest management assumes an unusually important place because a high percentage of the lands that are either too wet or too dry for agricultural use are relegated to forestry. The supply of moisture in soils largely controls the type of tree which can be grown and, thus, it influences the distribution of forests around the world. Soils that are abundantly supplied with minerals are completely unproductive without water, but even impoverished sands may support reasonably productive forests if supplied with adequate moisture. Abundant and well-distributed rainfall generally results in lush vegetation, best exemplified by tropical rain forests, but where summer droughts are frequent and severe, forests are replaced by grasslands and prairies. Semideserts and deserts result from still further decreases in precipitation.

Water is essential to the proper functioning of most soil and plant processes. Water affects directly or indirectly almost every plant process, and, to a large extent, the level of metabolic activity of cells and plants is related to their water content (Kramer, 1969). In addition to serving metabolic needs of the plant, water performs many functions in the soil. It is a solvent and medium of transport of plant nutrients, a source of hydrogen, a moderator of soil temperature and aeration, and a diluter of toxic substances in soils.

## SOIL WATER CHARACTERISTICS

Most areas of the United States are rather well supplied with water. During an average year, precipitation on the conterminous states is equivalent to 76 cm of rainfall, if evenly distributed over the landscape. Of our total water budget, 23 cm of precipitation (95.9 million hectare-meters) annually enter into the massed flow of our rivers. This leaves about 53 cm of rainfall (236.6 million hectare-meters) that infiltrates the soil and is eventually used in evapotranspiration from the vegetated lands on which it falls (Wadleigh, 1964).

All water that falls on land surfaces and is not lost by runoff enters the soil where it may be used in evapotranspiration, lost by percolation, or retained for a period of time in the soil pores. Soil pores form a continuous but complex system of varying-sized spaces, which usually constitutes 30 to 60 percent of the total soil volume. These pores may be filled entirely by water, as in a saturated soil, or largely by air, as in dry soils. Most soil pores will normally be partly filled with air and partly with water. Continuity of the water phase from pore to pore is highly important in regard to water and salt movement in the soil.

## Soil Water Energy

The retention and movement of water in soil and plants involve energy transfers of various types, generally categorized as free energy of water. Water molecules are attracted to each other in a polymerlike grouping, forming an open tetrahedral lattice structure. This asymmetrical arrangement results from the dipolar nature of the water molecule. The hydrogen atom of one molecule attracts the oxygen atom of an adjacent molecule in a kind of hydrogen bonding. Hydrogen bonding accounts for the forces of adhesion, cohesion, and surface tension that largely regulate the retention and movement of water in soils. *Adhesion* refers to the attractive forces between soil and water molecules. At the water-air interface, *surface tension* may be the only force retaining water in soils. It results from the greater attraction of water molecules for each other than for the air.

Water tends to move from a zone of high free energy to one of low free energy—from a wet soil to a dry soil. However, the amount of movement depends on the differences in the energy states (potential) between the two zones. Forces that affect the free energy of soil water are the attraction of the soil solids for water by adsorption and capillarity, called *matric* force, and the attraction of ions and other solutes for water result in *osmotic* forces, both of which tend to reduce the free energy of the soil solution. *Gravity* is a third force acting on soil water, tending to move it from a higher elevation to a lower level. Total potential of soil water, therefore, is the sum of matric, osmotic, and gravitational forces (plus other minor forces).

Gravitational potential is a positive force, but both matric and osmotic potentials are negative forces because they reduce the free energy level of water. The terms *suction* and *tension* are used as positive expressions of the forces of the two negative potentials that result from the attraction of soil for water.

The energy required to remove water from soil pores or from the attraction of soil particles can be measured by applying suction to a saturated soil sample placed over a permeable membrane, and is expressed either in atmospheres (or bars) of pressure or height of a water column needed to create such pressure. A *bar* is a unit of pressure equal to one million dynes per square centimeter. More commonly an atmosphere, or bar, identifies the suction created by a water column 1000 cm high.

In order to simplify the method of expressing tension of a water column, Schofield (1935) proposed a pF scale that is the logarithm of the column height in centimeters. That is, the tension that is equivalent to 10,000 cm of water would have a capillary potential of pF 4.0 (or approximately 10 atmospheres of pressure). In recent years bars have been more commonly used than pF to express the tension with which water is held in the soil.

Soil water can be conveniently grouped into three fractions on the basis of the force with which it is held in the soil: *gravitation* water drains away under the force of gravity through the large soil pore spaces (water in excess of 0.1 to 0.2 bar suction); *capillary* water is retained in the capillary pores and around soil particles by forces of cohesion and adhesion after gravitational water has moved out; *hygroscopic* water is retained very firmly as a thin film around individual soil particles after capillary water has been removed. It is essentially nonliquid and moves primarily in the vapor form. It is not available to higher plants.

## Moisture Constants

Investigators have long attempted to devise useful equilibrium points or constants in describing soil moisture. Such terms as *field capacity, moisture equivalent, hygroscopic coefficient,* and *permanent wilting point* have found their way into soils literature over the years. Most of these terms deal with hypothetical concepts and do not apply equally well for all soil conditions. Nevertheless, they have some usefulness in forest soil studies and, consequently, they will be reviewed briefly.

When water from a steady heavy rain passes through the forest floor and enters the mineral soil, air is displaced and the soil pores, both large and small, become filled with water. At this point, the upper part of the soil is saturated and is at its maximum retentive capacity. With the cessation of the rainfall, there is a relatively rapid downward movement of some of the water until, after a day or so, this movement will essentially cease and gravitational water will have drained away.

**Field Capacity**   This is a term used to describe the amount of water held in the soil after gravitational water has drained away and, for all practical purposes, the downward movement has stopped. Field capacity is an excellent concept for expressing the upper limit of plant available soil moisture. However, its usefulness is reduced by the difficulty of securing accurate determinations. Such factors as soil textural and structural properties, organic matter content, and depth to a water table significantly influence the values obtained. The presence of soil layers of different pore size markedly influences water flow and, hence, field capacity. Consequently, it is difficult to know how long the soil should be permitted to drain after saturation and before moisture content is determined. One or two days should be adequate under most conditions, resulting in a pF value of about 2.5, or $\frac{1}{3}$ bar, for silt loam soils, and a somewhat lower value for sands. One-tenth bar suction is often used to approximate field capacity in sands.

**Moisture Equivalent**   The amount of water held in a sample of sieved soil after it has been subjected to a centrifugal force of 1000 times gravity is termed moisture equivalent. It is sometimes used as an approximation of field capacity. It is a convenient determination, but because it is made on a disturbed sample it may have little relation to field capacity of undisturbed forest soils. The term is included mostly for its historical significance.

**Hygroscopic Coefficient**   This term is used to describe the amount of hygroscopically bound water in soils. It is sometimes used to mark the lower limit of available water. Hygroscopic coefficient is the moisture content of a soil that has lost its liquid water from even the smallest of micropores. The remaining water is absorbed on the soil particles, particularly the colloids, as a nonliquid and can move only in the vapor phase. Since it is usually determined by exposing a thin layer of dry soil to a saturated atmosphere for 24 hours, a *true* equilibrium is seldom reached. Furthermore, a soil that approaches an equilibrium point as a result of drying has more water than one that reaches equilibrium as a result of wetting, due to a hysteresis effect.

**Permanent Wilting Point**   The moisture content of a soil at which plants remain permanently wilted unless water is added to the soil is called permanent wilting point, or permanent wilting percentage. Just as field capacity has been widely used to refer to the upper limit of soil water storage for plant growth, the permanent wilting percentage is used to define the lower limit. The use of a test plant, such as a sunflower, is the most widely accepted method for determining this soil constant, but this awkward and time-consuming method is not available to most foresters.

The need for a more practical method for determining permanent wilting percentage in soils is obvious. The use of the hygroscopic coefficient or dividing the moisture equivalent value by 1.84 have been proposed, but they have not proven to be satisfactory. The amount of moisture retained at a tension of 15 bars exerted by a pressure membrane apparatus, as described by Richards and Richards (1957), is probably the best approximation of this moisture content, short of using a test plant.

These moisture constants are used in attempts to describe the *available water-holding capacity* of soils. The readily available soil water is considered to be the amount of moisture retained in the soil between field capacity and permanent wilting point. It is obtained by calculating the difference in moisture content of a soil that has been subjected to $\frac{1}{10}$ bar (or $\frac{1}{3}$ bar for fine-textured soils) and then 15 bars of tension. It is worth pointing out, however, that the capacity of deep-rooted trees to use water from the capillary fringe above a water table reduces the values of these moisture constants for some forest soils.

**Retention and Detention Storage**    Retention and detention are two constants used to describe soil water storage in watershed studies. Retention storage refers to water that can be held in soil capillary pores, expressed as centimeters of water per given depth of soil. It is termed retention storage because this water is retained in the soil and does not contribute directly to stream flow or ground water supplies. By contrast, water in the noncapillary pores is called detention storage because it is only temporarily held. Depending on permeability of horizons, detention water will either flow laterally to stream channels or downward to satisfy capillary moisture deficits and eventually to the water table (Hoover, 1949). Examples of retention and detention storage at different depths in an Ultisol are given in Table 8.1.

## SOIL WATER MEASUREMENTS

Measurements of soil moisture contents and water-holding characteristics of soils are often useful for management purposes, particularly for nurseries, seed orchards, and other special tree crops. Detailed discussions of the subject can be found in soil physics texts and other soils or plant physiology publications (Kramer, 1969; Brady, 1974).

Changes in water storage, commonly used to estimate evapotranspiration, are given by the equation

$$\Delta W = P - (O + U + E)$$

where $\Delta W$ represents the initial water content minus the final content during the period of measurement. (The $P$, $O$, and $U$ represent precipitation, runoff, and deep drainage beyond the rooting zone, respectively, and $E$ is evapotranspiration from plant and soil surfaces. The expression can be used for changes in soil water from large watersheds down to the soil volume supporting individual plants, but in all cases some measuring devices must be employed.

The use of large weighing or floating lysimeters (Patric, 1961) can be one of the most accurate means of determining $\Delta W$. However, the mixed nature of species composition, spatial distribution of the vegetation, depth and ramification of the root system, and size of plants make it difficult, if not impossible, to simulate natural forest conditions in the lysimeters. These factors place a limitation on the usefulness of lysimeters in forest soil moisture studies. Simpler methods for direct and indirect measurements of soil moisture content are generally used to calculate changes in soil water over time.

**Table 8.1** Retention and Detention Water Storages in a South Carolina Piedmont Soil (Ultisol) Under Undisturbed Forest Cover and Under Pine on Abandoned Farm Land (Hoover, 1949)

| Depth from Soil Surface | Undisturbed Forest | | Abandoned Farmland | |
|---|---|---|---|---|
| | Retention | Detention | Retention | Detention |
| | cm | | cm | |
| cm | | | | |
| 0–15 | 4.4 | 4.9 | 4.0 | 2.1 |
| 15–30 | 4.8 | 2.8 | 6.0 | 0.3 |
| 30–45 | 5.1 | 1.6 | 5.9 | 0.2 |
| 45–60 | 5.9 | 1.0 | 5.8 | 0.2 |

## Gravimetric Determination

This is the best-known method and, perhaps, the most accurate technique for measuring soil moisture, but it can be rather time consuming. By this procedure a weighed sample of wet soil is placed in an oven at 100 to 105°C and dried to a constant weight. The moisture lost by heating represents the soil moisture present in the moist sample. The percent soil moisture on a dry weight basis can be calculated by dividing the weight of soil water by weight of the dry soil and multiplying by 100.

## Tensiometers

These instruments result in direct measurement of the capillary potential as well as estimating soil water content. They consist essentially of a porous ceramic cup filled with water that is buried in the soil at a desired depth and connected by a water-filled tube to a manometer or vacuum gauge. The manometer indicates the pressure drop on the water in the cup, which is in equilibrium with the matric potential of the water in the soil. The tensiometer is an excellent measuring instrument in moist soils, but when soil moisture drops below about 0.8 bars suction it is not useful because of leakage (Richards, 1949).

## Resistance Blocks

These electronic instruments consist of a pair of electrodes embedded in a small block of nylon, fiberglass, or gypsum. The block is buried in the soil and connected by insulated leads to a resistance bridge. The blocks absorb moisture from moist soil until an equilibrium is reached. By calibrating the resistance readings with known soil moisture contents, one can approximate the amount of moisture in field soils. The blocks can be left in the soil for monitoring changes in soil moisture over rather long periods, although gypsum blocks are prone to disintegrate after a few months. Resistance blocks can be used to measure either moisture tension directly, or percentage moisture after calibration. Blocks are most sensitive at tensions from 1 to 15 bars suction and are not useful at high moisture contents (Bouyoucos, 1954).

## Neutron Scattering

This is an indirect method of measuring soil moisture content that has recently gained popularity. This method is based on the fact that hydrogen atoms have a much greater ability to slow and scatter fast neutrons than most other atoms, so that counting slow neutrons in the vicinity of a source of fast neutrons provides a means of estimating hydrogen content. Since water is the only significant source of hydrogen in most soils, the technique is a fairly accurate and convenient means of measuring water content in the soil profile. Where soil organic matter content or root density is high, organic hydrogen will affect the results, but the error may be acceptable in most mineral soils. However, the instrument cannot be used with accuracy in organic soils. The technique is one that can be used without repeated disturbance of the soil and it can be used in soils containing salts (Brady, 1974).

## Pressure Membrane and Tension Plate

Both instruments are useful for laboratory determinations of the relationship of matric potential to soil moisture content. The principles involved with the two instruments are quite similar. A soil core is placed firmly on a porous plate and, in the case of the pressure membrane, a pressure is created by introducing gas into the soil chamber. Water is forced out of the soil through the plate into a chamber maintained at atmospheric pressure. With the tension plate, a known suction is applied and the soil core is permitted to reach equilibrium with the tension on the plate. With both instruments, the soil moisture content is determined and related to the pressure or suction applied. The pressure membrane apparatus will measure much higher soil suction values than will the tension plate. The relations of soil moisture tension to soil moisture content for two soils of different textures are shown in Figure 8.1.

## Pressure Bomb

The pressure bomb is not really a device for measuring soil moisture but an instrument for measuring the internal water balance of plants as a means of integrating the several factors that influence soil water potential. Harms (1969) reported that water potential cannot be reliably estimated from soil moisture content and leaf water deficit alone because of the mutual relationship among the three variables. However, with information on soil moisture tension, as obtained from a moisture retention curve (Figure 8.1), the variations of residual leaf water deficit with variation of soil moisture content can be explained in terms of water potential.

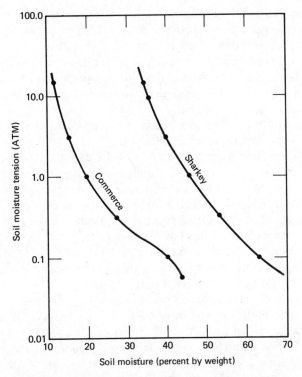

**Figure 8.1**  Relations of soil moisture tension to soil moisture content in Sharkey clay and Commerce silt loam (Bonner, 1968).

## WATER MOVEMENT IN UNSATURATED SOILS

Soil water is held largely by matric forces that bind water on the soil particle and, to a lesser extent, by osmotic forces developed by the salts dispersed in the soil solution. The availability of soil water to plants depends on its potential and on the hydraulic conductivity of the soil (Kramer, 1969).

Water movement when the soil pores are not completely saturated is of greater consequence to tree growth than that of saturated flow. Under unsaturated conditions, macropores are mostly filled with air and micropores with water and water movement is very slow. The rate of water movement through a soil, or its hydraulic conductivity, generally decreases with decreasing pore space. Movement of water through the soil is controlled chiefly by the gravitational potential in soils above field capacity and by the matric potential in soils drier than field capacity. Hydraulic conductivity decreases rapidly with decreasing water potential. Movement of liquid water is very slow in a dry soil and practically ceases at a water potential of approximately 15 bars. In other words, water moves only as a vapor in very dry soil. Differences in temperature between surface soil and deeper horizons result in measurable upward movement of water vapor in the winter and downward movement in the summer. The diffusion of water vapor from one area to another of lower moisture content occurs in relatively dry soils under the driving force of a vapor pressure gradient. However, it probably plays little part in water movement in forest soils, except during severe droughts.

There appears to be more movement of water in soil below field capacity than was once supposed (Kramer, 1969). This is particularly true of the upward movement of water from a shallow water table by capillary conductivity. However, capillary conductivity decreases rapidly as soil moisture content decreases, particularly in sandy soils. At 0.2 bar tension, for example, conductivity is 10,000 times that at 10 bars tension. At high tension, water in sands is held only at points of contact between the relatively large particles. Under these conditions there is no continuous water film and thus no opportunity for liquid movement. If water transfer takes place, it must move in the vapor state. Layers of sand in a profile of fine texture soil, therefore, can inhibit downward movement of water in a similar fashion to compact clay or silt pans (Brady, 1974).

*Stratification* of materials of different texture is common in soils. It results from horizon differentiation and from geological differences in materials and methods of transport. These can result in silt or clay pans or sand or gravel lenses. Layers of fine textured materials over coarse sands and gravel, as well as the reverse situations, are common in glacial outwash sands of the northern part of North America and to a lesser extent in the fluvial and marine deposits of sands in coastal plains. In any event, changes in texture from that of the overlying material result in conductivity differences that slow down movement. Textural layers can act as a moisture barrier until a relatively high moisture level is reached in the overlying soil. When a wetting front contacts a finer textured (smaller pores) material than that in which it is moving, the fine pores begin to fill rapidly because of their greater attraction for water. If the layered material has very fine pores as found in clay soils. resistance to movement may be so great as to markedly reduce flow. In the reverse situation, when a wetting front moving in a soil of relatively fine porosity contacts a layer of coarser textured material, the volume of pores capable of holding water at the tension which exists at the wetting front is reduced. Before the wetting front can advance in this layer, the moisture tension at that point must decrease until it is low enough to allow the pores to fill with water. Eventually the moisture tension becomes low enough that a significant number of large pores can fill, as seen in Figure 8.2. In this example, Miller and Gardner (1962) constructed layers of different-sized aggregates of a silt loam soil. Infiltration was not inhibited as much by the aggregate layers as by layers of sand of equivalent particle size because of the existence of small pores within the aggregates that allowed some water to move into the layer

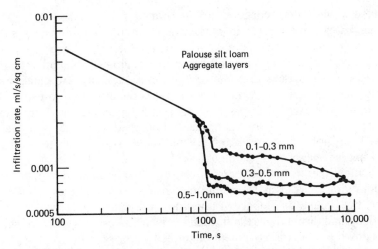

**Figure 8.2** The effect of subsurface aggregate layers on infiltration rates into a silt loam soil as a function of time for downward wetting. Layers were 0.5 cm thick at a depth of 8 cm (Miller and Gardner, 1962). Reproduced from *Soil Science Society of America Proceedings,* Volume 26, No. 2, Page 118, 1962 by permission of the Soil Science Society of America.

from the wetting front. Nonetheless, most any type of layering significantly influences rate of water infiltration and movement in soils and can result in a higher moisture content at "field capacity" than is justified.

## Factors Affecting Infiltration and Losses of Water
The intensity and total amount of annual precipitation, as well as the potential evapotranspiration, affect the amount of water entering the soil during the year.

**Infiltration**  The term infiltration is generally applied to the mode of entry of all water into the soil. The rate of water entering the soils is determined more specifically by initial water content, surface permeability, internal characteristics of the soil (such as pore space, degree of swelling of soil colloids, and organic matter content), intensity and duration of rainfall, and temperature of the soil and water (Kramer, 1969). Only when rainfall intensity exceeds the infiltration capacity of a soil can runoff occur. By virtue of the spongelike action of most forest floors and the high infiltration rate of the mineral soil below, there is little opportunity for surface runoff of water in mature forests. Overland flow is normally not a serious problem in undisturbed forests, even in steep mountain areas.

When rainfall does exceed infiltration capacity, the excess water accumulates on the surface and then flows overland toward stream channels. This unimpeded surface flow concentrates in defined stream channels and causes greater peak flows in a shorter time than water that infiltrates and passes through the soil before reappearing as streamflow. High-velocity overland flow can cause erosion damage. Soil compaction by harvesting equipment, disturbances by site preparation, and practices that reduce infiltration capacity and cause water to begin moving over the soil surface are of great concern to all forest managers. Special care is warranted on shallow soils and those that have inherent low infiltration rates. Soils that are wet because of perched water tables or because of their position along drainage ways are unusually susceptible to reduction in porosity and permeability by compaction.

The litter layer beneath a forest cover is especially important in maintaining rapid infiltration rates. This layer not only absorbs several times its own weight of water, but breaks the impact of raindrops, prevents agitation of the mineral soil particles and discourages the formation of surface crusts (Wooldridge, 1970). It slows down the lateral movement of surface water permitting a longer period for its infiltration. Infiltration rates through a crumblike mull humus and associated mineral soil are generally greater than through the more compacted mor humus and soil found beneath such cover. The incorporation of organic matter into mineral soils, artificially or by natural means, increases their permeability to water as a result of increased porosity. Forest soils have a high percentage of macropores through which large quantities of water can move—sometimes without appreciable wetting of the soil mass. Most macropores develop from old root channels or from burrows and tunnels made by insects, worms, or other animals. Some result from structural pores and cracks in the soil. As a consequence of the better structure, higher organic matter content, and the presence of channels, percolation rates in forest soils are considerably greater than in similar soils subjected to continued cultivation (Figure 8.3). Wood (1977) found that water infiltration rates were higher on 14 of 15 forest sites than in adjacent sites used for pastures, pineapple, or sugar cane production in Hawaii. In this study, lower bulk densities and greater porosities were found in forest-covered soils than in nonforested soils. These same properties influence the retention and detention storage capacities of soils, as shown in Table 8.1. Here the undisturbed forest soil had greater detention storage at all soil depths than the abandoned farm soil, but the latter had greater retention storage at the 15- to 45-cm depths.

The presence of stones increases the rate of water infiltration in soils because the differences in expansion and contraction between stones and the soil result in channels and macropores. However, stones reduce the retention storage capacity of soils, as shown in Figure 8.4 for 40 Oregon sites. A good example of a stony soil profile was shown in Figure 2.2.

If snow covers the soil before prolonged freezing weather in the fall, it may protect the soil water from freezing, thus favoring continued permeability during the winter. But if the soil freezes before a snow cover, a later snow cover will

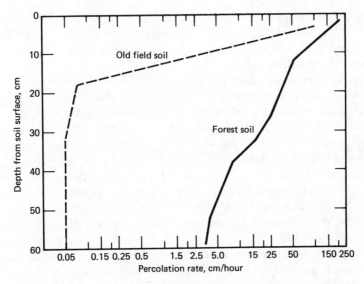

**Figure 8.3** Comparative percolation rates by soil depths for a forest soil and an adjacent old field soil in the South Carolina Piedmont (Hoover, 1949).

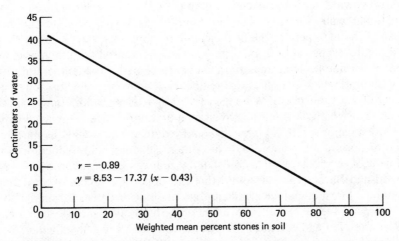

**Figure 8.4** Retention storage capacity in the surface 120 cm of 40 soil profiles representing four soil series in Oregon as a function of average stone content (Dyrness, 1969).

delay thawing in the spring, reduce soil in the spring, reduce soil infiltration and storage, and increase surface runoff. In the transition zone of the boreal forest and in the tundra, permafrost prevents water infiltration and results in mires even though total precipitation is rather low.

Infiltration from heavy rains generally decreases with steepness of slope, but the microrelief created by uprooting of trees in old stands, rock slides, slippages, and other causes provides catchment basins that favor infiltration on even the steepest slopes.

Burning of watersheds may result in a well-defined water-repellent layer beneath the surface ash (DeBano et al., 1970). This condition has been reported to cause excess soil erosion and water runoff, and to impede the reestablishment of vegetation on range land. It is thought that this heat-induced water repellency results from the vaporization of organic hydrophobic substances at the soil surface during the passage of fire and their subsequent condensation in the cooler underlying portions of soil. The substances responsible have been identified as basically aliphatic hydrocarbons derived from partially decomposed plant materials (Savage et al., 1972).

**Water Losses**  Once water infiltrates a soil its loss can result from gravity, evaporation, and transpiration. Losses by gravity may be quite low in regions of low rainfall or in warm climates where evapotranspiration rates are high. As moisture is added to the surface of a uniformly dry soil, the moisture content of the affected layer is raised to its capillary capacity and little effect is noted below this layer. With further additions of water, the lower boundary of the wet zone moves downward until the excess water passes into the ground water table. Gravitational water moves faster in sandy soils than in fine-textured soils because of the high percentage of noncapillary pores.

Moisture evaporation from surface soils is increased with increases in the percentage of fine-textured materials, moisture content, and temperature of the soil. Evaporational losses are also influenced by wind velocity and relative humidity of the atmosphere. Wind movement increases evaporation by speeding up the rate at which moist air above an evaporating surface is replaced by drier air. A forest canopy with *no litter layer* may reduce evaporation as much as 50 percent of that from a nonforested soil, due to shading and a reduction in wind movement. The presence of a forest floor reduces evaporation even further. Heyward and Barnette (1934) found that the upper soil layers of an unburned longleaf pine forest contained significantly more moisture than did similar soils in burned forests, due to the mulching effect of the litter layer in the former areas.

The retention and detention storage capacity of a soil also influence the loss of infiltrated water in a soil. A fine-textured soil has a higher retention capacity than sands and can store larger amounts of water following storm events. However, the effective depth of the soil (rooting depth) and the initial moisture content are factors that also influence the amount of rain water that can be retained in soils for later use by plants.


## GROUND WATER TABLE

The upper level of the zone of saturation in a soil is called the ground water table. Extending upward from the water table is a belt of moist soil known as the *capillary fringe*. In fine textured soils this zone of moisture may approach a height of a meter or more, but in sandy soils it seldom exceeds 25 to 30 cm. In some soils the height of the water table fluctuates considerably between wet and dry periods. In a forested soil in New England (Lyford, 1964) water tables were highest during late fall, winter, and early spring, gradually lower in late spring and early summer, and higher again in the fall. In the flatwoods of the United States southeastern coastal plain, the water table may be above the soil surface during wet seasons and fall to a depth of a meter or more during dry seasons. These fluctuations are due to variations in precipitation, transpiration, evaporation, and temperature.

Diurnal changes in the water-table levels have also been noted, due to differences in transpiration rates between day and night (Lyford, 1964). A rather consistent daytime drop was accompanied by a rise in the water table during the night. Diurnal changes are most pronounced in gently sloping watersheds with perched or shallow water-tables where the capillary fringe is in reach of tree roots.

Soil depth reflects the volume of growing space for tree roots above some restricting layer, and in wet areas this restricting layer often is a high water table. The water table may be a temporary (perched) zone of saturation above some impervious layer, following periods of high rainfall. The effects of such a water table on tree growth may be approximated by measuring depth to a fine-textured horizon or depth to a mottled horizon. These two variables account for a high percentage of the total variation in tree growth in coastal sands. The best growth rate of pine obtained where either the fine-textured horizon or mottling (an indication of poor aeration) occurs about 75 cm below the surface (Barnes and Ralston, 1955). The absence of a moisture-retaining fine-textured layer within the zone permeated by tree roots results in poor tree growth on deep sand ridges, while the lack of root-growing space in the low flat areas is often caused by a high water table.

Large seasonal fluctuation in depth to the water table are harmful to root development. Roots that develop during dry periods in the portion of the profile that is later flooded are killed under the induced anaerobic conditions. This periodic root pruning may create an imbalance in the root-top ratio. A reasonably high water table is not detrimental to tree growth as long as there is little fluctuation in its level. In an experiment in the southeastern coastal plain, slash pines grown for five years in plots where the water table was maintained at 45 cm by a system of subsurface irrigation and tile drains were 11 percent taller than trees in plots where the water table was maintained at 90 cm, but they were 60 percent taller than trees in plots where the water table was allowed to fluctuate normally (White and Pritchett, 1970).

The primary benefits derived from draining some nutrient-poor wet soils may result more from an increase in the nutrient supply than an increase in the soil oxygen supply. The improved nutrition is brought about by an increase in the volume of soil available for root exploitation, as well as an acceleration in mineralization of organic constituents. For example, fertilizer experiments in wet soils of the coastal plain indicated that increasing the fertility of undrained soils resulted in tree growth as rapid as that obtained on unfertilized, but drained, soils (Pritchett and Smith, 1974).

Trees obtain moisture from the water table or the capillary fringe within reach of their roots, even when the water table has been lowered to a considerable depth. It is apparently for this reason that the water tables of some wet soils are lowered to a greater extent where trees are grown than where grasses and other shallow rooted plants are grown. Wilde (1958) reviewed a number of reports of general rises in the soil water table following the removal of a forest cover on the plains of temperate and cold regions. The increase in the water-table level results from a reduction in interceptional and transpirational surfaces following the removal of the forest canopy (Trousdell and Hoover, 1955). Such an increase apparently does not hold true for mountainous and tropical areas. The higher ground water table under forests than in open areas in the tropics may be due to greater permeability of forest soils or to the greater evaporation from open areas during hot periods.

## SOIL WATER ABSORPTION BY TREE ROOTS

Tree roots absorb vast quantities of water to replace that lost by transpiration and that used in metabolic activities. Under favorable conditions this loss can be as much as 6 mm per day (75,000 liters per hectare) during early summer. Trees use 300 to 500 kg of water for each kilogram of dry matter produced. This water requirement may be much higher for trees on infertile soils than for trees growing on fertile soils. Tree roots can effectively exploit soil moisture even at relatively low soil-moisture contents. It appears that the mycelia of mycorrhizal fungi may play an important role in the extraction of soil moisture by pines and other extensively rooted species. Only a small proportion of the soil water lies in the immediate vicinity of the absorptive surfaces of roots of these species at any one time. Neither capillary movement of soil water to tree roots nor the rapid growth of roots into moist soil can account for the tremendous capacity for water absorption by trees at low soil moisture contents.

In spite of the efficiency of most root systems, the distribution and growth of trees are controlled to a great extent by water supply. Wherever trees grow their development is limited to some degree by either too little or too much water. Even in relatively humid areas of the temperate zone, forest soils may be recharged to field capacity only during the dormant season or for a brief time after very wet growing periods (Kozlowski, 1968). Most light rains are intercepted by the forest canopy or serve only to recharge the surface soil layers. Even though recharged, the upper layers of the soil containing the majority of the fine root system may be reduced to near-wilting points long before the remainder of the profile is dry. Consequently, soil water deficits during the growing season commonly are of long duration and can result in considerable losses in tree growth.

Tree growth is reduced by water deficits indirectly through interferences with physiological processes such as photosynthesis, nitrogen metabolism, salt absorption, and translocation, and directly by the effects of reduced cell turgor on cell enlargement and other processes more directly involved in growth (Kramer, 1969). To be sure, tree growth may be controlled more by moisture balance within the tissue than by soil moisture per se. This is because internal water deficits are controlled by relative rates of absorption through roots and loss of water by transpiration, which are only indirectly controlled by soil moisture content. The relation of soil water potential to transpiration, however, can be quite close as shown in Figure 8.5. During the day transpiration usually exceeds absorption whether the soil is relatively dry or moist. Although water deficits

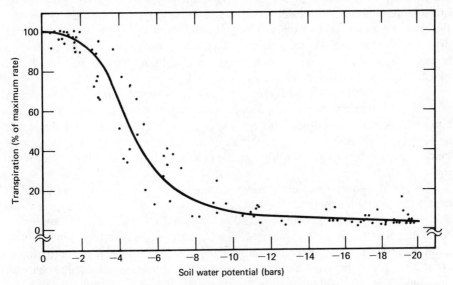

**Figure 8.5** The relation of transpiration rate of lodgepole pine to soil water potential. (Lopushinsky, 1975). Used by permission of USDA.

tend to develop in leaves during the day, they generally rehydrate during the night when absorption is greater than transpiration. Transpiration is largely regulated by the above-ground environment and leaf structure and only absorption is directly influenced by soil factors. Water absorption is affected by such soil factors as soil moisture tension, soil temperature, concentration of the soil solution, and soil aeration, as well as size and distribution of the root system.

Absorption of water by trees is apparently controlled by two groups of factors (Kramer, 1969). These factors affect the difference in water potential from soil to roots and the resistance to water movement through the soil and in the roots. Soil texture and hydraulic conductivity control water movement to root surfaces. Aeration, temperature, and degree of suberization of roots modify resistance to water movement into the roots. The driving force is the difference in water potential between the soil mass and the root surface and between the root surface and the xylem.

Often various parts of the tree root system are in a stratified soil possessing different levels of water potential, and they not only survive but appear to thrive with only part of their root system in soil above the permanent wilting percentage. The minimum soil water potential at which absorption can continue is limited by the minimum water potential that can be developed in the roots. Soil water will become limiting to rapidly transpiring plants at a higher soil water potential than that at which it becomes limiting to slowly transpiring plants.

The rate of root elongation is greatly affected by soil moisture levels during the growing season. Root growth is rarely affected by water deficits in the spring, but summer water stress causes slowing and cessation of expanding root tissues. For example, roots of radiata pine ceased elongation completely during warm, dry periods. Late in the season, if soil moisture recharges occur, roots may resume elongation in some species of trees even though shoots do not resume extension growth (Zahner, 1968).

Large root systems develop in tree seedlings when grown in soil maintained close to field capacity, but only sparse root systems may develop in soil that dries almost to permanent wilting before rewatering. Zahner (1968) pointed out that soil strength increases sharply as the soil dries and the resulting physical resistance to penetration by the root tip may be independent of the deficiency of water for absorption.

Roots of mature trees are known to develop profusely in zones of the soil profile that contain adequate supplies of available moisture. For example, Wilde (1958) presented examples of proliferation of small roots at the end of large sinker roots that encountered the capillary fringe above the ground water. On the other hand, excess water and poor aeration of the soil have as great an influence on gross root morphology as do deficient soil water or dense impermeable substrata. Soils with limited gas exchange, either because of high bulk density or excess water, support trees with superficial, shallow root systems.

## SOIL WATER AND STAND DEVELOPMENT

Soil water influences forest regeneration by controlling seed germination as well as survival and growth of young trees. Direct seeding of conifers on sandy soils may be feasible only where the water table is near enough to the surface to keep capillary moisture within the top 20 cm of soil during the germination period. Transplanted trees can become desiccated because of excessive transpiration at a time when roots grow too slowly to absorb sufficient water. The capacity of transplants to resume root growth rapidly is often critical to survival. Capillary movement of water from wet to drier regions in soils that are at or below field capacity is slow and if there is no capillary movement toward the roots, continuous root expansion is essential for growth and survival. Trees that survive transplanting often exhibit reduced growth for months or even years later due to unfavorable water balance. Soil moisture is equally critical to the developing stand. Kozlowski (1968) reported that the release of hardwoods upon cutting of a pine overstory resulted as much from improved moisture conditions as from favorable light conditions.

The resistance of trees to certain insects is apparently reduced under stress conditions brought about by soil moisture deficiencies. Flower initiation may sometimes be induced by moisture stress, but good soil moisture conditions are required for best flower production. Nutrient availability, uptake, and transport depend on adequate soil moisture. In fact, ample soil moisture is so essential to tree growth and stand development that site quality is largely determined by those soil physical properties that influence soil moisture relations. For this reason, soil physical factors and physiographic features that influence soil moisture storage and availability are the most frequently used factors in site quality classification (Carmean, 1975; Coile, 1952; Kozlowski, 1968).

# 9

# FOREST SOILS AND THE HYDROLOGICAL CYCLE

*Hydrological cycle* is a term used to describe the circulation of water in its various forms and storage systems throughout the world. About four-fifths of the world's surface is covered by the oceans. Radiant energy from the sun is absorbed by this vast area of salt water resulting in the evaporation of fresh water into the atmosphere. As water is changed into vapor, it is transported by the winds until it condenses as water droplets into clouds and is precipitated as rain, snow, or hail. Sometimes this water vapor is transported over extreme distances, but it is estimated (Pereira, 1973) that only about 25 percent of the water evaporated from the surface of the seas falls on land areas.

Some of the water falling on land areas runs off directly into streams and drainageways back into the sea, while some evaporates from soil and plant surfaces and from the surfaces of lakes and rivers. Perhaps the most important pathway from the standpoint of survival and growth of forests is that of infiltration into the soil. Part of the water entering the soil is stored and subsequently used by plants. The amount of water available for root exploitation is largely determined by soil texture and depth. Soil depth can range from a few centimeters in some rocky soils to several meters in deep, well-drained soils. Water that moves into the soil supplies the needs of plants, and a part of it seeps underground to supply springs and maintain the flow of rivers. Water absorbed by plant roots is transported through the plant and is eventually transpired from the foliage. Transpiration plays a prominant role in the growth and development of trees and other plants and it assumes an important role in the hydrologic cycle in all vegetated areas. Part of the moisture evaporated from soil, plant, and water surfaces condenses from the atmosphere onto the earth's surfaces as dew, but most returns as rain or snow and eventually makes its way back to the sea. This cycle is graphically illustrated in Figure 9.1.

Transpiration rates are largely determined by the radiant energy available for evaporation, the ability of the air to transport water vapor, and the capacity of the soil to supply water to roots. However, the types and amounts of plant cover also have a significant influence on transpiration rates. While humans can do little regarding the climatic effects on evapotranspiration, they can exert significant changes in evapotranspiration through manipulation of the types and density of plant cover.

One of the principal uses of forests in many areas of the world is for the protection of watersheds. Proper management of such watersheds, therefore, must take into consideration all forest practices that optimize yield and quality of water from these areas, as well as stabilize stream flow. Through management techniques, humans can influence the rate of infiltration of water into the soil, its storage and distribution, and its flow toward the oceans.

## WORLD WATER RESOURCES

It is generally agreed that approximately 97 percent of the world's total water is contained in the oceans, although the actual volume of water in these bodies is not accurately known (Pereira, 1973). Of the approximately 3 percent of the world supply of water that is fresh, some three-quarters is held frozen in the polar ice caps of the Arctic and Antarctic regions. These great storage reservoirs of fresh water do not remain constant. There were significant losses of water from the seas to glacial advances during the Pleistocene and there continue to be changes in the amounts of water held in glaciers and ice caps of the frigid zones. However, there appears to be little agreement on whether the glaciers are increasing or decreasing in size at this point in history.

The fresh water in lakes and rivers and in known underground reservoirs available for use is less than 0.05 percent of the world's total water supply (Pereira, 1973). Nonetheless, this small percentage amounts to about 500,000 cu km. In addition, there are large supplies of underground water that are too contaminated with salts to be immediately useful to land plants. Soil and ground water total about 0.6 percent of the world supply. The average flow of water into the sea by surface streams is about 924 cu m per second. This is sufficient on an annual basis to raise the level of the seas by about 10 cm. There are also substantial, but unmeasured, losses of fresh water into the sea by underground seepage along some 370,000 km of coastline.

The amount of water in the lower part of the atmosphere amounts to only about 2.5 cm (or about 0.001 percent of the world total) and its distribution above the earth's surface is very erratic. For example, air humidity over some desert areas can be particularly dry for extended periods. Atmospheric circulation is

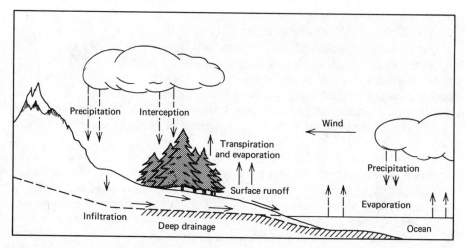

**Figure 9.1**  Schematic presentation of the hydrologic cycle.

due mainly to temperature differences between the polar and equatorial regions. When the wind systems cause air masses to converge they are forced upwards into colder conditions and precipitation usually occurs. While it is probable that up to 90 percent of the precipitation over land masses is of marine origin, evaporation from land surfaces can have a significant influence on local weather conditions.

## SOILS, FORESTS, AND THE WEATHER

It is recognized that a forest cover can have significant direct effects on soil water regimes through increased losses by interception, root absorption, and evaporation. It has also been proposed that large forested areas can influence the climate of the region through increased precipitation and stream flow when compared to nonforested areas on similar soils.

The mechanisms that supposedly explain the greater rainfall in forested areas are (1) greater air turbulence generated by a forest compared to uniform low-growing forms of land cover—especially in flat areas and where the tree cover is not continuous; (2) interception by trees of atmospheric moisture in the form of fog, mist, or haze; and (3) increased atmospheric water content of forested areas, presumed to result from greater evaporation rates. The increased water loss by the latter mechanism is well established (Stanhill, 1970), but there is considerable disagreement as to its size and its significance in increasing rainfall. In fact, there is little valid data on the relative effects of any of the above mechanisms on rainfall in forested areas as compared to other forms of land use.

In early Russian calculations, reported by Stanhill (1970), a 20 percent afforestation of some areas caused an 8 percent increase in annual precipitation in those areas due to increased turbulence. The greater evaportranspiration from such areas was calculated rather empirically to have led to a 5 percent increase in summer rainfall. Later research from the same area suggested a much smaller effect (10 percent afforestation producing a 1.5 percent increase in rainfall), and most of this increase was attributed to the enrichment of atmospheric water content rather than to turbulence.

Pereira (1973) reviewed later research in western Russia that indicated a strong positive trend of increasing annual water yield when the proportion of forest cover increased from 10 to over 80 percent of the land area of large river basins. The Russian scientists concluded that the extra water yield in the forested catchment was largely due to increased snow storage.

In especially favorable mountainous regions, increases in precipitation may occur through horizontal interception of atmospheric moisture. This is a very local phenomenon such as found in forest belts along the coast of some northern Pacific regions (Penman, 1963). There have been reports of particularly heavy drip of as much as 25 cm per year as a result of condensation of mist on tree foliage in the San Francisco Bay area and along the coast of Japan (Matsui, 1956). Under freezing conditions such interceptions may be increased even more.

There appears to be no solid support that transpiration from forests increases the rainfall of the forested area. Penman (1963) reviewed the evidence and meteorological arguments rather thoroughly and concluded that, although vegetation does affect the disposal of precipitation, there is no evidence that it can affect the amount of precipitation received. The limited evidence suggests that for large areas any increase in rainfall brought about by the presence of forests are smaller than the errors in rainfall measurements.

## FORESTED WATERSHEDS

The presence of water in soils is not only essential to the growth of forests, but improved water yield and quality are becoming increasingly important management objectives of many forested watersheds. Without minimizing the traditional role of watershed management in preserving and ameliorating the soil and timber resources, progressive forest management is often tempered by the effects of such practices on increasing or decreasing water supplies from those areas. A cursory examination of the water balance equation may help explain the role of management on water yield and quality from these forested areas.

### Water Balance Equation

A simplified version of the water balance equation, shown graphically as part of the hydrologic cycle in Figure 9-1, was presented in Chapter 8. In this equation[1]

[1]$\Delta W = P - (O + U + E)$.

$\Delta W$ represents the change in soil water storage during a specified period and $P$, $O$, $U$, and $E$ are the precipitation, runoff, deep drainage, and evaporation from plant and soil surfaces. Runoff, $O$, can be considered to include both surface and subsurface flow, while $U$ is defined as the amount of water passing beyond the root zone as deep drainage and contributing to the groundwater flow (Kramer, 1969).

Obtaining quantitative measurements of certain components of the water balance equation is difficult and subject to gross errors. Precipitation and runoff are relatively accurately measured (Figure 9.2), but deep drainage and evaporation-transpiration must often be estimated. If one ignores, or else measures, drainage from a watershed, then $E = \Delta W$, during periods between rains. Therefore, $E$ can sometimes be determined by measuring changes in soil water storage under a plant community. Measuring $\Delta W$ is probably most accurately accomplished by using weighing lysimeters. However, the spatial distribution and size of vegetation and the restriction on root development of plants grown in containers make it difficult to simulate a natural environment inside the lysimeter.

Changes in soil moisture content can be measured directly on soil samples of known weight or volume. The water content is expressed as grams of water per gram of oven-dry soil or grams of water per cubic centimeter of oven-dry soil.

**Figure 9.2** A monitored watershed in Vermont.

Representative samples are difficult to obtain because of considerable vertical and horizontal variability in soil water distributions. Intensive sampling with a soil tube or auger may disturb the ground vegetation and it generally requires considerable time and labor.

Indirect methods of measuring soil moisture content, such as the neutron scattering, gamma ray absorption, electrical capacitance, thermal conductivity, and tensiometeric techniques, permit installation of a number of sensing elements in a study area so that repeated measurements can be made at the same point. These indirect methods result in considerable labor saving after the initial installation and calibration, but they generally have a high initial cost and require considerable skill to properly operate.

## Interception by Tree Canopies

*Interception* refers to precipitation retained by the aerial part of vegetation and either absorbed by it or evaporated. The loss through interception is measured by gauging the precipitation that falls through the canopy (throughfall) plus rainfall that runs down tree trunks (stemflow) and then subtracting the sum of these from precipitation measured in clearings or areas outside the forest (Stanhill, 1970).

Stemflow is a relatively small percentage of total precipitation, but it can be quite important in some instances. For example, stemflow from a 30-cm diameter hickory tree amounted to as much as 113 liters during a 2.5-cm rainfall. Stemflow in a beech *(Fagus sylvatica)* forest in Austria was reported to be as much as 21 percent of precipitation in rains of 2.5 cm or more (Kittridge, 1948). It can be essentially zero in small storms and generally amounts to no more than 1 to 5 percent in heavier storms. Stemflow serves to funnel a portion of the intercepted water into the soil at the base of the tree. This often results in higher soil moisture levels directly under the tree than found at the edge of the tree canopy. It may also result in increased leaching and podzolization of soil under the stem (Buol et al., 1973).

The evaporation of intercepted rainfall is often regarded as a separate pathway for water loss additional to, and independent of, losses by leaf transpiration and soil surface evaporation. More recently, some scientists have proposed that evaporation of intercepted water is compensated for by an equivalent decrease in transpiration because little transpiration takes place at 100 percent relative humidity. However, there is evidence that more than an equivalent amount of transpiration water may be lost by interception. Transpiration water loss is less per unit energy received than interception losses because of greater resistance in the system (Stanhill, 1970). It appears that interception losses may or may not be compensated for by reductions in transpiration depending on local conditions. Johnson (1964) found that, in a plantation of radiata pine in which the soil water was near the permanent wilting percentage, rain appreciably reduced the leaf water deficit, even when a plastic cover on the soil surface prevented the rain from reaching tree roots. In other research reviewed by Rutter (1968), the above-ground parts of *Picea abies* were found to absorb water and tritiated water was reported to enter leaves of *Pinus halepensis,* but little indication was given of the quantities of water involved and the techniques of measuring isotopic exchange were not perfected. Rutter (1968) calculated that the change in leaf water deficit of *Pinus sylvestris* was only about 6 percent of the intercepted water. He suggested that the assumption that all intercepted water is evaporated is not seriously in error. However, he pointed out that absorption of intercepted water may be of physiological significance to trees in dry soil when they are wetted by dew or by rainfall that is too light to penetrate the soil.

Monteith's equation (Monteith, 1965) may be used to calculate the rate of evaporation from a rain-wetted canopy ($E_0$) relative to the rate of transpiration from the same canopy in a dry condition ($T$). Assuming that the radiation balance of the two surfaces is the same, the ratio $E_0/T$ depends on the air temperature and the relative stomatal and aerodynamic resistances of the forest canopy to the diffusion of water vapor (Stanhill, 1970). Rutter (1968) reported that an average $E_0$ was four times as great as $T$ in a forest of *Pinus sylvestris.* In other words, one-fourth of the intercepted water was equivalent to transpiration that would otherwise have occurred in the same atmospheric conditions, while three-fourths was evaporation that would not have occurred in absence of interception.

The hydrological potential for intercepted water loss depends on both the length of time that the canopy remains wet (as influenced by the distribution of rainfall periods and its relation to evaporating conditions) and the amount of intercepted precipitation that the canopy can hold. Zinke (1967) reported that most estimates of canopy saturation capacity are in the range of 0.5 to 2.00 mm. Ovington (1954) compared adjacent 0.1 ha plots of 12 species in Kent, England, and showed interception by dry foliage to vary from 21 to 34 percent of precipitation in deciduous trees and 36 to 54 percent in coniferous trees.

Swank et al. (1972) found interception losses from 5-, 10-, 20-, and 30-year-old loblolly pine averaged 14, 22, 18, and 18 percent of the annual 1.37 m precipitation in the Piedmont of South Carolina. On the average, the loss of water intercepted annually by loblolly pine stands appeared to be about 10 cm greater than interception losses from a number of hardwood sites. Some of the results from this study are summarized in Table 9.1.

In a summary of research on rainfall interception by several North American conifer species, Helvey (1971) noted that canopy interception losses are greatest in the spruce-fir-hemlock type, intermediate in pine, and least in broad-leaved deciduous forests. He stated that surface area index is an important common denominator for extending experimental results to ungauged areas, but that attempts to relate interception values with basal area have not been successful, except in a heavily thinned stand. Fir and hemlock species not only prevent more water from reaching the forest floor, but they also transpire more water than pine species (Helvey, 1971). Apparently transpirational rates of Douglas-fir and grand

**Table 9.1**  Annual Interception Losses Computed for a Region with Mean Rainfall of 137 cm Delivered in 75 Storms (Swank et al., 1972).[a]

| Species | Age | Annual Interception Loss | |
|---|---|---|---|
| | | cm | % of Precipitation |
| Loblolly pine | 5 | 18.5 | 14 |
| Loblolly pine | 20 | 24.4 | 18 |
| White pine | 10 | 20.6 | 15 |
| White pine | 60 | 36.1 | 26 |
| Mixed hardwood | Mature | 15.2 | 11 |

[a]*Journal of Soil and Water Conservation* **27**(4): 163. Copyright by the Soil Conservation Society of America.

fir are less affected by moderate soil drying than are transpiration rates of ponderosa and lodgepole pines because of species differences in stomatal control. Lopushinsky (1969) reported that the firs were less sensitive to increasing leaf moisture stress than the pines.

The practical significance of the greater interception losses from needle-leaved trees was pointed up by Swank et al. (1972) in their calculations that the current practice of conversion of Piedmont hardwood forests of the southeastern United States to loblolly pine could reduce annual streamflow by as much as 10 hectare cm for each hectare converted. The interception of snow by conifers is significantly greater than by defoliated hardwoods. In cool climates, the interception of snow by canopies of coniferous forests can amount to 13 to 27 percent of seasonal fall. Thus conversions of hardwood stands to conifers under these conditions may produce even greater reductions in streamflow than those in warmer climates.

Rainfall interception by groundcover vegetation after the over-story vegetation has become saturated can be substantial and is generally related to leaf area of the groundcover. Field capacity of the litter (water held against drainage) averages about 215 percent by weight and annual losses range from 2 to 17 percent of gross rainfall (Helvey, 1971). The loss of water intercepted by litter can be considered as a function of accumulated weight of the litter per unit area, water retention characteristics of the litter, and its wetting frequency and drying rate. Swank et al. (1972) report that the average annual interception losses in loblolly pine stands ranging from 10 to 30 years old varied form 5.3 to 6.1 cm or 3 to 4 percent of the 1.37 m of rainfall in South Carolina.

## Transpiration by Tree Canopies

Transpiration can be defined as the loss of water from plants in the form of vapor. Although basically an evaporation process, transpiration is modified by plant structure and stomatal behavior operating in conjunction with the physical principles governing evaporation. Trees are admirably constructed to lose water by transpiration to the atmosphere. The extensive leaf area of forests makes up a vast evaporating surface. Kozlowski (1968) pointed out that a hectare of forest often possesses an aggregate leaf surface area of more than 5 ha. Under ideal conditions with free access to unlimited water supplies, trees absorb and transpire tremendous amounts of water. Willows and cottonwoods along stream banks may transpire as much as 100 cm of water in one growing season. However, the amounts actually transpired vary greatly with soil water availability. It has been estimated that mature hardwoods of the southern Appalachians transpire about 50 cm of the annual rainfall of 2 m and that mixed forests of the United States lose as much as 75,000 liters of water per hectare per day (Kozlowski, 1968).

Transpiration largely controls the rate of absorption and ascent of sap in trees through the production of an energy gradient that causes movement of water into and through the plant. Transpiration can cause temporary leaf water deficits, and when drying soil causes absorption to lag behind water loss, permanent water deficit may cause injury and death by desiccation. Kramer (1969) pointed out that more plants are injured or killed as a result of transpiration exceeding water absorption than any other cause. He ventured that if transpiration could be reduced, plants would thrive over large areas that are now semidesert or desert. Kramer further suggests that some of the functions ascribed to transpiration, such as assisting in ascent of sap, absorption of mineral nutrients, and cooling of leaves, could be accomplished in the absence of, or with a reduction in, transpiration, and that in many respects transpiration is little more than an unavoidable evil. It is doubtful that all plant physiologists would agree with this conclusion.

Not all water absorbed by trees is transpired; some is combined with $CO_2$ in photosynthesis and some is retained within the plant body without being chemically changed. Assuming an annual dry matter production of forests of about 20,000 kg per hectare, Ovington (1962) calculated an annual chemical combination equivalent to about 1.2 mm depth of water. He also reported that the liquid water content of a number of plantations of different species, just under 50 years old, was between 100,000 and 200,000 kg per hectare.

**Measuring Evapotranspirational Losses**    Measurements of water losses by evapotranspiration *(ET)* are extremely difficult and expensive to carry out and, as a consequence, they are often calculated rather than directly measured.

To be sure, transpirational losses from potted plants can be determined directly by serial weighing of the plants and their containers and determining losses during limited periods of growth. The soil surface must be covered, usually with plastic, to prevent evaporation therefrom. Note that measurements involving plants enclosed in transparent coverings are subject to error because the environment of plants in the container is different from that of plants in the open. Furthermore, measurements of water losses from cut twigs or leaves are subject to large errors and caution must be exercised in extending transpirational rates for single plants (or plant parts) to stands of plants growing in the open under different environmental conditions (Kramer, 1969).

Evaporation (evapotranspiration) may be measured directly by means of weighing lysimeters, but these large containers also have shortcomings, as mentioned earlier in this chapter. Evapotranspirational losses from plant communities are more commonly calculated by means of a vaporization term in the water balance equation. If one knows, or can measure, the amount of precipitation *(P)* above the forest stand, the quantity of leakage water (runoff and deep drainage, *O, U)* and the change in the soil water storage ($\Delta W$) in the rooting zone during the measurement period, then the evapotranspiration from a closed watershed can be calculated:

$$E = P - O - U - \Delta W$$

The use of such gauged watersheds as shown in Figure 9.2 can provide much information on soil water storage and movement, as well as evapotranspiration losses. However, one encounters considerable difficulty in obtaining a suitable watershed where both deep and shallow drainage can be accurately measured.

Vaporization losses can be estimated from the *radiation balance* resulting from the difference between fluxes directed toward and away from a forest stand. Direct solar radiation, diffuse sky radiation, and thermal atmospheric radiation flow toward the stand, while the reflected fractions of the incoming components, as well as thermal terrestrial radiation, flows away from the stand. While the net radiation of surfaces with different reflectance characteristics may differ markedly, for surfaces covered with vegetation the variations in net radiation (for a given amount of solar radiation) are not great. Albedo values (the ratio of reflected light to the total amount falling on a surface) for most cropped surfaces average about 20 percent (Pruitt, 1971). The net exchange of long-wave radiation is also affected by surface temperature, cloudiness, and the amount of water vapor in the air. The effect of climate and season on net radiation, therefore, results in the need for considerable empirical adjustment in any evaporation prediction formulas that are based largely on solar radiation.

The *energy balance* equation, based on the principles of the conservation of energy, applies over any time period from minutes to years. The equation

$$R_n = G + H + LE$$

can be used to approximate evapotranspiration, although some components such as photosynthesis are often neglected (Pruitt, 1971). In the above energy balance equation, $R_n$ = net exchange radiation; $G$ = net rate at which the heat content of the soil and water is changing; $H$ = sensible heat transfer; and $LE$ = latent heat flux due to evaporation or condensation.

It was on the energy balance equation that the concept of potential evapotranspiration was developed. An early misconception on potential *ET* was that it should be about equal to net radiation, even on a short-term basis. While net radiation $(R_n)$ of a surface is the single most important factor affecting $ET$, the $LE$ factor can exceed it under some climatic conditions. In order to make the term more limiting, a revised definition of potential $ET$ was described by Pruitt (1971) as "the rate of evapotranspiration, primarily determined by the weather, from an extended surface of a uniform, short green crop, actively growing and completely covering the ground, and not short of water."

The energy balance equation can be a useful tool for calculating $ET$, if one recognizes that it has limitations, such as the effects of stomatal resistance, degree and stage of maturity of the plant cover, depth of plant rooting, and moisture status of the soil.

**Factors Affecting Transpiration**    Approximately two-thirds of all rain falling in the continental United States is returned to the atmosphere by evapotranspiration processes. The actual amount of water loss through these processes is influenced by various climatic and environmental factors. The primary source of energy required in evaporation of water is radiant energy coming directly or indirectly from the sun. Thus, the rate of transpiration depends on the supply of this energy, which maintains the concentration gradient constituting the driving force, and on the resistance to diffusion in the vapor pathway. Most of the water vapor escapes through stomata, but some passes through the epidermis and its cuticular covering and through lenticels in the bark of branches and twigs.

The chief environmental factors affecting transpiration are light intensity, vapor pressure of the air, wind, and soil water supply to plant roots. Plant factors include the extent and efficiency of the absorbing roots; area, arrangement, and structure of the leaves, and their stomatal behavior.

Evapotranspiration can be significantly altered by manipulation of some plant and site factors, although it is not always practical to do so. Increases in available soil moisture by increasing rooting depth (moisture storage capacity), by drainage, or by irrigation can increase evapotranspiration. Factors that affect vigorous tree growth, such as site preparation and fertilization, may influence the rate of $ET$. Altering the vegetative cover of a watershed has been reported to influence the water available to streamflow (Swank and Douglass, 1974; Hewlett and Hibbert, 1961). Presumably under conditions where advection occurs, plant communities with open, well-ventilated structures can be expected to transpire more than dense, even communities. Furthermore, differences in depth of rooting, length of growing season, stand morphology, and degree of stomatal control over transpiration may result in differences in water loss among species. Perhaps the most dramatic effect on water yield is achieved by removal of the tree canopy by harvesting.

# EFFECTS OF ALTERING THE FOREST COVER ON WATER YIELD

Several investigators have noted a rise in the soil water table due to reduction in evapotranspiration following the removal of the forest canopy in wet areas

(Trousdell and Hoover, 1955; Wilde, 1958). In coastal flats the water table may become so high that it interferes with subsequent reforestation operations. Hibbert (1967) reviewed results from 39 studies dealing with the effects of altering forest covers on water yields. First-year response to complete forest reduction varied from 34 mm to more than 450 mm of increased streamflow (Figure 9.3). Hibbert pointed out that most complete cuts produce less than 300 mm of extra water during the first year after treatment. Figure 9.3 also shows that, in well-watered regions, streamflow responses were about proportional to reduction in forest cover. A practical limit of yield increase appears to be about 4.5 mm per year for each percent reduction in forest cover in humid areas. Of course, the amount of increase in water yield depends on annual precipitation as well as the type of overstory vegetation removed, as shown for three watersheds in Arizona in Figure 9.4.

As the forest regrew following cutting, increases in streamflow declined as a result of increased transpiration losses (Figure 9.5). Streamflow declines are greater in areas that are restocked with conifers than in those restocked to hardwoods. This results from greater annual transpiration losses from well-stocked coniferous forests than from deciduous forests due largely to the reduction of transpirational surfaces from the latter forests, during the winter months. In a study in the Appalachian Mountains a monitored watershed was clear-cut and maintained in a low coppice-herb condition by annual recutting for 7 years.

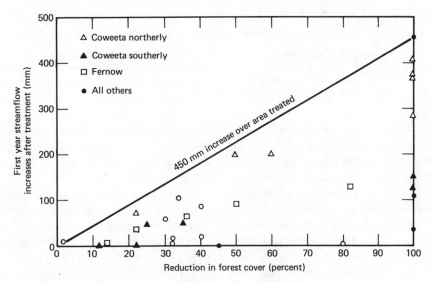

**Figure 9.3** First-year streamflow increases after treatment versus reduction of forest cover of 30 water-yield experiments. Line depicts practical maximum increases of 450 mm (Hibbert, 1967). Reprinted with permission from *International Symposium on Forest Hydrology* (Sopper and Lull, ed.); *Forest Treatment Effects on Water Yield.;* 1967; Pergamon Press, Ltd.

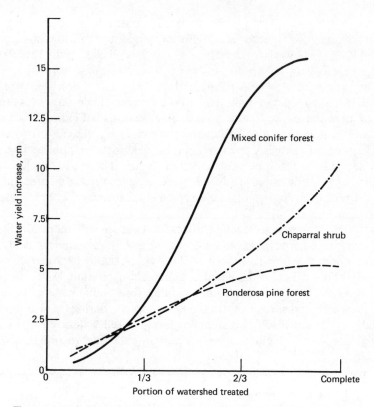

**Figure 9.4** Relationships between average water yield increase and extent of clearing of three types of overstory on watersheds in Arizona (Ffolliott and Thorud, 1974). P. F. Ffolliott and D. B. Thorud, College of Agriculture, University of Arizona, Tucson.

Figure 9.6 shows the average monthly flow before harvesting the hardwood cover and the average monthly increase in flow with repeated cuttings. About 60 percent of the 20 cm of increased flow came in the July through November period, and the remainder during the winter months.

Swank and Douglass (1974) reported that 15 years after two experimental watersheds in the southern Appalachians had been converted from a mature deciduous hardwood cover to white pine, annual streamflow was reduced about 20 cm, or 20 percent below that expected for the hardwood cover. Streamflow was reduced during every month, with the greatest monthly reduction (1.5 to 3.5 cm) occurring in the dormant and early growing season. Other experiments in the same humid areas (Hewlett and Hibbert, 1961) indicated that conversion of mature forests to low-growing vegetation increased streamflow by the equivalent of 12.5 to 40 cm per year the first year after cutting; the increases being greater on north-facing than on south-facing slopes.

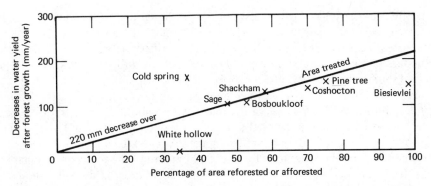

**Figure 9.5** Decreases in annual water yield as a result of reforestation of eight watersheds (Hibbert, 1967). Reprinted with permission from *International Symposium on Forest Hydrology* (Sopper and Lull, ed.); *Forest Treatment Effects on Water Yield;* 1967; Pergamon Press, Ltd.

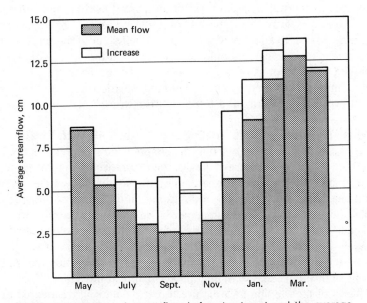

**Figure 9.6** Timing of mean flow before treatment and the average increase in flow produced by a Coweeta, North Carolina, hardwood watershed that was clear-cut and recut annually for seven years (Douglass and Swank, 1972).

**153**

Effect of the understory on evapotranspirational losses appears to be influenced by several environmental factors. Penman (1963) reported no major differences in soil water content that could be ascribed to differences in densities of understory of several hardwood and coniferous forest stands in Europe. Rutter (1968) reviewed the effects of forest understory in a number of United States studies on water loss. At the Coweeta watershed, the removal of *Kalmia latifolia* and *Rhododendron maximon* that covered about 75 percent of the floor under mixed hardwoods increased streamflow by 10 cm in the first year and by 1.6 cm after 6 years, as the understory became reestablished. In this study, the evergreen understory accounted for about 13 percent of the total evapotranspiration. Rutter pointed out that the removal of a deciduous hardwood understory from a 50-year-old stand of loblolly-shortleaf pine stand in Arkansas resulted in no significant reduction in soil water depletion during the spring when the soil was generally moist. However, by midsummer the depletion rates were 25 percent greater in areas where the understory had not been removed than in areas where it had been eradicated.

The effects of cutting either forest dominants or the understory on soil moisture losses are related to changes in the canopy density and structure. Forest thinnings expose lower branches of the remaining trees and the understory to greater solar radiation and air movement. Light thinnings may be made with only small reductions in total radiation absorbed at leaf surfaces, and the empty spaces in the upper canopy resulting from these thinnings may soon be filled by expanded growth of the surrounding trees. Large gaps created by thinning may not be invaded by trees for many years and if there is no significant understory vegetation, water should be conserved in the areas from which trees are removed. The amount of water consumed by herbaceous communities in those gaps may be reduced by their sheltered position in relation to wind and their shallower rooting compared to neighboring trees.

## POSSIBILITIES FOR SOIL AND SITE MANAGEMENT

The correct conclusion to draw from this chapter and Chapter 8 is that, in most areas of the world, soil moisture is the dominant environmental factor controlling the distribution and growth of forest communities. In a large measure, moisture conditions are inherent to the soil, topography, and climate of the area and they can not be easily altered by management. Nevertheless, certain facets of the hydrological cycle can be significantly influenced through watershed management. When one considers that in the temperate forest areas of North America more than one-third of the annual rainfall is lost as transpiration, possibilities for water control through ecosystem manipulation appear intriguing. Some examples of the effects of clearfelling of a mature forest or altering the degree or type of cover have been reviewed. Total runoff increases of 30 percent or more during the first years following clearing of a watershed are possible (Hornbeck et al.,

1970). Possibilities exist for decreasing water yields through forest management as well as increasing yields.

In addition to altering transpirational losses, there are considerable potentials for influencing the hydrological cycle by soil and site manipulations. These include changes in the soil infiltration rate, water storage, and surface runoff. Dams and other structures can be used to collect and store water and to slow the movement of water in drainage ways and streams on its way to the sea. The improper construction of roads, excessive disturbance of the soil and ground cover during harvest, soil compaction by logs and heavy equipment, and intensive site preparation techniques are all operations that potentially influence infiltration and runoff. The type and density of plants that cover the soil can influence infiltration and rate of evaporation. Plant cover is also a nonstructural approach to problems of surface water movement, soil erosion, and flood control.

Fire protection of watersheds managed primarily for water supplies is a particularly urgent problem. A wild fire not only destroys trees and reduces transpiration but may consume much of the litter that protects the ground. Pereira (1973) reported that the flow patterns of streams in the Snowy Mountains of Australia were changed abruptly following a fire. Storms that would have been expected to produce a flow of 60 to 80 cu m per second produced a peak of 370 cm m with a suspended sediment increase of 100-fold in comparison to soil content prior to the fire. Similar damage has been reported when fire destroyed the dry chaparral shrub in the San Dimas catchment of southern California. Fire in those areas may also induce hydrophobicity in the soils, further reducing their infiltration capacity.

The rise in the water table associated with the destruction or removal of a forest cover contributes to storm flow by decreasing the potential for water storage in the soil. Rain falling on a saturated soil runs directly to channels without attenuation in the soil profile. In steep areas the increased water movement poses erosion problems, and contributes to lake siltation, and possible eutrophication. In level areas, a higher water table following clear-cutting has prevented growth of new trees and the recovery of the ecosystem to its original state.

Whether the objectives of watershed managers are to increase or to decrease the rate of water cycling, they have at hand a number of tools for attacking the problem. It is becoming increasingly apparent that water quality is of even greater importance to society than water yield. Forest soils and associated litter layers are excellent filters, as well as sponges, and water that passes through this system is relatively pure. However, forest disturbances of various kinds can speed up the movement of water from the system and, in effect, short circuit the filtering action. While disturbances are inevitable, in most instances they need not contribute to poor water quality.

# 10

## SOIL AND ROOTS

Dr. Hans Jenny, a well-known soil scientist, once lamented that "trees and flowers excite poets and painters, but no one serenades the humble root, the hidden half of plants." The paucity of information on roots is particularly true for trees, because of the great difficulties encountered in extracting these large underground organs without destroying or modifying them. The extent and gross morphology of soil-grown root systems of a number of species have been described (Kramer, 1969; Bilan, 1971; Fayle, 1975), and a number of other studies have dealt with growth responses to variations in soil and site properties (Haines and Pritchett, 1964; Kaufman, 1968; Lyford and Wilson, 1966). However, studies on physiological processes have largely been confined to roots grown in nutrient solutions and to excised roots. Such approaches can be rewarding, but care has to be exercised in extrapolating results of these studies to field conditions.

Plant roots supply the connecting link between the plant and the soil. In a similar manner, the study of tree root systems forms the bridge between forestry and soil science, and the results of such studies are especially pertinent to forest soil science. Roots provide anchorage for the tree and serve the vital functions of absorption and translocation of water and nutrients. They exert a significant influence on soil profile development and, upon dying, roots contribute to soil organic matter content. It should not be surprising that growth and distribution of roots are influenced by essentially the same environmental factors that affect growth of the above-ground tree. Not only do variations in the chemical, physical, and biological properties of the soil have profound effects on tree roots, but the influence of such factors as light intensity, air temperature, and wind may be reflected as much in root growth as in shoot growth. This is illustrated (Figure 10.1) by the close relationship between root weight and branch weight in radiata pine in New Zealand (Will, 1966).

## FORM AND EXTENT OF ROOTS

When grown under favorable soil conditions, each tree species tends to develop a distinctive root system that is especially noticeable during early development. This characteristic pattern of root development often persists throughout the life of the tree, but it may become appreciably modified in later years by unfavorable site conditions. Soil texture, compaction, available moisture, impeding layers, and nutrition are factors that can influence the pattern, depth, and extent of root development. Density of stand or competition among individual trees also has a significant effect on the extension of lateral roots, but under favorable conditions it is not uncommon for trees to possess lateral roots that extend two to three times the radius of the crown (Zimmerman and Brown, 1971).

## Root Form

Root systems can be characterized on the basis of (1) rooting habit, which relates to the form direction and distribution of the larger, framework roots; and (2) root intensity, which pertains to the form, distribution, and number of small roots. Although the habit, or form, of a root system is influenced by local site conditions, it tends to be under genetic control. Most tree root systems can be conveniently grouped into tap, heart, or sinker (flat) root forms. Tap-rooted trees are characterized by a strong downward growing main root, which may branch to some degree, as in most *Carya, Juglans, Quercus, Pinus,* and *Abies*. Numerous strong roots radiating diagonally from the base of the tree, without a strong tap root, is characteristic of heart root trees, as generally found in *Larix, Betula, Carpinus,* and *Tilia*. The sinker root habit is dominated by strong laterals from which vertical sinkers grow straight down, as in *Populus, Fraxinus,* and some *Picea* (Bilan, 1971). Rooting habits are graphically illustrated in Figure 10.2.

    The rooting habit of a tree has considerable influence on the type of habitat in which it will thrive. Root form may well determine whether a species is capable of fully exploiting a site and competing successfully with neighboring species or becoming suppressed and eliminated. Root systems of species that are under strong genetic control tend to retain their rooting habit regardless of soil conditions to which they are subjected. Consequently they only grow well over a rather limited range of site conditions. On the other hand, a few species such as red maple can adapt their juvenile root system to a variety of environments and, thus, become established and grow on both wet and dry sites (Figure 10.3).

    Trees that develop strong tap roots are capable of penetrating the soil to great depths for support and moisture. For example, many pines, such as longleaf pine, have well developed tap roots and are capable of surviving on deep, relatively dry sandy sites. This well-developed tap root form is also shared by a number of deciduous trees, exemplified by burr oak and black walnut. Many tap-rooted trees also have extensive laterals and sinkers that permit them to

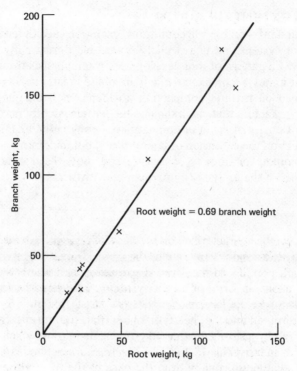

**Figure 10.1** Relationship between root weight and branch weight in 18-year-old radiata pine (Will, 1966). Used with permission by For. Res. Inst., New Zealand For. Serv.

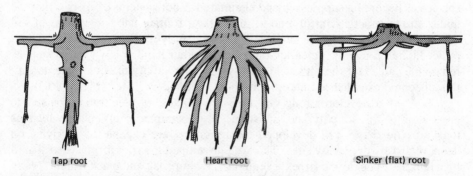

**Figure 10.2** Schematic presentation of tree rooting habits.

**Figure 10.3**  Partial exposure of a red maple root system on a moderately well-drained stony soil in Massachusetts.

survive on shallow soils and those with fluctuating water tables. Although roots do not usually persist in a zone of permanent water saturation, tap roots of pitch pine have developed below the water table (McQuilkin, 1935). Cypress (*Taxodium*) and some other species common to swamps have special root adaptations that permit them to grow in saturated zones. If young tap roots are injured, several descending woody roots may occur at the base of the tree in place of a single tap root. Tap roots of slash pine are often fan shaped from the capillary zone to below the mean water table. These fan-shaped roots were reported by Schultz (1972) to be two-ranked, profuse, and often dichotomously branched. Boggie (1972) reported that where the water table was maintained at the surface of deep peat, root development of *Pinus contorta* was confined to a multitude of short roots from the base of the stems, giving a brushlike appearance. Laterals developed only on trees where *Sphagnum* hummocks had raised the soil surface above the mean water level.

Tree species with inherent shallow, or flat, root systems and those systems under weak genetic control have particular advantages on shallow soils. Black spruce is an example of a species that can be found growing over a wide range of soil profile conditions from peats with high or fluctuating water tables to deep sands. However, trees with shallow root systems may have a real disadvantage compared to other species on deep, penetratable soils, for they are poorly equipped to exploit these conditions and are more subject to wind throw than tap-rooted trees.

Species possessing a heart root form with a number of lateral and oblique roots arising from the root collar, thrive best on deep, permeable soils, but they are also capable of exploiting fissures in fractured bedrock to a greater extent than other types of root systems.

## Root Intensity

While a tree's rooting habit may dictate the actual volume of soil occupied by its root system, the number and distribution of small roots determines the intensity with which the occupied soil volume is used. In this respect, root systems can be thought of as *intensive* or *extensive* systems, regardless of rooting habit. Intensive systems are those in which a relatively small volume of soil is penetrated by a large number of roots, while an extensive system exploits a larger volume of soil but with fewer roots per unit area. The intensity of fine roots in the surface soil is, perhaps, as much a function of soil nutrient and moisture supplies, temperature, and aeration as it is of genetic control.

The bulk of the root system of most trees is within less than a meter of the soil surface and the majority of small roots lie in the upper 20 cm of the surface soil. Schultz (1972) reported that 50 percent of the total root surface area of slash pine was in the surface 30 cm of the soil while only 6 percent was below 135 cm. Roots of less than 2 mm in diameter made up 50 percent of the total root surface area.

The lateral roots of conifers (Figure 10.4) often extend to great distances, and in spite of the high concentration of small roots in the surface soils, pines are among the least branched (most extensive) of the important forest species. They may possess up to 20 or more first-order lateral roots somewhat evenly spaced around a tap root and extending horizontally for a distance of 15 m or more. However, they apparently rely heavily on fungal mycelia and associated mycorrhizae for nutrient absorption in the forest floor and surface soil. The advantage of an extensive lateral root system on a heterogeneous site is that it permits a tree more flexibility in utilizing favorable microsite conditions. In a study of the extent of root systems of northern hardwoods, Stout (1956) noted that the mean root-crown ratio was about 4.5:1. He concluded that under each unit area of ground surface, in closed stands, at least four trees were competing for the available space, water, and nutrients.

## Biotic Influences on Root Development in Soils

There may be rather large differences among tree species, and provenances within species, in root developmental responses to site conditions (Kozlowski, 1971). For example, initiation of root growth in the spring may vary as much as a month among species on any given site. Some species begin top growth more

**Figure 10.4** Lateral roots of a white spruce extending more than 20 m (parent tree with white spot in background) at 0 to 10 cm below the forest floor. (Root is painted white for contrast.)

than 20 days earlier than root growth, while others have active roots as much as 20 days before top growth starts. These factors should be taken into consideration in scheduling soil management operations. In provenance trials, larch from a high elevation, where the soil warmed late, began root growth later than larch from other locations (Kostler et al., 1968). Scots pine grown on different soils from four seed sources exhibited features of the tap root and horizontal roots characteristic of the sources, although root systems of trees grown on the better sites were stockier and the total cross-sectional area of the horizontal roots at the stump were greater than were the more slender roots of trees grown on poorer sites. Steinbeck and McAlpine (1966) reported significant variation in the respiratory rate of roots of four black willow clones but none among clones of yellow-poplar, red maple, or weeping willow. There were significant differences among species. Yellow-poplar had the lowest respiratory rate (the most intolerant to flooding), while red maple was intermediate, and willow had the highest respiratory rate.

Fayle (1975) reported that the maximum rooting depth of red pine was reached in the first decade and maximum horizontal extension of roots was virtually completed between 15 and 20 years. The main horizontal roots of red pine seldom exceeded 11 m in length. McQuilken (1935) found that laterals replaced the tap root as the most prominent root in 8- to 12-year-old shortleaf pine and pitch pine, and elongation of the horizontal system continued to lengths of 7.6 to 9.1 m at about age 30 years. After that time the rate of elongation decreased but the intensity of rooting within the soil volume increased.

Early prominence of the horizontal roots was reported by McMinn (1963) for Douglas-fir. Long pioneer roots were characteristic to about age 10 years; older trees had a more compact, densely branched system that occupied the soil in the vicinity of the bole. This exploitation of the soil by secondary roots was as intensive as that of hardwoods and was greater than that of spruce and pine.

Kalela (1957) found the greatest number of active root tips on Scots pine at 35 years, an intermediate number at 65 years, and the smallest number on trees 110 years old. Kalela also found a greater number of root tips and greater length of small-diameter laterals in the early part of the growing season than at other times. It appears that a high percentage of the small-diameter laterals on primary laterals are ephemeral. The cause of their disappearance is not at all clear. Age in itself may be unimportant, because some nonwoody laterals live for many years (Lyford, 1975). Excessive dryness can cause small roots to die, and insects and diseases continually destroy small roots—particularly in the F and H layers of the forest floor. Formation of replacement roots and forks occurs on most second- and third-order laterals but not on many third- and fourth-order nonwoody roots.

It is generally agreed that dominant trees have vigorous lateral roots with well-developed, fine root systems, while codominant trees may have only a third as much total root length and 40 percent as much root mass. Total root length of suppressed trees may be only about 10 percent and root mass about 8 percent of dominants. Horton (1958) found a good relationship between height of lodgepole pine and extent of the horizontal root systems. This relationship between dominance and root extent would suggest that larger trees in a stand absorb a major share of available nutrients and respond more vigorously to fertilizer applications than intermediate and suppressed trees.

## SOIL CONDITIONS AND ROOT GROWTH

The physical, chemical, and biological properties of soils have profound effects on the rate of root growth and development and, to a considerable extent, on the habit and intensity of root systems of established trees. Consequently, variations in root systems among individuals of a species grown in different soils may be as great as that among different species on the same soil.

## Physical Impedance

Most pines have an inherent ability to produce extensive root systems, especially on deep sandy soils. However, the presence of compacted soil layers, bedrock, and fragipans can sometimes limit the penetration of even strong tap roots. Species that normally develop tap roots, such as Douglas-fir, pines, and aspens, have been observed to develop shallow root systems where bedrock or cobbles were near the soil surface. Development of a deep tap root and extensive laterals is particularly evident in longleaf pine, but high water tables and restricting layers can result in a tap root of less than a meter in length. This condition may also result from impedance by a cemented spodic horizon of a Spodosol (Groundwater Podzol) in some coastal plains.

Over 70 percent of the root weight of two-year-old loblolly pines was concentrated in the upper 15 cm of well-drained sandy soil, and these roots were much finer textured than those at lower depths (Bilan, 1971). However, soil texture influences distribution of lateral roots even in well-drained soils. Pines generally have less root mass in fine-textured soils than in sands, and roots in the former soils tend to be coarser, shorter, and less branched. Texture and structure may influence rooting through physical impedance and by their influence on soil aeration. A critical limit of 9 to 10 percent air space at field capacity was found for the 25 to 75 cm layer for citrus roots (Patt et al., 1966). However, trees differ in tolerance to reduced aeration. Loblolly pine apparently has a greater tolerance to poor aeration than shortleaf pine, and red pine is among the most sensitive to reduced conditions of all conifers.

Soil compaction may affect both the size and distribution of root systems of planted trees, regardless of soil texture. Seedlings of Douglas-fir, Sitka spruce, western hemlock, western redcedar, lodgepole pine, Pacific silver fir, and red alder were grown in soil columns compacted to bulk densities of 1.32, 1.45, and 1.59 g/cc. In two years, the roots of lodgepole pine, Douglas-fir, red alder, and Pacific silver fir penetrated soil densities that were found to prohibit the growth of Sitka spruce, western hemlock, and western redcedar roots (Minore et al., 1969). Machine logging may so compact soil that the following generation experiences reduced growth. Youngberg (1959) found that the growth of Douglas-fir seedlings was significantly greater in cutover area than in compacted soil of roadways and berms.

It has been observed that roots of machine planted pines tend to grow along the slit in the direction of planting (Haines and Pritchett, 1964). This tendency for roots to grow down the row may result, in part, from the original orientation of the roots in the slit during planting, but compaction of the sides of the slit and ease of root penetration in the direction of the opening is probably also important. Schultz (1972) reported that blowdown resulting from root deformities of hand-planted trees (balled or J-shaped tap root) may be as serious as L-shaped tap roots associated with machine planting.

## Soil Moisture

Moisture probably has a greater influence on root development and distribution than any other soil factor. Often, slow tree growth rate associated with shallow soils is not due primarily to lack of space for root development but rather to the limitation in water (and nutrient) supply associated with shallow rooting. Lyford and Wilson (1966) suggested that red maple roots grow rather well over a broad range of soil texture and fertility conditions if the soil is maintained at near-optimum moisture levels. On the other hand, rooting habits change appreciably when maples are grown under very moist, poorly aerated conditions. Kozlowski (1968) reported that large root systems develop in tree seedlings grown in soil maintained close to field capacity, in contrast to sparse root systems in soils allowed to dry to near the wilting point before rewatering. It has been noted that small roots die almost immediately in local dry areas (Lyford and Wilson, 1966). Lorio et al. (1972) found that mature loblolly pine trees on wet, flat sites were nearly devoid of fine roots and mycorrhizal roots compared to neighboring trees on mounds. Decreases in surface area of mycorrhizal roots occurred during drying periods, probably in response to temporary improvement in soil aeration. They proposed that loblolly pines may obtain much of their water through roots other than mycorrhizal and unsuberized root tips in dry soils. It is doubtful that any significant amount of root growth takes place when the soil moisture drops to near the wilting point (Kozlowski, 1968). Because root growth may slow or stop anytime the demand imposed on the soil exceeds its capacity to supply moisture, growth may be exceedingly slow during the dry summer months that would otherwise be favorable for rapid root growth.

Even among species not tolerant to excess moisture there is considerable variation in their response to changes in soil moisture. Steinbrenner and Rediske (1964) found that roots of Douglas-fir seedlings concentrated near the surface of well-watered soils but penetrated to considerable depth when surface soil moisture was below optimum. This feature was also shown for red pine (Figure 10.5). On the other hand, ponderosa pine developed deep root systems regardless of the moisture regime. Bilan (1968) reported that loblolly pine produced a more extensive, but superficial, root system in a mulched soil than in an unmulched soil. Mulching apparently produced more favorable moisture and temperature conditions for root development in the surface soil.

Optimum soil moisture for root development depends on soil texture and temperature, among other factors, but soil moisture near "field capacity" is optimum for most species. Roots of most trees grow best in moist, well-*aerated*

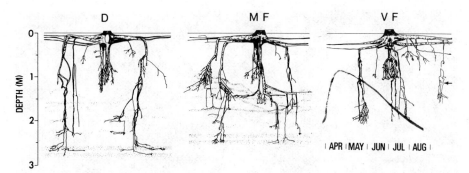

**Figure 10.5**  Vertical root systems of red pine on dry (D), moderately fresh (M F), and very fresh (V F) sites. Layers of finer-textured material are shown for D and M F soils and the location of a moderately cemented layer in the V F soil (←). The typical seasonal fluctuation in water-table level is indicated for the V F soil (Fayle, 1975). Reproduced by permission of the National Research Council of Canada from the *Canadian J. For. Res., 5*: 109–121, 1975.

soils, and they generally proliferate in layers affording the greatest moisture supply only if well aerated. Water saturation of the soil results in a deficiency of oxygen and an accumulation of carbon dioxide. Such conditions usually result in reduced root growth and, eventually, in root mortality. Some species such as the swamp species of baldcypress and water tupelo, and certain species of willow and alder, are capable of obtaining oxygen and growing in water-saturated soils. However, roots of most tree species will not long survive in saturated soils. Small lateral (feeder) roots of slash pine grow above the forest floor into moist, but aerated, clumps of grass during periods when the water table approaches the soil surface (Figure 10.6).

   Large fluctuations in depth to the water table, as found in such soils as those of coastal wetlands, tend to restrict the penetration and development of tree roots. Roots that develop when the water table is low are later killed when the water table rises for extended periods. This periodic root pruning tends to create an imbalance in top:root ratio and may later produce toxic substances from decaying root tissue. White et al. (1971) reported that stabilizing the water table at 46 and 92 cm below the surface of a coastal plain flatwood soil (Spodosol) increased slash pine root biomass by 69 percent and 43 percent, respectively, over root biomass of trees on soils with fluctuating water tables, or else the roots were repeatedly pruned. On the other hand, the spodic horizon offered little resistance to root penetration once water table levels were lowered and controlled.

   Boggie (1972) found that some water control was necessary for the initial establishment of trees in peat soils, but that once vigorous tree growth was started, drying as a result of evapotranspiration was progressive; that is, improved tree growth was accompanied by further drying of the peat. Pritchett

**Figure 10.6** Slash pine small lateral root with ectomycorrhizae and fungal mycelia growing in clump of grass above forest floor on wet site.


and Smith (1974) reported similar conditions to exist in the wet savanna soils of some coastal areas.

## Soil Temperature

Soil temperature affects many aspects of root growth and distribution both directly and indirectly. The minimum, optimum, and maximum soil temperatures for best tree growth vary with species and environmental conditions. The minimum soil temperatures for root growth range from slightly above 0° to 7°C; the optimum from 10° to 30°C; and the maximum from 25° to 30°C (Lyr and Hoffman, 1967). Barney (1951) reported that roots of loblolly pine seedlings grew most rapidly at 20° to 25°C, while growth at both 5° and 35°C was less than 10 percent of the maximum. The optimum temperature for red maple root growth was reported to be about 12° to 15°C (Lyford and Wilson, 1966). Root growth in cool climate species begins and ceases at lower temperature than in warm climate or tropical species.

Roots may not experience a true dormancy, such as that of buds, but a type of quiescence may be induced by environmental factors (Zimmermann and Brown, 1971). Roots of many warm climate and fast-growing species, such as radiata and many eucalypts, apparently never completely cease growth during winter months, although their rate of growth may be slowed by decreases in soil temperature. The root growth of southern pine is slowest during winter months, but roots seldom completely cease growth during most of this period (Kaufman, 1968). Loblolly pine roots continued to elongate until average weekly minimum air temperatures fell below −2°C (Bilan, 1967). Roots resumed growth in the spring when the daily minimum no longer fell below −1°C (Figure 10.7).

Lyford and Wilson (1966) found that day-to-day variations in growth rate of red maple root tips were more closely related to soil temperature than to air temperature fluctuations. They grew parts of lateral root systems of large maple trees in sheltered trays and found that the rate of root growth in unheated trays closely paralleled the variations in daily air temperature outside the shelter. However, when the trays were heated, the red maple roots continued to grow in spite of low outside temperature. During the winter, the temperature of the soil where the trees were located outside shelter (rhizotron) was less than 1°C. While some growth probably resulted from the starch stored in the roots, it was assumed that translocation of metabolites took place despite the low outside soil temperature.

Cool night temperatures are important in root regeneration of outplanted tree seedlings. Ponderosa pine root regeneration was significantly increased after exposure to at least 90 nights of less than 6°C air temperature (Krugman and Stone, 1966). Loblolly pine seedlings also regenerate new root tips more rapidly during periods of cool night temperatures. Soil temperature may also influence the severity of attacks from soil-borne organisms, and affect the morphogenesis of root systems of some species.

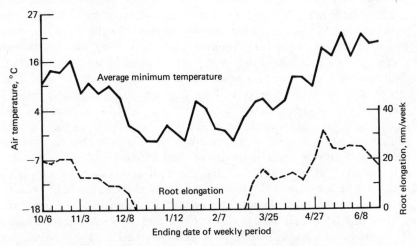

**Figure 10.7** Weekly mean minimum air temperature and root elongation of one-year-old loblolly pine (Bilan, 1967). Used with permission.

## Soil Chemical Conditions and Root Development

Acidity and nutrient deficiencies, or imbalances, are the chemical conditions of the soil most likely to restrict plant root growth and development in humid regions. Low soil pH may inhibit roots due to toxicity of the hydrogen ion itself. More frequently, root inhibition results from toxicity of aluminum, the solubility of which increases with increasing soil acidity. Root tolerance to acidity differs widely among plant species, but it appears that most woody species are relatively tolerant to acid conditions.

Nutrient availability in the soil affects both the growth rate and distribution of roots. Pines generally have an inherent capacity to produce extensive root systems, especially on deep, coarse-textured soils. They are able to survive on dry and infertile sands because of their ability to obtain moisture and nutrients from a large volume of soil. Apparently they are capable of adapting to a variety of edaphic conditions as a result of roots concentrating in soil layers most favorable to growth and development.

Fertilizer applications stimulate growth and development of hardwood and conifer roots in deficient soils. For example, Kohmann (1972) reported that nitrogen applications had a positive effect on the quantity of fine roots of Scots pine (*Pinus silvestris*) after two years, but this effect had largely disappeared after nine growing seasons following fertilization. Kohmann also reported that, while small applications of nitrogen stimulated root growth in the forest floor humus, large amounts of nitrogen may have a negative effect on root development. Roots in the upper mineral soil layer tolerated higher rates of nitrogen than did roots in the forest floor.

In a field study in which water tables of a flatwood soil were controlled at varying heights, White et al. (1971) reported that total root biomass of five-year-old slash pine was 58 percent greater in fertilized plots. Furthermore, the greatest increase in root biomass resulting from fertilizer use was obtained on those plots where the roots were restricted to a small volume of soil by a high water table.

Soil nutrient levels may also affect the ratio of top-to-root biomass. The dry weights of slash and loblolly pine seedling roots were approximately one-half as great in phosphorus deficient wet coastal savanna soils as in a similar soil to which 80 ppm phosphorus had been added (Pritchett, 1972). In these pot tests, the addition of nitrogen alone suppressed root development, but the addition of nitrogen or phosphorus fertilizer treatments had no significant effect on top:root ratio. It has generally been assumed that the application of nitrogen to deficient soils results in an increase in plant top:root ratio as a consequence of an increase in growth hormones that promote top growth at the expense of root growth (Wilkinson and Ohlrogge, 1964). However, it appears that top:root ratios of woody plants may be less affected by soil nutrient levels than are top:root ratios of nonwoody plants.

Hoyle (1971) reported that deficiencies of calcium and nitrogen prevented yellow birch primary root development in the lower substrate of New England Spodosols (podzols). Aluminum was toxic to birch root development, but the degree of toxicity was influenced by the level of other elements in the substrate. Greatest reductions in root growth associated with high aluminum levels were where magnesium or sulfur were low. In most cases, forest humus adequately supplied with nutrients in the upper layers did not compensate for nutrient deficiencies in the lower substrate. The assumption is often made that deep placement of fertilizer and/or lime in soils that contain spodic or other restricting horizons will encourage deep root penetration and the exploitation of a larger volume of soil. Such practices are no doubt beneficial for some crops under certain conditions but not for all soils according to Robertson et al. (1975).

Band, or localized, placement of fertilizer is often used as a means of reducing phosphorus fixation in some soils and of preventing excessive growth of competing weedy vegetation. High concentrations of fertilizer salts may also result in rapid proliferation of roots adjacent to the fertilizer band. Root proliferation is primarily the result of increased growth of smaller roots, since first-order roots are apparently not affected by fertilizer placement. Furthermore, it is likely that root proliferation in fertilized zones results from elongation and increased branching. Branching may occur after root tip meristems are killed by high salt concentrations.

## CONTRIBUTION OF ROOTS TO SOIL PROPERTIES

It seems apparent from this discussion that tree roots grow in a complexity of forms and to varying degrees of intensity, governed to a large extent by genetic forces but, nonetheless, modified by local conditions of soil and site. The influence of these latter factors on the nature and abundance of roots is not well understood because of the difficulty of studying them under field conditions. It is well known, however, that roots play a major role in the life of trees by providing physical support and serving as organs for the absorption and transport of water and nutrients. Other functions of roots that are often overlooked, but of tremendous importance, are those related to stability and development of the soil.

Tree roots provide a significant stabilizing force in mountainous areas where the soil is subject to erosion or mass movement. Fine roots and fungal mycelia serve as cohesive binders for surface soils and, where larger roots penetrate the surface horizons, they can anchor the soil mantle to the substrate (Swanston and Dyrness, 1973). It has been noted that the number of landslides from cut-over areas increases within three to five years after logging. This increase is attributed to a reduction in soil shear strength caused by the decay of tree roots following logging. The presence of living tree roots to anchor shallow soils to the underlying subsoil is particularly important in small drainages where winter storms can cause a sharp rise in groundwater level. Clear felling further contributes to a decline in soil retention by roots through greater exposure of the soil surface and

**Figure 10.8**  Wind-thrown black spruce in boreal forest showing only slight disturbance of soil profile.

increased soil moisture levels following the removal of the intercepting and transpiring tree canopy.

The form and extent of the tree root system also influences the amount of soil disturbance resulting from windthrow. Soil mixing and the development of a mound and pit microrelief is brought about by uprooting of individual trees during storm periods, with the greatest soil modifications wrought by trees with deep tap or heart roots. A tree with a shallow root system may disturb a large area of soil when it is windthrown, but generally only the surface layer is affected, as shown in Figure 10.8.

Roots make a considerable contribution to the organic matter content of the mineral soil. White and Pritchett (1970) calculated the root biomass of a 5-year-old slash pine stand to be as much as 14.8 kg/ha, which represented approximately 20 percent of the total tree biomass. Will (1966) estimated an annual

**Table 10.1** Chemical Composition of 18-year-old Radiata Pine Roots in New Zealand (Will, 1966) and 16-year-old Loblolly Pine in North Carolina (Wells et al., 1975)[a]

| Root Component | N | P | K | Ca | Mg |
|---|---|---|---|---|---|
| | Percent (o.d.) | | | | |
| | | | Radiata Pine | | |
| <3 mm in diameter | 0.22 | 0.18 | 0.44 | 0.14 | 0.13 |
| 3–12 mm in diameter | 0.11 | 0.09 | 0.28 | 0.13 | 0.11 |
| 13–50 in diameter | 0.15 | 0.08 | 0.28 | 0.10 | 0.09 |
| 15.2 cm dia. root: Bark | 0.20 | 0.02 | 0.075 | 0.27 | 0.06 |
| Outer wood | 0.095 | 0.015 | 0.080 | 0.02 | 0.03 |
| Middle wood | 0.070 | 0.007 | 0.044 | 0.01 | 0.02 |
| Inner wood | 0.065 | 0.006 | 0.033 | 0.01 | 0.03 |
| | | | Loblolly Pine | | |
| *Laterals* | | | | | |
| <3 mm in diameter | 0.58 | 0.098 | 0.146 | 0.37 | 0.10 |
| 3–10 mm in diameter | 0.31 | 0.099 | 0.372 | 0.35 | 0.15 |
| 10–20 mm in diameter | 0.23 | 0.094 | 0.328 | 0.33 | 0.13 |
| 20–40 mm in diameter | 0.21 | 0.022 | 0.148 | 0.05 | 0.03 |
| >40 mm in diameter | 0.10 | 0.012 | 0.068 | 0.05 | 0.02 |
| *Taproot* | | | | | |
| Wood | 0.06 | 0.013 | 0.106 | 0.04 | 0.06 |
| Bark | 0.22 | 0.095 | 0.783 | 0.09 | 0.09 |

[a] Used with permission.

production of more than 3000 kg/ha of roots by an 18-year-old radiata pine stand in New Zealand, exclusive of fine roots that are produced and die back each year and those roots below a depth of 91 cm. Tree roots not only add to the soil store of organic matter, but they contribute slowly released organically bound elements that play an important role in nutrient cycling. For example, the 5-year-old slash pine root system mentioned above contained approximately 53, 11, 12, and 23 kg/ha, respectively, of nitrogen, phosphorus, potassium, and calcium, part of which is annually recycled back to the above-ground tree as older roots are replaced. The chemical composition of roots of 18-year-old radiata pine and 16-year-old loblolly pine are shown in Table 10.1. Fisher, Role, and Eastburn (1975) reported that a 20-year-old loblolly pine plantation in southern Illinois produced 7.5 tons of roots per hectare annually. Although the forest produced three times more root biomass and more litter per year than an adjacent old field dominated by *Andropogon virginicus* and *Solidago altissima,* the old field soil (Ochreptic Typic Fragiudalf) contained more organic matter. They attributed the higher decomposition rates of forest organic matter (litter of the forest floor and roots in the mineral soil) to larger microfloral and mesofaunal populations than in the old field. This activity plus the greater root biomass apparently accounted for the lower bulk density in the forest soil than that in the old field soil.

Roots also contribute to soil weathering by exploiting fissures in bedrock, developing channels in compacted soil for the movement of water and fine-textured soil materials, and supplying a favorable environment for microbes within the rhizosphere of individual roots. Exudates of roots encourage the proliferation of organisms that may be capable of such functions as dinitrogen fixation, mineral solubilization, and nutrient mineralization (Richards and Voigt, 1965; Fisher and Stone, 1969).

# 11

# MYCORRHIZAE: FORMS AND FUNCTIONS

Mycorrhizae are specialized rootlike organs formed as a result of the symbiotic association of certain fungi with the roots of higher plants. Specific fungi grow upon and vigorously invade portions of the root in that area of the root system that is primarily responsible for nutrient absorption (Hacskaylo, 1967). In 1885, Frank, a German forest pathologist, coined the term "mycorrhiza" meaning fungus-root to denote these particular associations of roots and fungi. A number of excellent surveys of the subject have been published, including the early work by Hatch (1937) and a more recent treatment by Harley (1969). Without mycorrhizae, most of our important tree species could not long survive against the dynamic, fiercely competitive biological communities that inhabit forest soils. Furthermore, the mycorrhizal condition is the rule, not the exception, in nature. Roots of most cultivated and noncultivated plants are infected with mycorrhizal fungi. The morphology of mycorrhizae varies among plant species, and each species tends to have characteristic groups of fungi capable of producing mycorrhizae. However, Melin (1963) pointed out that many different Basidiomycetes are able to form mycorrhizae with the same tree species. For example, more than 40 species have been proven to form mycorrhizae with *Pinus silvestris*. Even a single tree may be associated simultaneously with many fungal species and more than one species of fungi has been isolated from a small segment of lateral root, and even from individual mycorrhiza.

## TYPES OF MYCORRHIZAE

On the basis of the spatial interrelation of threadlike fungal hyphae and root cells, mycorrhizae are often divided into two general classes: (1) *ectomycorrhizae*, in which the fungal hyphae occur in the intercellular spaces of the cortical cells and form a compact mantle around the short roots; and (2) *endomycorrhizae,* in which hyphae occur intracellularly as well as extracellularly in the root cortex, but do not form a fungal mantle. Although Frank's classification into the two broad groups is still widely used (Hacskaylo, 1967), a third type, *ectendomycorrhizae,* is also recognized. It possesses some of the features of both ectomycorrhizae and endomycorrhizae.

## Ectomycorrhizae

Ectomycorrhizae have a much more limited distribution among plant species and a greater morphological uniformity than the endomycorrhizae. Ectomycorrhizae are restricted almost entirely to trees, but they occur naturally on many forest species. More than 2000 species of ectomycorrhizal fungi are estimated to exist on trees in North America. Most are Basidiomycetes, but certain of the Ascomycetes also form mycorrhizae. The fruiting bodies of these fungi produce spores that are readily and widely disseminated by wind and water. Ectomycorrhizae are characteristic of the families Pinaceae, Fagaceae, and Betulaceae, the principal tree species of cool and temperate forests. *Eucalyptus* and some tropical hardwood species are generally ectomycorrhizal, while such angiosperm families as Salicaceae, Juglandaceae, Tilliaceae, and Myrtaceae may be either ectomycorrhizal or endomycorrhizal, depending on soil conditions.

Ectomycorrhizal infection is initiated from spores or hyphae (propagules) of the fungal symbionts in the rhizosphere of feeder roots. Initial contact between hyphae of a mycorrhizal fungus and a compatible short root may originate from spores germinated in the vicinity of the roots, by extension through the soil of hyphae from either residual mycelia or established mycorrhizae, or progression of hyphae through adjacent internal root tissue (Hacskaylo, 1971). Growth of mycorrhizal fungi on the surfaces of short roots is stimulated by exudates from the roots. These exudates contain at least one growth promoting metabolite designated the M-factor. Fungal mycelia grow over the feeder root surfaces and form an external mantle or sheath (Figure 11.1). Following mantle development, hyphae grow intercellularly, forming a network of hyphae (Hartig net) around root cortical cells. Entrance of ectomycorrhizal fungi into tree roots may involve the secretion of pectolytic enzymes by the fungi, which dissolve the middle lamella and thus permit the hyphae to grow through the intercellular regions of the root cortex. Physical or chemical properties of roots restrict the hyphae of all mycorrhizal fungi to the cortex delimited by the endodermis and meristematic cells of the root tip. The exact mechanism of this resistance to hyphae penetration by other parts of the plant is unknown. The Hartig net, which may completely replace the middle lamellae between cortical cells, is the major distinguishing feature of ectomycorrhizae (Kormanik, Bryan, and Schultz, 1977).

The growing, absorbing root of the species of forest trees that become enveloped by ectomycorrhizal fungi are subjected to growth regulatory sub-

**Figure 11.1**   Ectomycorrhizae on slash pine. Note fungal mantle on short roots (courtesy D. H. Hubbell).

stances produced by the fungi. Auxin produced by the fungi modifies subsequent root growth, retarding elongation of short roots and frequently initiating dichotomous branching in pine (Hacskaylo, 1971), as graphically illustrated in Figure 11-2. Such roots are usually less than 0.5 cm in length, have no root cap, and possess a monarch stele. The cortical cells are oriented somewhat differently from those in nonmycorrhizal roots, and the presence of intercellular hyphae causes a swollen appearance. Ectomycorrhizae may appear as simple unforked feeder roots, or as bifurcated, multiforked coralloids (nodule-shaped modifications of feeder roots) as shown in Figure 11.3. Hyphae on short roots often radiate from the fungus mantle into the soil thereby greatly increasing the absorbing potential of the roots.

## Endomycorrhizae

The endomycorrhizal fungi are the most widespread and important root symbionts. They are found throughout the world in both agricultural and forest soils. They occur on most families of angiosperms and gymnosperms, including perhaps all agronomic and horticultural crops. Among the forest tree genera with this type of mycorrhizae are *Acer, Alnus, Fraxinus, Juglans, Liquidamber, Liriodendron, Platanus, Populus, Robinia, Salix,* and *Ulmus* (Gerdemann and Trappe, 1975).

Endomycorrhizae may be divided into two subgroups: (1) those produced

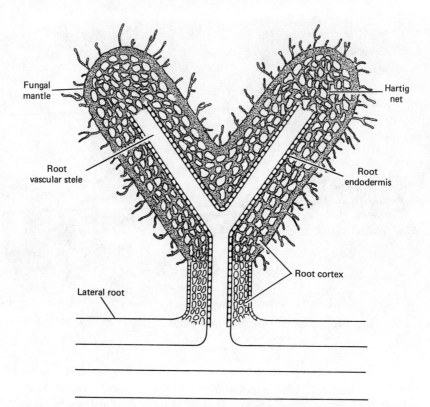

**Figure 11.2** Graphic presentation of an ectomycorrhiza (not to scale). Marx (1966).

by septate fungi, and (2) those produced by nonseptate fungi (Gerdemann and Trappe, 1975). Endomycorrhizae of trees are produced most frequently by nonseptate fungi of the family Endogonaceae (*Endogone* species). They generally form an extensive network of hyphae on feeder roots and extend as far as 1 cm from the root into the soil, but they do not develop the dense fungal sheath characteristic of ectomycorrhizae. Endomycorrhizal fungi form large, conspicuous, thick-walled spores in the rhizosphere and on the root surface and, sometimes, in the feeder root cortical tissue (Kormanik et al., 1977). The fungal hyphae penetrate epidermal cell walls of the root and then grow into cortical cells where they develop specialized absorbing structures called arbuscules in the cytoplasmic matrix. In some instances, the fungus completely colonizes the cortical region of the root, but it does not invade the endodermis, stele, or meristem (Gray, 1971). Thin-walled, spherical vesicles may also be produced in cortical cells, leading to the term "vesicular arbuscular" to denote this type of

**Figure 11.3** Clumps of ectomycorrhizae on conifer root, showing bifurcated short root.

endomycorrhizae. Endomycorrhizal infection does not result in major morphological changes in roots and, therefore, cannot be detected by the unaided eye. The anatomy of an endomycorrhizae is shown in Figure 11.4.

The fungi that form endomycorrhizae are mainly Phycomycetes. They do not produce large, above-ground fruiting bodies or wind-disseminated spores as do most ectomycorrhizal fungi. Some of them produce large azygospores and chlamydospores on or in roots, while others may produce sporocarps containing many spores in roots. Although these fungi depend mainly on root contact, moving water, or soil fauna for dissemination, they are nonetheless widespread in soils (Kormanik et al., 1977).

## Ectendomycorrhizae

This type of mycorrhizae has features of both ectomycorrhizae and endomycorrhizae. Ectendomycorrhizae are apparently limited in occurrence and, on forest trees, they generally appear on roots normally colonized by ectomycorrhizal fungi. Very little is known about the fungi involved. Wilcox (1971), in a review of the morphology of ectendomycorrhizae in red pine, pointed out that these infections in pine are almost exclusively confined to one- to three-year-old seedlings. He noted that ectendomycorrhizal fungi usually did not enter the new root growth on transplanted seedlings, which were immediately infected by an

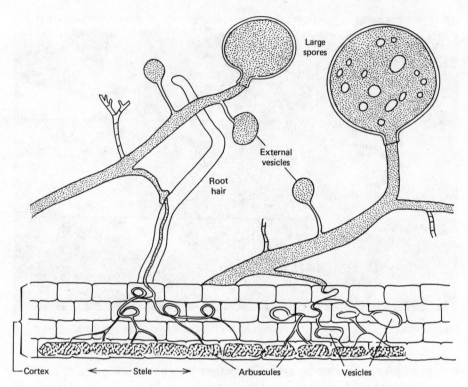

**Figure 11.4**  Graphic presentation of an endomycorrhiza (not to scale) (Nicolson, 1967). T. H. Nicolson, Vesicular—Arbuscular Mycorrhiza—a universal plant symbiosis. *In* Science Progress. 55:561–181, 1967. Blackwell Scientific Publications Ltd., Oxford, England.

ectomycorrhizal fungus. Some soil factor appeared to give the latter indigenous fungi a competitive advantage over the ectendomycorrhizal fungi. It is suggested that the same fungus may produce ectendomycorrhizae in seedlings growing in nursery soils and ectomycorrhizae when growing under forest conditions.

## SOIL FACTORS AFFECTING MYCORRHIZAL DEVELOPMENT

Environmental factors may influence mycorrhizal development by affecting either the tree roots or the fungal symbionts. After the formation of a receptive tree root, the main factors influencing susceptibility of the root to mycorrhizal infection appear to be photosynthetic potential and soil fertility (Marx, 1977). High light intensity and low to moderate soil fertility enhance mycorrhizal development, while the opposite conditions may reduce or even prevent mycorrhizal development. These factors may influence the biochemical status of the root by controlling the level of reducing sugars, or they may affect the synthesis of new (susceptible) feeder roots (Figure 11.5).

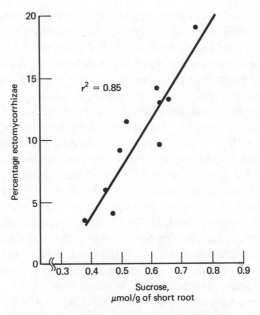

**Figure 11.5** Relationship of sucrose concentrations in short roots of loblolly pine seedlings maintained under 10 levels of soil fertility to percentage ectomycorrhizae formed after innoculation with *Pisolithus tinclorius* (Marx, Hatch, and Mendicino, 1977). Reproduced by permission of the National Research Council of Canada from the *Canadian J. Bot.*, **55**: 1569–1574, 1977.

When high concentrations of readily available nitrogen and phosphorus are absorbed from soils and translocated upward to the source of the photosynthate, soluble carbohydrates are assimilated rapidly during formation of new protoplasm and cell walls in the shoot. Consequently, the quantity of soluble carbohydrates translocated to and accumulated in the roots and the quantity of the M-factor secreted would be low. It appears that the formation of ectomycorrhizae is not favored under these conditions. Furthermore, low light intensity, which suppresses shoot growth, results in a similar low amount of soluble carbohydrates in the roots and a similar suppression of mycorrhizal formation. Mikola (1973) reported that ectomycorrhizal formation can be stimulated on white pine seedlings by applications of phosphate fertilizers to soils containing a low population of relatively inactive mycorrhizal fungi. It appears possible that growth suppression of *Pinus* spp. following nitrogen applications to phosphorus-deficient soils results from a reduction in mycorrhizal development (and phosphorus absorption) due to the high nitrogen:phosphorus ratio in host plant tissue (Figure 11.6).

The effects of soil fertility and fertilizer additions on the development of endomycorrhizae appear to vary with the original fertility of the soil before fertilizers are added, and the nutrient content of the host plant. If the level of soil organic matter affects vesicular-arbuscular development, as suggested by Hayman (1975), this might help explain the occasional presence of endomycorrhizae on nursery seedlings of species that are normally ectomycorrhizal in the forest.

There is no evidence that light has any *direct* effect on mycorrhizal development in soils. However, temperature can have a profound effect on the growth of certain mycorrhizae. Optimal temperatures for mycelial growth lie between 18°C and 27°C for the majority of species. For many ectomycorrhizae, growth ceases above 35°C and below 5°C, as found for *Thelephora terrestris* (Marx et al., 1970) and as illustrated for *Rhizopogon luteolus* in Figure 11.7. Others may have wider temperature tolerance for it has been pointed out that alpine and arctic timber lines are formed by ectomycorrhizal tree species (Moser, 1967). This suggests that symbiosis with fungi enables trees, during the short growing season, to obtain sufficient nutrients to produce drought and frost resistant young shoots and thus survive the unfavorable period of the year. Such ectomycorrhizal fungi as *Pisolithus tinctorius,* which develop at soil temperatures of 34°C or higher, offer advantages in forestation of adverse sites (Marx et al., 1970).

Apparently all mycorrhizal fungi are obligate aerobes, and it is presumed

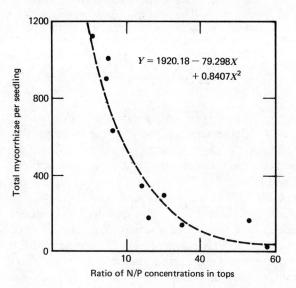

$$Y = 1920.18 - 79.298X + 0.8407X^2$$

**Figure 11.6** The relationship of number of ectomycorrhizae to ratio of nitrogen: phosphorus concentrations in loblolly pine seedling tops. (Pritchett, 1972). Used with permission.

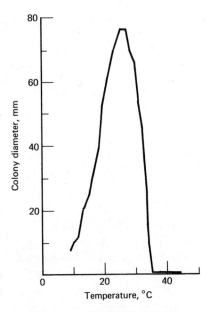

**Figure 11.7** Relationship of temperature to growth of *Rhizopogon luteolus* in 11 days (average of two strains) (Theodorou and Bowen, 1971). Used with permission.

that mycelial growth decreases with lowered oxygen tensions. It is also likely that the requirements of mycorrhizal fungi for nutrient elements are not greatly different from those of the host plants, although little research has been conducted on this point. It is generally conceded that formation of ectomycorrhizae on tree roots is greatest under acid conditions. Richards (1965) concluded that poor mycorrhizal formation in alkaline soils was due to nitrate inhibition of mycorrhizal infection rather than alkalinity per se. However, Theodorou and Bowen (1969) reported that alkaline conditions in the rhizosphere severely inhibit the growth of some types of mycorrhizal fungi apart from the possible effect of nitrate inhibition of infection. They further stated that nitrate inhibition of mycorrhizae formation under acid conditions is mainly due to a paucity of infection and not due to poor fungal growth. The effect of soil pH on mycorrhizal infection and growth of radiata pine is shown in Table 11.1.

The presence of antagonistic soil microorganisms can influence survival of the symbiont as well as root growth of the host plant. Fungicides used in plant disease control can inhibit mycorrhizal fungi under some conditions or they may stimulate mycorrhizal development by reducing microbial competition. Eradicat-

**Table 11.1**  Effect of Soil pH on Growth, Mycorrhizal Infection, and type of Infection of *Pinus radiata* Seedlings (Theodorou and Bowen, 1969)[a]

| Soil pH (at beginning) | Seedling Weight g | Mycorrhiza per Seedling % | Mycorrhizae Type | |
|---|---|---|---|---|
| | | | Brown % | White % |
| 4.5 | 1.16 | 41 | 43 | 57 |
| 6.2 | 1.61 | 53 | 67 | 33 |
| 8.0 | 0.33 | 18 | 100 | 0 |
| LSD ($P = 0.05$) | 0.15 | 15 | | |

[a]Used with permission.

ing ectomycorrhizal fungi from nursery soils by fumigation is not a problem in most areas because these fungi produce wind-disseminated spores that soon recolonize the soil. Artificial reinoculation of fumigated nurseries, however, may be necessary in cold climates. Fungitoxicants applied to young seedlings in nurseries may have an inhibitory effect on the development of vesicular-arbuscular mycorrhizae. Alteration of the host metabolism as a result of applications of systemic fungitoxicants can suppress fungal invaders, but the effect appears to be short lived.

## BENEFITS DERIVED FROM MYCORRHIZAE

The dependence of most species of forest trees on mycorrhizae to initiate and support healthy growth has been most strikingly illustrated by the problem encountered in introducing trees in areas devoid of the symbionts. Kessell (1927) failed to establish *Pinus radiata* in Western Australia nurseries that lacked mycorrhizal fungi. The seedlings grew normally only after soil from a healthy pine stand was added to the beds and ectomycorrhizae formed. Similar results have been obtained where exotic species have been used in the Philippines, Rhodesia, New Zealand, and Puerto Rico (Vosso, 1971). Other areas where ectomycorrhizal trees and their symbiotic fungi do not occur naturally, as reviewed by Marx (1977), include former agricultural soils of Poland, oak shelterbelts on the steppes of Russia, and the former treeless plains of the United States.

The observed benefit of mycorrhizae in the growth and development of trees has been ascribed to several factors. Among these factors are (1) increase in nutrient and water absorption by virtue of an increased absorptive surface area resulting from the formation of bifurcated or coralloid short roots and as provided by mycelia permeating the soil in the vicinity of short roots; (2) increase in nutrient mobilization through biological weathering; and (3) increase in feeder root longevity by providing a biological deterrent to root infection by soil pathogens. There is general agreement that mycorrhizae increase the capacity of infected plants to absorb nutrients and this can be especially important on infertile and adverse sites, such as mine spoils. Carbon compounds synthesized in the green tissue of the host not only nourish the host itself but serve as a source of carbon for fungal mycelia. In turn, soil-derived nutrients absorbed by the mycelia in the soil pass on into host tissue. Ectomycorrhizae are able to absorb and accumulate in the fungus mantle various elements, such as nitrogen, phosphorus, potassium, and calcium, and then translocate these elements to host plant tissue (Harley, 1969). Fungal hyphae completely permeate the F and H horizons of the forest floor and most minerals mobilized in this zone are absorbed before they reach the mineral soil. The importance of mycorrhizae to nutrient absorption is illustrated in Figure 11.8.

The fungi of ectomycorrhizae may be able to break down certain complex minerals and organic substances in the soil and make essential elements from these materials available to their host plant (Voigt, 1965a). However, the importance of the solubilization of minerals by mycorrhizal fungi to the nutrition of the host plant is not well understood, and there seems to be little evidence that *Endogone* mycorrhizae can exploit less soluble forms of phosphates. Neither is there good evidence that mycorrhizal fungi are themselves directly involved in dinitrogen fixation, but there is some indication the mycorrhizal system does

**Figure 11.8** Effect of mycorrhizal fungi on the growth of six-month-old Monterey pine seedlings (A) fertile prairie soil, (B) fertile prairie soil plus 0.2 percent by weight of Plainfield sand from a forest, and, (C) Plainfield sand alone (courtesy S. A. Wilde).

somehow stimulate fixation in the rhizosphere (Voigt, 1971), probably by stimulating microbial activity as a result of root or fungal exudates.

Mycorrhizal fungi have been reported to afford protection to delicate root tissue from attack by pathogenic fungi (Zak, 1964). This protection apparently results from the fungal mantle serving as a physical barrier to infection. Even without the mantle, root cortex cells surrounded by the Hartig net also are resistant to pathogens (Marx, 1972). In addition to a physical barrier there is an antibiotic mechanism of resistance. This resistance derives from a chemical substance identified as diatretyne nitrile. Normally susceptible roots adjacent to ectomycorrhizal roots are often resistant to attack by pathogenic fungi. These nonmycorrhizal roots are apparently protected by translocation or diffusion of an antibiotic from mycorrhizal roots (Marx, 1972). It has been suggested by Zak (1964) that the fungus may aid tree growth by protecting absorbing roots from soil phytotoxins.

There is some evidence that the symbiotic fungus may supply the higher plant with more than inorganic nutrients from the soil. They may also provide the host plant with growth hormones, including auxins, cytokinins, gibberellins, and growth regulating B vitamins (Kormanik et al., 1977).

## SILVICULTURAL IMPLICATIONS

The significance of mycorrhizal associations for forest trees was probably first noted when experimental plantings of exotic pines in various parts of the world invariably failed until suitable mycorrhizal fungi were introduced. Mikola (1973) speculated that in other countries and treeless areas where exotic pines were successfully introduced without intentionally importing mycorrhizal fungi, it is likely that fungi were inadvertently imported in potted seedlings or other materials.

In naturally regenerated forests there appear to be few possibilities of improving stand condition through mycorrhizal manipulations because of the widespread occurrence of the fungi. On the other hand, in the introduction of exotic species, in afforestation of adverse sites, and even in the establishment of some routine plantation forests, there may be a practical application for mycorrhizal symbiosis.

In introducing exotic pines to the tropics, nursery soils have generally been inoculated with ectomycorrhizal fungi. The same procedures have been followed in establishing new nurseries in temperate zone prairies or areas that have long been in agricultural use. Because ectomycorrhizal infection is easily spread by airborne spores, it is generally not necessary to artificially inoculate new pine nurseries in forested areas. In these instances, attention is directed toward correcting any nursery soil problems, such as acidity, fertility, and organic matter content, so as to favor introduction of fungi.

## Methods of Inoculation

Three types of materials have been used for mycorrhizal inoculation in nurseries: (1) soil from natural forests or nurseries; (2) mycorrhizal seedlings; and (3) pure cultures of mycorrhizal fungi. Soils have been the most widely used inocula because of the simplicity of operation. A thin layer of soil taken from an inoculated area is spread on the nursery bed and mixed with underlying soil or mixed with the growing medium for container-grown seedlings. Soil is expensive to transport and, consequently, forest floor litter, if readily available, is sometimes used as a mulch for the nursery beds. While these are generally satisfactory methods of introducing mycorrhizal fungi, the spreading of weed seeds and pathogenic organisms is also possible with the use of either soil or litter materials.

Vigorous mycorrhizal pine seedlings have been planted in seed beds at 1- to 2-m intervals as a source of infection for adjacent seedlings. This method has been used for *Pinus merkusii* in Indonesia, but is uncommon in large-scale nursery practice (Mikola, 1973).

The use of pure cultures of mycorrhizal fungi appears to hold the most promise as a method of inoculating nursery soils. Although commonly used experimentally (Marx, 1977), this technique has been rarely used in large-scale nursery operations because of the difficulties in mass producing quality inoculum. Commercial production has recently been initiated and an expansion in the use of inoculum can be expected in some areas.

## Cultures for Nursery Inoculation

The fact that mycorrhizal fungi differ in their environmental requirements poses the possibility of nursery inoculation with selections of species or strains that perform exceptionally well under certain difficult site conditions. The probability of obtaining a growth response of seedlings to supplemental inoculation with spores of selected fungi is largely determined by the prevalence and vigor of native strains and the relative persistence and efficiency of the introduced fungi in stimulating growth. Theodorou and Bowen (1970) reported improved survival and growth of young trees inoculated in the nursery with *Rhizopogon luteolus* and *Suillus granulatus*. After 32 months the growth rates of inoculated and uninoculated trees were about the same; however, height differences exhibited up to that time were maintained and apparently all trees were mycorrhizal.

Pure culture inoculation is used on an operational basis in Austria for the afforestation of high elevations with *Pinus cembra* (Mikola, 1973). *Boletus plorans* is not only a suitable symbiont for the pine near timberline, but it grows reasonably well in pure culture. Because the species is often lacking in nursery soils of the valley and in alpine areas above timberline, it is introduced in the nursery artificially so that seedlings may have an efficient symbiont on their roots at time of transplanting.

Marx (1977) reported that *Pisolithus tinctorius* was the most prevalent ectomycorrhizal fungus on roots of pine and spruce in strip-mined coal spoils in the northeastern United States. This fungus was also found on pines growing on low fertility, sheet-eroded clay soils in various parts of the United States. *P. tinctorius* increased the tolerance of pines to high temperature and improved tree survival and growth on these adverse sites. Other species of ectomycorrhizal fungi appeared on roots and produced basidiocarps primarily after litter had accumulated under the older seedlings on these sites, presumably due to the high soil temperatures in bare spoils. Soil temperatures between 35°C and 65°C were recorded in wastes at a depth of 6 to 7 cm, and pine seedlings with *P. tinctorius* ectomycorrhizae survived and grew as well at 40°C as at 24°C, but seedlings with *Thelephora terrestris* (or without mycorrhizae) survived poorly and did not grow at 40°C (Marx, 1977).

On the assumption that *P. tinctorius* became dominant on adverse sites because of its tolerance to high soil temperatures that restricted early establishment of other symbiotic fungi, Marx (1977) inoculated a number of fumigated nursery soils, using pure cultures of vegetative mycelial inoculum of this ectomycorrhizal fungus, in attempts to produce superior seedlings for difficult sites. Virginia and loblolly pine seedlings were grown with either *P. tinctorius* or *T. terrestris* in nurseries and outplanted on coal mine spoil banks in Kentucky and Virginia. Seedlings with *Pisolithus* mycorrhizae had better survival and made significantly greater growth than seedlings with *Thelephora* mycorrhizae on these acid (pH 3.9) spoils after two years (Table 11.2). The same pine species were planted on severely eroded sites in the Copper Basin in Tennessee. Seedling survival was excellent for both mycorrhizal treatments, but after two years *Pisolithus* ectomycorrhizae significantly improved growth of both pine species by 88 to 92 percent over that of seedlings with *Thelephora* mycorrhizae. Equally interesting results were obtained with loblolly pine seedlings grown with *T. terrestris, P. tinctorius,* or *Cenococum graniforme* mycorrhizae in the nursery and transplanted to strip-mine clay spoil and kaolin spoil banks in Georgia.

**Table 11.2**  Survival and Growth of Virginia and Loblolly Pines after Two Years on Coal Spoils. Seedlings Were Infested in a Nursery with *Pisolithus tinctorius* or *Thelephora terrestris* Ectomycorrhizal Fungi (Marx, 1977)[b]

| Mycorrhizal Condition at Planting | Virginia Pine | | | Loblolly Pine | | |
|---|---|---|---|---|---|---|
| | Survival % | Height cm | Volume cm³/tree | Survival % | Height cm | Volume cm³/tree |
| *P. tinctorius* | 49[a] | 49[a] | 130[a] | 90 | 101[a] | 962[a] |
| *T. terrestris* | 1 | 19 | 3 | 79 | 81 | 379 |

[a]Denotes significance at 0.05 level.
[b]TAPPI Conference Papers—Annual Meeting 1977.

**Table 11.3**   Survival and Growth of Slash Pine Seedlings with *Pisolithus tinctorius* or
Natural Ectomycorrhizae after Two Years on Wet Site, with and without
Fertilizer (Marx, 1977)[a]

| Mycorrhizae from Nursery Formed by | Fertilizer | | Survival | Height | Volume |
|---|---|---|---|---|---|
| | N | P | | | |
| | kg/ha | | % | cm | cm³/tree |
| *Pisolithus* | 0 | 0 | 83 | 72 | 209 |
| *tinctorius* | 90 | 90 | 77 | 150 | 2215 |
| Natural | 0 | 0 | 73 | 56 | 89 |
| Inoculum | 90 | 90 | 73 | 150 | 2188 |

[a] TAPPI Conference Papers—Annual Meeting 1977.

Marx (1977) pointed out that producing seedlings in the nursery with specific ectomycorrhizae has potential benefits for forestation of adverse sites. Many disturbed sites have soil factors that exert selective pressures on symbiotic fungi, and fungi that can tolerate these factors are better adapted and should be used in producing seedlings for those aeas. However, it was difficult to maintain the integrity of the specific ectomycorrhizal associations under the conditions tested. The ubiquitous *Pisolithus* recolonized roots of control seedlings from air-borne sources so rapidly that valid growth comparisons between seedlings initially treated with different pure cultures were sometimes difficult to obtain. For example, slash pine seedlings grown in fumigated nursery soil inoculated with a culture of *Pisolithus tinctorius* and transplanted to a phosphorus-deficient wet coastal site made better growth than naturally inoculated seedlings without fertilizer, but there were no differences in growth where fertilizer was added after two years (Table 11.3). Nevertheless, the technique of infesting fumigated nursery soils with mass-produced pure cultures of selected ectomycorrhizal fungi appears to hold considerable promise for adverse sites where other fungi may not readily colonize roots of planted trees.

Pure cultures of endomycorrhizal fungi have also been tested in fumigated nurseries for the production of hardwood seedlings. Soil fumigation can effectively reduce populations of endomycorrhizal fungi in surface soils and, since spores of these fungi are not readily wind disseminated, recolonization of the fumigated soil is slow and erratic. Even in well-fertilized nursery soils, nonmycorrhizal sweetgum seedlings were unable to obtain sufficient nutrients for good growth in a Georgia trial (Marx, 1977). However, seedlings treated with a culture of *Glomus mosseae* endomycorrhizae were several times larger than nonmycorrhizal seedlings after six months, and seedlings in control plots infested with naturally occurring endomycorrhizae in a sweetgum forest soil were largest of all. These tests indicated that poor early growth of hardwood seedlings in fumigated nursery soil is often followed by acceptable growth beginning in midseason, apparently due to root penetration and subsequent endomycorrhizal development in the zone below effective soil fumigation. Nonetheless, certain endomycorrhizal fungi may stimulate hardwood seedling growth more than others in the nursery, and this suggests the possibility of selection among endomycorrhizal fungi for those species ecologically adapted to specific sites with high soil temperatures or other adverse conditions.

# 12

# NUTRIENT CYCLING IN FOREST ECOSYSTEMS

A unique characteristic of most forest ecosystems is the development of a distinct forest floor resulting from the periodic return through litterfall of leaves, branches, bark, fruit, and sometimes entire trees. This litterfall contains a large proportion of the nutrients extracted by the trees from the soil, with only a relatively small percentage retained in the growing biomass. The dead vegetation on the floor, in turn, decomposes, liberating minerals for reuse by the growing stand. Other nutrients may be brought in from the atmosphere or made available by biological fixation or weathering of parent rock, while some nutrients are lost through crop harvests, burning, and removal in surface or ground water. Thus, it is through this dynamic and rather complex system of geological, chemical, and biological cycling that the soil organic matter and nutrient supplies are replenished and maintained, thereby ensuring continued productivity of the site.

The importance of nutrient cycling was recognized in the early history of forestry. Ebermayer (1876) assessed annual litterfalls and growth increments in Bavarian hardwood and coniferous forests and analyzed samples for their nutrient contents. He was able to demonstrate that nutrients are not only withdrawn from the soil by trees, but that significant quantities are returned annually to the forest floor in litterfall. Not long after Ebermayer's work, the gradual deterioration of site productivity as a consequence of interrupted nutrient cycling was dramatically demonstrated in forests where litter had been removed repeatedly for animal bedding; a rather common practice in Europe during the last century.

Nutrient cycling received renewed attention when use of commercial fertilizers became a viable alternative in intensive forest management. It became important to know whether applied nutrients are cycled rapidly and reused many times, or whether they are leached or become immobilized in a physiologically less active component of the system. Nutrient cycling undoubtedly has an influence on the long-term effectiveness of any fertilizer treatment.

Mineral cycling has been studied in varying degrees of detail for a number of forest ecosystem types in recent years. Duvigneaud and Denaeyer-DeSmet (1970) summarized the result of Belgian research in deciduous forests and interpreted their data in the context of mineral cycling in European deciduous forests. Remezov (1959) reviewed accomplishment in the U.S.S.R., and more recently Rodin and Bazilevick (1967) reported on biological mineral cycling in the better-known ecosystems of the world. They recognized a zonal distribution of biomass, nutrient content, and circulation parallel to that of plant formation zones. A number of rather detailed studies have been made in Great Britain on cycling in hardwood forests (Ovington, 1962; Carlisle et al., 1967), and Young (1967) edited a symposium that included reports on coniferous forests in the Pacific Northwest and mixed deciduous-coniferous forests in the east and south-eastern regions of the United States. Some of the most comprehensive information on mineral cycling has been generated by the United Nations International Biological Program, aimed at establishing reference values for the productivity of the world's major plant communities.

The transfer of minerals in and out and among the various nutrient pools is a continuous process and can be measured on a daily, seasonal, or annual basis. They have been most widely measured on an annual basis. Remezov (1959) recognized two major nutrient cycles in forest ecosystems: (1) an external *geochemical* system and (2) an internal *biological* cycle. The former "open" system concerns the import-export relationship of nutrients into and out of the ecosystem, while the latter "closed" system involves plant-soil exchanges of nutrients. These two major cycles will serve as a basis for the discussions that follow. Admittedly, the discussion is somewhat an oversimplification of a complex phenomenon, and few scientists agree as to the importance of its various components. In fact, Switzer and Nelson (1972) proposed that Remezov's biological cycle be split into two parts: one cycle dealing with translocation of nutrients within trees, and the other cycle involving the soil-plant relationship. They called the internal transfer of nutrients within the standing tree mass the *biochemical* cycle and the transfer between the soil and the tree biomass the *biogeochemical* cycle.

## GEOCHEMICAL NUTRIENT CYCLING

The geochemical cycle involves the transfer of mineral elements into or out of the ecosystem. This cycle imports from such sources as dust and precipitation, weathering of parent rock, biological fixation of nitrogen, and fertilization; while outputs include leaching and erosional losses in drainage water, volatile losses from fire and denitrification, and removal in harvests. The amounts of nutrients gained or lost annually by an ecosystem are influenced by such factors as soil properties, climatic conditions, type of vegetation, and location of ecosystem in relation to the sea and industrial areas. These fluxes tend to reach an equilibrium in a mature forest, unless disturbed by man or acts of nature.

## Nutrient Inputs (Gains)

**Atmospheric Inputs.**   Inputs by precipitation and dust vary by location, as well as season of the year, depending largely on dust load and lightning activity. The forest canopy is an efficient agent in capturing airborne dust particles from industrial and agricultural sources. Duvigneaud and Denaeyer-DeSmet (1970) pointed out the difficulty of assessing the relative amounts of nutrients washed as dust from foliage and branches as part of an external cycle, as compared to nutrients leached from leaves and twigs by throughfall and stemflow. Nevertheless, estimates of leaching losses can be made and they will be treated later in the discussion on internal cycling.

The atmosphere is a particularly important long-term source of nitrogen. Through electrical discharge, molecular nitrogen is converted to ammonium, nitrate, or various nitrogen oxides that dissolve in atmosphere humidity and reach the soil in precipitation. Large quantities of combined nitrogen are introduced into the atmosphere by industrial pollution and automobile exhaust in some regions. The National Precipitation Network (Wolaver, 1972) reported that additions of ammonium plus nitrate nitrogen along the eastern seaboard of the United States varied from 0.3 to 5.0 kg per hectare per year. A more realistic figure for northern Europe and eastern United States might be between 4 and 10 kg per hectare per year, with a sizable proportion present as organic nitrogen (Wollum and Davey, 1975). The main source of combined atmospheric nitrogen probably originates from soil and ocean surfaces, according to these authors. Ovington (1968) calculated an average input of minerals into ecosystems by precipitation in temperate regions ranged as follows: nitrogen, 0.2 to 0.6; potassium, 1 to 10; calcium, 3 to 19; and magnesium 4 to 11 kg per hectare per year. Switzer and Nelson (1972) reported values for precipitation additions in 20-year-old loblolly pine in Mississippi as about middle of the range given by Ovington (1968), except for a higher nitrogen value. These and other data from temperate regions are shown in Table 12.1. Although one finds a fairly wide range of input values from precipitation and dry fallout among locations, average values are meaningful worldwide and of interest in cycling studies. For example, the average input for nitrogen from all sites was about 9 kg per hectare per year. The inputs for phosphorus, potassium, calcium, and magnesium averaged about 0.3, 3.2, 8.0, and 2.8 kg per hectare per year, respectively. These values are in line with other reports on precipitation inputs summarized by Weetman and Webber

(1972) and Woodwell and Whittaker (1967). Potassium, calcium, and magnesium additions to forest ecosystems via the atmosphere originate largely as aerosols over oceans and agricultural lands. Under drought conditions, a significant part of the atmospheric input of nitrogen and other nutrients can come from dust particles. These additions are usually larger in coastal areas than inland (Table 12.1). The average annual nutrient inputs from atmospheric sources may not appear large, but taken in the context of a forest rotation, they can be significant.

**Biological Nitrogen Fixation.**   Fixation of atmospheric nitrogen by microorganisms is probably the most important pathway for this element to enter the forest ecosystem. Nitrogen may be fixed by both free-living organisms and by symbiotic relationships between microorganisms and higher plants.

   *Nonsymbiotic* dinitrogen fixation by blue-green algae and certain free-living bacteria (*Clostridium* and *Beijerinckia* species) is not believed to account for more than a few kilograms of nitrogen per hectare annually in most forest soils (Youngberg and Wollum, 1970; Wollum and Davey, 1975). Forested soils are generally too acid for appreciable activity by free-living bacteria. An exception is a report by Brouzes and Knowles (1969) which indicated that appreciable quantities of nitrogen could be fixed by *Clostridium* spp. in a strongly acid, water-saturated black spruce *(Picea mariana)* forest floor. Members of all three families of photosynthetic bacteria are capable of dinitrogen fixation under anaerobic condition only. Blue-green algae are active mainly in areas of higher light intensity than found under most forest canopies. However, blue-green algae that occur in lichens colonizing crowns of Douglas-fir stands contribute several kilograms of nitrogen per hectare per year under some conditions (Denison, 1973). Richards and Voigt (1965) reported annual accessions of nitrogen in coniferous forest ecosystems that were too large to be explained by the normal activities of the known agents of fixation. However, they were not able to identify the organisms responsible for these increases. Wollum and Davey (1975) suggest that dinitrogen fixation in the rhizosphere of numerous plants could be significant under anaerobic conditions, apparently involving species of *Clostridium.*

   *Symbiotic* nitrogen fixation is carried on by *Rhizobium* in concert with members of the family *Leguminosae,* and perhaps by *Streptomyces* and other unidentified organisms in conjunction with a large number of nonleguminous plants. Many leguminous plants are capable of fixing large amounts of nitrogen (in excess of 100 kg nitrogen per hectare per year), but as they have high light and nutrient requirements they normally are grown under agricultural conditions and do not thrive in a forest environment. A large number of leguminous species are believed to occur naturally in forests, especially in disturbed areas, but little is known of their fixing capacity.

   Some legumes have been successfully introduced into problem areas as a

**Table 12.1**  Input of Nutrients in Precipitation in kg per Hectare per Year

| Location | Quantity of Precipitation | Quantity of Nutrients | | | | | Source |
|---|---|---|---|---|---|---|---|
| | mm/year | N | P | K | Ca | Mg | |
| | | kg/ha/year | | | | | |
| Wisconsin | — | 13.1 | 0.3–0.4 | 1.0–4.0 | 2.0–7.0 | 0.5–1.1 | Boyle and Ek, 1972 |
| Great Britain | 1717 | 8.7 | 0.3 | 2.8 | 6.7 | 6.1 | Carlisle et al., 1967 |
| Belgium | — | 6.0 | — | 2.9 | 9.1 | 2.3 | Duvigneaud and Denaeyer-DeSmet, 1967 |
| Great Britain | — | — | — | 2.8 | 4.2 | 10.7 | Ovington and Madgwick, 1959 |
| Germany | 624 | 20 | 0.1 | 4.6 | 19 | — | Neumann, 1966 |
| Nigeria | 1850 | 14.0 | 0.4 | 17.5 | 12.7 | 11.3 | Nye, 1961 |
| Mississippi | 1270 | 13.3 | 0.3 | 4.0 | 5.0 | 1.0 | Switzer and Nelson, 1972 |
| USSR | 204 | 5.6 | 0.5 | 7.7 | 15.4 | 2.5 | Remezov and Pogrebnyak, 1969 |
| Sweden | 420–648 | 1.4–5.2 | — | 0.6–3.7 | 2.6–13.9 | 0.6–2.6 | Tamm, 1958 |
| N. Carolina | 1169 | 3.5 | 0.3 | 0.9 | 3.4 | 0.7 | Wells et al., 1972 |

source of nitrogen, in preparations for afforestation (Gadgil, 1971a). However, neither native nor planted legumes play an important part in nitrogen accretion in temperate-region forests at the present time. Lupines have been used with some success in the amelioration of degraded forest soils in Germany, but this success has depended on cultivation and fertilization to secure the legume establishment. On the other hand, symbiotic fixation by nonlegumes may be important under some conditions, particularly where they serve as pioneer species. For example, on sites with significant numbers of alder, sweet fern, or wax myrtle, annual fixation may amount to as much as 50 kg nitrogen per hectare or more (Young-berg and Wollum, 1970). Nitrogen accretion by *Alnus glutinosa* was estimated at 56 kg per hectare per year by Silver and Mague (1970). The same authors estimated that wax myrtle *(Myrica cerifera)* may fix as much as 15 kg nitrogen per hectare per year in coastal areas of the southeastern United States. A nonleguminous nitrogen fixing plant common to many cool climate forests is *Comptonia* (Figure 12.1).

**Weathering of Parent Rock.**    The geological weathering of parent material is thought to be one of the most important means of replenishing nutrient reserves in most forest ecosystems. In fact, very little is known of the actual contributions

**Figure 12.1** Symbiotic dinitrogen fixation by nonleguminous plants can add several kilo-grams per hectare per year to forest ecosystems, usually as pioneers such as *Comptonia peregrina.*

from this source because of the difficulties in obtaining reliable measurements. Among factors that influence the rate of input from this source are the nature of the weatherable rock, climatic conditions, topography, and vegetation.

Nitrogen inputs from weatherable rock are generally too small to be considered in balance sheets, except under some special conditions, such as coal mine spoil banks (Cornwell and Stone, 1968). On the other hand, phosphorus appears to be supplied from parent material in ample quantities for good tree growth on such soils as those derived from residual material over igneous rock, deep glacial till, and fluvioglacial outwash material. In fact, soils with an impoverished phosphorus cycle are largely confined to those derived from marine deposits of sands along coastal plains, organic soils, or ancient soils with high sesquioxide contents.

Most forest soils contain sufficient primary and secondary minerals as components of the parent material to insure an adequate cycle of cations through normal weathering processes. Potassium is sparingly supplied to forest ecosystems from feldspars and muscovites, but is more richly supplied where biotite is abundant in the parent material. Calcium and magnesium silicates weather more rapidly than potassium-containing primary silicates, except biotite. These elements may also occur as carbonates in deeper portions of the soil profile. As a result, significant amounts of calcium and magnesium are continuously released into many forest ecosystems, often in sufficient quantities so that losses by leaching are sizable. Average inputs for several temperate forest ecosystems are given in Table 12.2.

**Fertilizer Contribution to Cycling.** The application of commercial fertilizer is an accepted means of increasing rates of nutrient cycling and tree growth in nutrient deficient stands. Where one or more nutrient is in limited supply, trees are unable to fix energy at a satisfactory rate, nor utilize other nutrient elements to the extent that they are available. Nitrogen may become a limiting factor to

**Table 12.2** Estimated Cation Inputs to Temperate Forest Ecosystems from Weathering of Parent Rock

| Location | K | Ca | Mg | Na | Source |
|---|---|---|---|---|---|
| | kg/ha/year | | | | |
| Brookhaven Forest (Long Island, N.Y.) | 11.1 | 24.3 | 8.3 | 6.7 | Woodwell and Whittaker, 1967 |
| Hubbard Brook Forest (New Hampshire) | 4.0 | 8.0 | 8.0 | — | Likens et al., 1970 |
| Cedar River Forest (Washington) | 15.2 | 17.4 | — | — | Cole et al., 1967 |

tree growth in boreal and other forest ecosystems, because this element is largely immobilized in the materials accumulated on the forest floor (Weetman, 1962). The addition of fertilizer may hasten the decomposition of this organic matter and speed up the cycling process.

In phosphorus-deficient wet savanna soils of the lower coastal plain, slash pine often grows in open stands with very sparse crowns. The additions of phosphorus and nitrogen fertilizers relieve the deficiency and alter the functioning of the nutrient cycling. For example, during the first 15 years of a slash pine stand development, energy fixed in the surface biomass increased from 42.5 tons per hectare to over 224 tons per hectare, and nutrient cycling through the system was similarly accelerated (Pritchett and Smith, 1974). While only 45 kg nitrogen per hectare were added in fertilizer, nearly 560 kg were in the above-ground biomass. Of this quantity, nearly 224 kg were in the forest floor. Nearly five times as much potassium was cycling in the fertilized stand as was cycling in the unfertilized stand—representing considerably more than that applied in the fertilizer. Similar quantitative enhancement of other nutrients, such as calcium and magnesium, also occurred. Suppressed trees in the unfertilized areas were apparently unable to effectively absorb calcium and potassium and other elements from lower soil layers and add them to the nutrient cycle. For example, calcium is known to be helpful in the establishment of native legumes. Increased calcium circulation in response to added fertilizers may explain why nitrogen in the fertilized system was increased, presumably resulting from fixation by both symbiotic and nonsymbiotic organisms. It is probable that if organisms are unable to absorb, retain, and subsequently recycle nutrients released by the soil, these nutrients may be lost from the system to the ground water.

## Nutrient Outputs (Losses)

**Leaching and Surface Runoff.**    Nutrient losses in drainage water from undisturbed forest ecosystems are generally conceded to be minimal. Gessel and Cole (1965) reported that while 4.8 kg nitrogen per hectare per year leached from the forest floor into the upper 3 cm of mineral soil, only 0.6 kg nitrogen per hectare per year leached beyond the rooting zone of a mature Douglas-fir stand in Washington. In the same lysimeter experiment, 10.1 kg of potassium and 16.6 kg of calcium were leached from the forest floor but only 1.0 and 4.5 kg per hectare per year, respectively, were leached beyond 100 cm.

Phosphorus output from a deciduous hardwood forest in Minnesota averaged 0.06 kg per hectare per year in surface runoff with an additional 0.03 kg per hectare lost in sediment carried in the runoff water. The sum of these two outputs was slightly less than the 0.11 kg phosphorus per hectare per year input from precipitation (Singer and Rust, 1975).

Outflow from a number of watersheds has been monitored to check losses of nutrients from forested areas. Likens et al. (1970) reported nitrate concentrations in stream water in the northern hardwood region at Hubbard Brook watershed of about 0.1 ppm in summer months (June–September) when available nitrogen was utilized by growing vegetation and heterotrophic activity in the soil. After growth declined with leaf fall and diminished biological activity in November, available nitrates leached through the soil, with a peak in concentration of about 2 ppm in March or April. These small losses were offset by inputs of nitrates in precipitation. Average leaching losses of nitrogen, calcium, magnesium, and potassium amounted to 2.3, 11.4, 3.2, and 2.0 kg per hectare per year from this undisturbed watershed (Table 12.3). Leaching losses of nutrients from an undisturbed hardwood stand in the central Appalachians were considerably less than those reported for Hubbard Brook (Aubertin and Patric, 1972). However, losses of cations to groundwater or streamflow can amount to as much as 25 and 50 kg per hectare per year for sodium and calcium on some sites, as shown in Table 12.3.

**Harvesting.** Human activity has often upset the near equilibrium in nutrient cycling frequently reached in mature forests. An early example was the practice of gathering litter from the floor of coniferous forests by Bavarian farmers. The damaging effects on tree growth of nutrient losses by repeated litter removal became clearly evident and it was eventually prohibited by law. However, it was estimated that as much as 600 kg nitrogen per hectare were removed from some forests of Germany before the practice was eliminated (Kreutzer, 1972).

The amount of nutrient removal in conventionally harvested crops depends on such factors as species, age, stocking, and site quality, but equated to an annual basis the losses probably amount to no more than 5 to 10 kg per hectare of nitrogen, potassium, and calcium and only a kilogram or so of phosphorus and magnesium (Rennie, 1955). Other average values for nutrients removed in harvested crops are given in Table 12.4.

**Table 12.3**  Precipitation Inputs and Losses to Groundwater or Streamflow from Three Watersheds

| | N | P | K | Ca | Mg | Na |
|---|---|---|---|---|---|---|
| | | | kg/ha/yr | | | |
| | Oregon (Douglas-fir)—Fredriksen, 1970 | | | | | |
| Input | 0.9 | 0.3 | 0.1 | 2.3 | 1.3 | 2.3 |
| Output | 0.4 | 0.5 | 2.2 | 50.3 | 12.4 | 25.7 |
| | New Hampshire (Northern hardwood)—Likens et al., 1970 | | | | | |
| Input | 7.2 | — | 0.7 | 2.7 | 0.6 | — |
| Output | 2.3 | — | 2.0 | 11.4 | 3.2 | — |
| | New York (Oak-pine)—Woodwell and Whittaker, 1967 | | | | | |
| Input | — | — | 2.4 | 3.3 | 2.1 | 17.0 |
| Output | — | — | 3.9 | 9.6 | 7.3 | 23.2 |

**Table 12.4**  Nutrients Removed in Conventional Bolewood Harvest

| Species | N | P | K | Ca | Mg | Source |
|---|---|---|---|---|---|---|
| | | | kg/ha | | | |
| Northern hardwood (45–50 yr) | 120 | 12 | 60 | 241 | 24 | Boyle and Ek, 1972 |
| Oak (47 yr) | 151 | 11 | 118 | 173 | 23 | Ovington, 1962 |
| Beech (37 yr) | 128 | 16 | 94 | 79 | 28 | Ovington, 1962 |
| Douglas-fir (36 yr) | 125 | 19 | 96 | 117 | — | Cole et al., 1967 |
| Loblolly pine (16 yr) | 115 | 15 | 89 | 112 | 29 | Wells et al., 1976 |
| Spruce-fir (uneven-aged) | 79 | 11 | 47 | 150 | 14 | Weetman and Webber, 1972 |

Disturbing the soil during harvesting and subsequent site preparation may hasten the decomposition of forest floor materials and result in nutrient loss by leaching and erosion. This increased loss is quite small following normal harvest operations (Gessel and Cole, 1965), but it may be significant if the soil is unduly disturbed. For example, nitrate concentrations in Hubbard Brook stream water was 41-fold higher from a clearcut area, where all vegetation was left to decompose on the soil surface and regrowth was inhibited for two years by periodic application of herbicides, than from adjacent undisturbed forest watersheds. There were also severalfold increases in bases in the stream water. While the magnitude of the nitrogen losses for the treated watershed can, no doubt, be explained by the large amount of organic material left on the site and the suppression of ground cover regrowth, rather large losses from the undisturbed watershed appear to be unique to this northern hardwood forest ecosystem. Studies from other areas have shown losses to groundwater or streamflow following normal timber harvests to be minimal (Aubertin and Patric, 1972; Gessel and Cole, 1965; Douglass and Swank, 1972).

**Volatile Losses.**   Nitrogen may be volatilized from soils as ammonia by nonbiological means under alkaline conditions; as nitrous oxides by chemical decomposition of nitrite under acid conditions; as $N_2$ by nonenzymatic reaction of nitrous acid with ammonium or amino acids; and as $N_2$ and nitrous oxide by microbial denitrification (Alexander, 1977). However, volatile losses from forest soils are likely to be insignificant because conditions generally prevailing in these soils do not favor the formation of gaseous forms of nitrogen.

Volatilization of free ammonia is insignificant below pH 7.0 and most forest soils are much too acid for losses from this route. While nitrite may decompose under the acid conditions found in many forest soils (below pH 5.5), there must first be oxidation of ammonium to nitrite. It is generally conceded that there is little ammonium oxidation to nitrite or nitrate in acid surroundings (Alexander, 1977). Furthermore, in the absence of nitrate there is little opportunity for denitrification in spite of the energy source and anaerobic conditions found in wet sites. Overall, it appears that none of the reactions that result in gaseous loss of nitrogen are common under forest conditions.

On the other hand, the gaseous losses of nitrogen and sulfur during forest fires can be significant. For example, from 20 to 80 kg nitrogen per hectare may be lost during a prescribed burn for the purpose of fuel reduction. The loss during a wildfire may be many times this amount. Burning not only disrupts the nutrient cycle, but on steep terrain, burning may increase the likelihood of erosional and leaching losses. Erosion is not a problem on nearly level landscapes, such as encountered in coastal plains, and the loss of nitrogen from burning may be largely offset by increased nitrogen fixation by symbiotic and nonsymbiotic organisms as a result of decreased soil acidity on these sites.

### Short-term Nutrient Balances

Annual nutrient balances based on estimated inputs and outputs have been prepared for a number of forest ecosystems (Weetman and Webber, 1972; Boyle and Ek, 1972). The nutrient budgets for the three watersheds in Table 12.3 indicate a net gain of nitrogen, small net losses of potassium, somewhat larger net loss of magnesium, and sizeable losses of sodium and calcium. However, the cation losses from these forested watersheds are probably fully or almost fully compensated by inputs from weathering of the parent rock (Table 12.2). The studies reported on by Gessel et al. (1973) for the Cedar River watershed and by Wells and Jorgensen (1975) for a loblolly pine plantation also show small net gains for nitrogen and minor net losses for phosphorus, potassium, and calcium.

These findings generally support the contention that nutrient pools of terrestrial ecosystems increase as succession progresses and that there is a near equilibrium in nutrient fluxes as mature (climax) conditions are approached. On the other hand, this balance can be significantly altered by such human intervention as harvesting, burning, or fertilizing.

## BIOLOGICAL NUTRIENT CYCLING

Biological cycling involves the transfer of nutrients between the forest floor-soil and the associated plant and animal communites. In forest ecosystems it may also include the internal transfer of nutrients among organs within the tree. The major steps within closed cycles, therefore, include (1) uptake, (2) retention, (3) restitution, and (4) internal transfers.

### Nutrient Uptake by Trees

Nutrient absorption by trees is influenced by type and age of the forest cover, and the soil and climatic conditions of the community. The annual uptake by most forest species is in the same order as that of many agricultural crops, but because a major portion of the absorbed nutrients are returned to the forest floor, or translocated within the tree, relatively small amounts are retained each year in an annual accretion of biomass. Data in Table 12.5 indicate a wide range in the amounts of nutrients absorbed among the different species, with the greatest uptake by a mixed stand of hardwoods (oak and ash). The lowest uptake rates were for mature Scots pine stands, while that of a 20-year-old loblolly pine stand was only slightly higher in nitrogen and potassium and slightly lower in phosphorus uptake.

Average nitrogen uptake is approximately 10 times that of phosphorus and three times that of potassium for the species shown in Table 12.5. Although not shown in this table, calcium uptake may be considerably higher than that of nitrogen by most hardwoods, but lower than nitrogen uptake by most conifers. Depending on the supply in the soil, calcium can be slightly lower to many times higher than that of potassium, but the rate of calcium uptake does not appear to be closely related to the rate of tree growth. Although an essential element, calcium has rarely been shown to be a growth limiting factor in forest ecosystems.

The effect of soil properties, particularly the nutrient-supplying power of soils, on nutrient uptake by trees is well recognized. An example of the effects of parent material is shown in Table 12.5, illustrating that European beech growing on dioritic soil absorbed 65–80 percent more phosphorus and potassium than the same species on granitic soils. The stage of stand development also has an effect on the annual uptake of nutrients. There is an annual increase in nutrient uptake in the early years of stand development. Remezov (1959) reported a sharp increase in mineral uptake when productivity, or annual increment, was at a maximum. However, as plantations mature, the rate of nutrient uptake decreases. This decrease may be accompanied by a reduction in litterfall (Wells and Jorgensen, 1975). Once canopy closure is reached the gross annual uptake of nutrients appears to be relatively constant. Ovington and Madgwick (1959a) reported only small differences in uptake in natural birch stands between the ages of 6 and 55 years on fens in England. Similar results were obtained by Mina (1955) working with 25- to 212-year-old oaks on mineral soils of Russia.

## Nutrient Retention and Distribution

Net annual nutrient accumulation may be considered as the difference between total nutrient uptake and that returned to the soil in the form of dead roots, litter, and canopy leachings. It is usually calculated from data obtained from measurements of periodic biomass changes, together with the chemical composition of the tissue making up the increments. Since nutrient accumulation usually follows biomass expansion, it increases linearly or exponentially during periods of rapid early growth, and at a diminishing rate as the stand reaches maturity. Wells and

Table 12.5   Annual Uptake, Retention, and Restitution of Some Nutrients in Forest Ecosystems

| Forest Cover | Nitrogen (kg/ha/year) | | | Phosphorus (kg/ha/year) | | | Potassium (kg/ha/year) | | | Source |
|---|---|---|---|---|---|---|---|---|---|---|
| | Uptake | Retention | Return | Uptake | Retention | Return | Uptake | Retention | Return | |
| Scots pine (mature) | 45 | 10 | 35 | 4 | 1 | 3 | 6 | 2 | 4 | Duvigneaud and Denaeyer-DeSmet, 1970 |
| European beech (mature) | 50 | 10 | 40 | 12 | 2 | 10 | 14 | 4 | 10 | Switzer and Nelson, 1972 |
| Loblolly pine (20 years) | 42 | 5 | 37 | 2 | — | — | 25 | — | — | Ehwald, 1957 |
| Scots pine (100 years) | 34 | 12 | 22 | 3 | 1 | 2 | 7 | 2 | 5 | Gessel et al., 1973 |
| Douglas-fir (37 years) | 39 | 24 | 15 | 7 | 6 | 1 | 29 | 14 | 15 | Ehwald, 1957 |
| Norway spruce (100 years) | 61 | 21 | 40 | 5 | 2 | 3 | 12 | 7 | 5 | Duvigneaud and Denaeyer-DeSmet, 1962 |
| Red oak-European ash (140 years) | 123 | 44 | 79 | 9 | 4 | 5 | 75 | 21 | 45 | Klausing, 1956 |
| European beech (115–125 years) (dioritic soil) | — | — | — | 9 | 6 | 3 | 8 | 4 | 4 | |
| (granitic soil) | — | — | — | 6 | 3 | 3 | 5 | 2 | 3 | |

Jorgensen (1975) reported that the maximum accumulation period appears to be somewhere between 10 and 15 years of age for rapidly growing trees in North Carolina. In a loblolly stand in Mississippi (Switzer et al., 1968), nitrogen accumulation leveled off after 40 years, phosphorus and potassium accumulation after 30 years, and calcium accumulation after 50 years. Total nitrogen accumulation at age 60 years was approximately 250 kg per hectare in this stand. The annual and total dry matter accumulation in the stems of this loblolly stand is shown graphically in Figure 12.2.

In early growth stages, most of the nutrients are contained in the foliage. For example, Switzer et al. (1968) showed that nitrogen increased rapidly in the stand foliage mass until the twelfth year and that it declined slightly thereafter. In contrast, nitrogen in the stem and branchwood steadily increased up to 60 years of age, although the rate of increase noticeably diminished after age 40. At 20

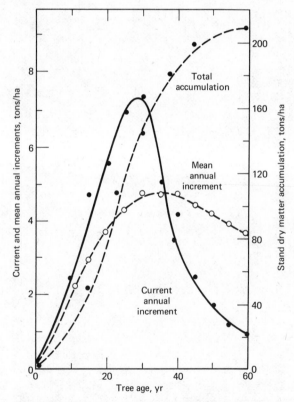

**Figure 12.2**  Annual and total dry matter accumulations in stems of a loblolly pine stand during 60 years (Switzer et al., 1968).

years, slightly more than one-half of the nitrogen in the standing biomass of the loblolly pine stand was contained in stemwood. This ratio did not change greatly as the stand grew older.

Total nutrient accumulation in rapid growing species can be rather large, as illustrated by the 15-year-old fertilized slash pine and the slightly older loblolly and radiata pines and Douglas-fir in Table 12.6. Species grown in long rotations, such as the northern conifers, may accumulate even larger quantities of nutrients in the tree biomass, but the greatest accumulations are, perhaps, found in the stands of mixed hardwoods growing on good sites, as exemplified by the oak-ash stand in Table 12.6.

The ratio of nutrients accumulated in the component parts of the tree varies with species and managment practices, as well as age of stand. For example, in a black spruce *(Picea mariana)* stand and a mixed balsam fir *(Abies balsamea)* red spruce *(Picea rubens)* stand, examined by Weetman and Webber (1972) in eastern Canada, nitrogen in the stemwood comprised less than 25 percent of the total contained in the above-ground biomass. In a second-growth Douglas-fir stand (Gessel et al., 1973), nitrogen in the stemwood amounted to 28 percent of the total nitrogen in the above-ground biomass. However, nitrogen in the bole-wood and bark to a 10 cm top, represented more than 50 percent of the nitrogen in the above-ground biomass of a 15-year-old slash pine stand in Florida (Pritchett and Smith, 1974). It appears that there is generally a greater proportion of the total biomass in the foliage and branches of northern conifers than in southern pines.

## Nutrient Returns

A major portion of the nutrients taken up annually into the above-ground components of trees is returned to the soil in litterfall and canopy wash. The percentage restitution varies somewhat with species, site, and stand age. For example, a higher percentage of the absorbed nutrients appears to be returned to the soil in vigorously growing stands than in nutrient-deficient stands and hardwood stands generally return more nutrients than conifers, but the reasons for the wide variations in retentions and returns are not clear. In the ecosystems shown in Table 12.5, from 38 to 88 percent of the nitrogen taken up into the aerial parts of the trees was returned. The highest percentage of nitrogen returned was in the loblolly pine stand and the lowest percentage return was in the Douglas-fir stand. Phosphorus retention is generally higher and the amounts returned to the forest floor are lower then that of nitrogen, with a very high retention in the Douglas-fir stand. The average percent of nutrients recycled in all ecosystems shown in Table 12.5 was 68 percent of the nitrogen, 55 percent of the phosphorus, and 59 percent of the potassium. The amount of litterfall from understory vegetation is small, but its contribution to total nutrient returns can be significant because of its high nutrient content.

**Table 12.6** Nutrient Accumulation in the Above-ground Biomass of Several Forest Ecosystems

| Forest Cover | Age | N | P | K | Ca | Mg | Source |
|---|---|---|---|---|---|---|---|
| | | | | kg/ha | | | |
| Slash pine[a] | 15 | 345 | 24 | 137 | 226 | 53 | Pritchett and Smith, 1974 |
| Loblolly pine | 16 | 257 | 31 | 165 | 187 | 46 | Wells and Jorgensen, 1975 |
| Loblolly pine | 25 | 190 | 20 | 115 | 100 | 32 | Switzer et al., 1968 |
| Radiata pine | 26 | 224 | 28 | 224 | 129 | — | Orman and Will, 1960 |
| Douglas-fir | 37 | 320 | — | — | — | — | Gessel et al., 1973 |
| Scots pine | 100 | 424 | 55 | 307 | 352 | 77 | Remezov and Pogrebnyak, 1969 |
| Norway spruce | 100 | 477 | 86 | 294 | 557 | 65 | Remezov and Pogrebnyak, 1969 |
| Red oak-ash | 115–160 | 947 | 63 | 493 | 1338 | 126 | Duvigneaud and Denaeyer-DeSmet, 1970 |

[a]Fertilized at time of planting.

Although litterfall is the major pathway of nutrient flow from standing biomass to the soil, considerable portions of nitrogen, phosphorus, calcium, and magnesium are leached from the tree canopy and returned to the soil in throughfall and stemflow, as shown in Table 12.7. In the case of potassium, the greatest portions are often returned to the soil in canopy wash.

In addition to the accumulation of nutrients in the tree biomass, there is a parallel accumulation of nutrients in the forest floor as the stand approaches maturity. Wells and Jorgensen (1975) found that the forest floor of a 16-year-old loblolly pine plantation had quantities of nitrogen and phosphorus approximately equal to that of the tree components. However, the tree biomass contained three times as much calcium and magnesium and eight times as much potassium as did the forest floor. Switzer and Nelson (1972) reported that, during the initial 20 years of development of a loblolly pine plantation ecosystem, the total biomass increased from about 10 tons (mainly herbaceous) to about 90 tons per hectare, of which some 16 tons were in plant debris on the soil surface. Increases in the nutrient content of the forest floor of this plantation are shown in Table 12.8 by 5-year intervals.

Organic matter in the forest floor of loblolly pine plantations approached an approximate equilibrium at 30 to 35 tons per hectare at about 30 years of age in the coastal plain of the United States (Wells and Jorgensen, 1975). From ages 13 to 16 years, the accumulation rate was nearly 4 tons per hectare per year. Total nutrient and biomass accumulations in the forest floor are greatly influenced by climatic conditions, with accumulations generally increasing with decreases in mean annual temperature (Figure 4.3).

There is also a long-term accumulation of nutrients in the surface of most mineral soils under forests. This accretion results largely from the uptake of nutrients from weathered parent rock and the subsequent return of these nutrients to the forest floor in the litter and canopy wash. Part of the humus formed from the decomposition of forest floor litter is washed downward or else mixed into the surface mineral soils by mesofauna. Decomposing plant roots and remains of microorganisms also play a major role in this accumulation of 2 to 10 tons per hectare of nitrogen found in the surface horizon of most forest soils.

Wells and Jorgensen (1975) found a significant decrease in nutrient content of mineral soils in loblolly pine plantations during the period between the fifth and fifteenth year. They reported a decline in nitrogen content in the top 60 cm of soil from 2392 kg to 2010 kg per hectare during the 10 years. They found that the accumulation of elements in the vegetative components of the ecosystem approximately equaled the decrease in the mineral soil. An estimation of accumulation in the trees and forest floor of the 16-year-old plantation showed an annual rate of 39, 5, 16, 21, and 6 kg per hectare per year for nitrogen, phosphorus, potassium, calcium, and magnesium, respectively, balanced against a 38, 0.34, 4.8, 35, and 10 kg per hectare per year decrease in to 0 to 60 cm of

**Table 12.7** The Return of Nutrients to the Forest Floor in Litterfall and Canopy Wash
(Cole, Gessel, and Dice, 1967)[a]

| Method | N | P | K | Ca | Mg | Ecosystem |
|---|---|---|---|---|---|---|
| | | | kg/ha/year | | | |
| Litterfall | 58.2 | 7.8 | 16.0 | 29.2 | 6.9 | Loblolly pine |
| Throughfall + | | | | | | (Wells and Jorgensen, 1975)[a] |
| stemflow | 9.6 | 0.5 | 12.3 | 6.0 | 2.0 | |
| Litterfall | 13.6 | 0.2 | 2.7 | 11.1 | — | Douglas-fir |
| Throughfall | 1.5 | 0.3 | 10.7 | 3.5 | — | (Cole et al., 1967)[a] |
| Stemflow | 0.2 | 0.1 | 1.6 | 1.1 | — | |

[a]Used with permission.

mineral soil. As trees grow larger they are capable of extracting a larger proportion of their nutrients from the lower part of the soil profile, but it is likely that the greater part of their nutrient needs are still met through nutrient cycling. As a result, the A horizon reaches an equilibrium that is maintained by additions from the forest floor and withdrawal by roots, according to the authors.

Woodwell and Whittaker (1967) proposed that in terrestrial ecosystems the inventory of nutrients retained within the system increases as succession progresses. In the early stages of succession when the ecosystem structure is simple and the inventory is small, inputs exceed losses and there is a rapid gain in biomass. In later stages, in ecosystems approaching climax for instance, the inventory approaches a relatively stable maximum, with losses equaling inputs. They point out, however, the simplicity of the concept tends to camouflage the total quantities, relative proportions, pathways of movement, and shifts in nutrient requirements with changes in the structure and diversity of the commu-

**Table 12.8** Nutrient Contents of the Forest Floor of a Loblolly Pine Plantation During the First 20 Years by 5-year Intervals (Switzer and Nelson, 1972)[a]

| Plantation age | N | P | K | Ca | Mg | S |
|---|---|---|---|---|---|---|
| year | | | | kg/ha | | |
| 5 | 15 | 1.1 | 5 | 16 | 2.3 | 2.7 |
| 10 | 75 | 6.9 | 12 | 59 | 10.5 | 8.1 |
| 15 | 108 | 8.2 | 14 | 73 | 14.2 | 9.8 |
| 20 | 124 | 9.1 | 16 | 80 | 15.5 | 10.7 |

[a]Used by permission of Soil Science Society of America.

nity. They further state that the nutrient budget of an ecosystem is as important a diagnostic criterion as species composition, morphology, metabolism, and production, but that it is sometimes more difficult to measure.

Woodwell and Whittaker (1967) measured the cation budget in a late successional oak-pine ecosystem with a standing crop of dry matter (excluding humus) of about 101 tons per hectare supported by a net primary production (current annual growth) of 12.6 tons per hectare per year. In a climax community it is assumed that all of the net primary production would be consumed by heterotrophic organisms (or lost), but in their successional forest the production was partitioned between an annual contribution to the standing crop of 5.6 tons and to the support of communities of heterotrophs (7.0 tons). The latter was divided into (1) consumption by herbivores, estimated from the area of leaves consumed to be about 320 kg per hectare, and (2) annual litter fall consisting of 2640 kg per hectare of leaves, 690 kg per hectare of branches and bark, and 220 kg per hectare of fruits. In addition there was a contribution of about 3110 kg of roots and 60 kg of herbs to the humus layers. Thus, with a net primary production of 12.6 tons, minus the respiration of heterotrophs (7.0 tons), the authors arrived at a net ecosystem production of 5.6 tons and a gross production of 26.6 tons per hectare pre year. The net primary production contained about 37.5, 41.5, 10.1, and 1.0 kg per hectare of potassium, calcium, magnesium, and sodium, respectively, or 17 percent of the biomass inventory.

## Internal Transfer

Seasonal variation in mineral composition of tree components is well recognized (Duvigneaud and Denaeyer-DeSmet, 1970). These authors showed that in deciduous forests potassium and calcium concentrations in xylem sap varied during the vegetative cycle. Spring sap had higher concentrations than summer and winter sap, with another peak during autumnal yellowing. There is a general translocation of nutrients from senescent organs to actively growing regions of the tree. Young leaves are always richer in nitrogen, phosphorus, and potassium, but poorer in calcium, than mature leaves. During active growth, nitrogen, phosphorus, and potassium concentrations steadily decrease but remain constant when leaves are completely developed, until a decline during autumnal yellowing.

Switzer and Nelson (1972) and Wells and Jorgensen (1975) reported that 25 to 39 percent of the nitrogen requirements of 17- to 20-year-old loblolly pine plantations were met by internal cycling. A similar portion of the potassium requirement was also met in this fashion, but internal transfer of calcium and magnesium did not appear to play a major role in meeting the tree requirements for these elements. Switzer and Nelson reported that a major part of the phosphorus requirement was met by internal transfer, but Wells and Jorgensen did not support this conclusion.

Switzer and Nelson (1972) suggested that internal transfers may explain the long-term growth responses often obtained from fertilizing forests, especially with phosphorus and potassium. They concluded that the internal transfer relationships were important enough to consider as a separate cycle within the ecosystem and proposed that it be called the *biochemical* cycle. Failure to recognize or consider the internal transfer of nutrients can produce misleading conclusions about the nutritional functioning of these ecosystems. However, it appears that it can be more appropriately considered as one of the steps in the biological cycle.

In spite of large variations in the amounts of geochemical and biological cycling among ecosystems, there can be little doubt that these cyclic transfers of nutrients play a major role in the continued productivity of forest soils. A good understanding of these dynamic processes is crucial to the development of effective forest management programs.

## SILVICULTURAL IMPLICATIONS OF NUTRIENT CYCLING

Because of the conservative nature of nutrient cycling, mature forest stands make only minimal demands on the soil for nutrients. Although the annual uptake of nutrients by a vigorous stand of trees approaches that of an agricultural crop, on an areal basis, a large percentage of the absorbed nutrients are returned to the forest floor as leachings from the forest canopy or as litter fall. The slowly decomposing litter releases nutrients for reuse by the forest stand, so that only small additional amounts of nutrients are needed to sustain tree growth and bolewood production.

The input of nutrients from atmospheric sources, dinitrogen fixation by soil organisms, and geological weathering of parent rock is sufficiently rapid to supply those nutrient demands not met by biological cycling in undisturbed stands. However, disturbances of forests by both natural and human activities are apparently common occurrences, even in natural forests. Major disturbances can have dramatic effects on nutrient cycling. Whether they produce long-term effects on tree growth and development depends on the nature of the disturbance and on the resilience of the system; that is, the capacity of the soil to replenish the nutrient supply. Nitrogen is the critical nutrient in most forest ecosystems. An example of nitrogen cycling, based on data from pine stands in coastal sandy soils, but applicable to other similar ecosystems, is graphically presented in Figure 12.3.

Although the 2225 kg nitrogen present in the soil and forest floor is almost entirely in humus (organic) complexes, almost 90 percent of this material is in the mineral (Al) horizon (soil reserve). This latter reserve is a relatively stable fraction that has accumulated over thousands of years and is not readily mobilized. The most active fraction of the soil reserve is that of the forest floor, specifically the F and H layers. Small feeder roots of trees and mycelia of mycorrhizal fungi permeate this zone, absorbing directly a major part of the 60 to 80 kg nitrogen per hectare per year requirement of the forest, as it becomes available from the forest floor. The reserves in the forest floor are continuously replenished by the 30 to 40 kg of nitrogen added each year in litterfall and canopy leaching. The system is so efficient that it is estimated that less than a kilogram per hectare is lost each year to deep percolation or runoff waters. This is more than compensated by the 2 to 6 kg added from atmospheric sources and dinitrogen fixation. Table 12.9 shows estimated transfer of nitrogen and the nutrients in a 16-year-old loblolly pine plantation in the North Carolina piedmont (Wells and Jorgensen, 1975).

Fire can have a major impact on the nitrogen reserves of the forest floor, because a large percentage of that element in the litter layer (100 to 200 kg per hectare) is volatilized, even in a controlled fire. However, conditons for microbial fixation of nitrogen may be enhanced because of the reduction of surface soil acidity by the ash layer, and this fixation may partly compensate for the volatile losses. Long-term studies do not show major decreases in ecosystem nitrogen due to occasional control burns. Control burning can hasten microbial activity and nutrient cycling in the boreal forest and areas where thick humus layers immobilize a large part of the nitrogen reserves.

Other small losses of nitrogen from the forest ecosystem can result from

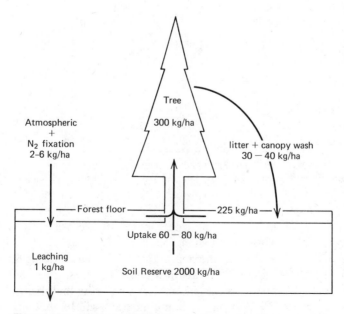

**Figure 12.3** Nitrogen distribution and annual cycling in coastal plain pine forest ecosystems.

**Table 12.9** Estimated Transfer Rates for Nitrogen, Phosphorus, Potassium, Calcium, and Magnesium in a Plantation of Loblolly Pine at Age 16 Years (Wells and Jorgensen, 1975).

| Component | N | P | K | Ca | Mg |
|---|---|---|---|---|---|
| | | | (kg/ha/hr) | | |
| **REQUIREMENTS OF TREES** | | | | | |
| New needles | 55.0 | 6.3 | 31.9 | 8.1 | 4.8 |
| One-year-old needles | 0 | 0 | 0 | 8.1 | 0 |
| Initiated branches | 4.3 | 0.7 | 5.2 | 1.7 | 0.7 |
| One-year-old branches | 2.3 | 0.3 | −0.6 | 2.5 | 0.4 |
| Stem | 5.3 | 0.6 | 4.4 | 3.0 | 1.2 |
| Roots < 1 cm in diameter | 48.7 | 12.3 | 20.0 | 34.4 | 9.2 |
| Roots > 1 cm in diameter | 1.5 | 0.4 | 2.4 | 0.7 | 0.7 |
| Total requirements | 117.1 | 20.6 | 64.5 | 58.5 | 17.0 |
| **TRANSFER WITHIN TREES** | | | | | |
| From two-year-old needles | 17.0 | 0 | 18.0 | 0 | 0 |
| **TRANSFER TO FOREST FLOOR** | | | | | |
| Litterfall | 58.2 | 7.8 | 16.0 | 29.2 | 6.9 |
| Throughfall + stemflow | 9.6 | 0.5 | 12.3 | 6.0 | 2.0 |
| Total to forest floor | 67.8 | 8.3 | 28.3 | 35.2 | 8.9 |
| **TRANSFER TO MINERAL SOIL** | | | | | |
| Forest floor to mineral soil | 25.2 | 4.0 | 20.6 | 24.1 | 4.7 |
| Roots < 1 cm in diameter to soil | 48.7 | 12.3 | 20.0 | 34.4 | 9.2 |
| Total to mineral soil | 73.9 | 16.3 | 40.6 | 58.5 | 13.9 |
| **ADDITIONAL TRANSFER** | | | | | |
| Requirements in excess of transfer to mineral soil | 26.2 | 4.3 | 4.7 | 0 | 3.1 |
| Soil depletion | 38 | 0.34 | 48 | 35 | 10 |
| Loss in ground water | 0.70 | 0.03 | 1.56 | 1.26 | 0.88 |

Excerpt from *Forest Soils and Forest Land Management,* edited by B. Bernier and C. H. Winget, 1975, published by Les Presses de l'Université Laval and reproduced with their permission.

grazing by domestic livestock and from accelerated erosion as a result of improper road construction and logging methods. However, the most notable interruption in the nutrient cycle of managed forests is that resulting from routine harvests. In a conventional harvest, about one-half of the nitrogen in the above-ground tree components is removed in the bolewood and bark, while the remainder is added to the litter layer. Some nitrogen will be mobilized and lost from the forest floor materials as a result of the removal of the forest canopy and disturbance by harvesting and site preparation operations. This happens before the regeneration develops an effective nutrient sink. However, the losses from this route should be minimal, unless the disturbance is excessive and no effect is made to reestablish a vegetative cover, as happened in the Hubbard Brook test (Likens et al., 1970).

Some increase of mineralization from nutrient reserves is needed for the development of the new forest, until the forest floor again approaches equilibrium in weight and nutrient content. Afterward, biological cycling supplies a large portion of the elements available to the trees. Soil and site conditions largely determine the time required for the forest floor to accumulate and the size of the compartments in which the nutrient elements are held. Rate of nutrient cycling, therefore, is primarily a function of and the major contributor to forest soil fertility.

# 13

## SOIL PROPERTIES AND SITE PRODUCTIVITY

The reproduction, survival, and growth of trees on a particular site represents an integrated response to a complex of many fluctuating and interacting environmental factors. Certainly soil is one of the essential components of this complex of related factors and the factor to which we will devote the greatest attention, but in order to keep the effects of soil properties on site productivity in perspective, other biological and physical site factors will also be reviewed. More than one factor often limits tree growth on the same site and the various limiting factors may interact in different ways. There are often positive interactions among factors, and, in some instances, one factor may substitute for another factor, making it essential that all features be considered when assessing site productivity.

The capacity of a tree species to thrive and successfully compete on a particular site is influenced by both internal (physiological makeup of the tree) and external (environmental) factors. The integration of these combined properties determine *forest productivity*. Of particular concern are the environmental or site factors of which the soil is a major component. These external factors largely determine *site quality*, the capacity of the land to produce trees. Site quality is based on the physiography, climate, soil, and other features of the environment that are not easily altered. It is essentially a fixed quantity. In contrast, *site productivity* can be considered to be the sum of site quality plus management input. Productivity is subject to varying degrees of alteration by manipulation of growing stock or modification of the site.

Site factors that influence productivity can be conveniently grouped for discussion into biotic (biological) and abiotic (physical) components (Remezov and Pogrebnyak, 1969).

## BIOTIC FACTORS

There are a number of biological factors that contribute to the overall productivity of soil and site. Although these variables are more transitory than the abiotic factors, failure to recognize their influence on tree growth can contribute significant errors to the measurement of site productivity. Some biotic components of particular importance in this regard are stand density, genetic variability of the stand, competing vegetation, and disease and insect problems.

### Stand Density

Stand density is particularly important to the growth and survival of intolerant species. While the tree population affects growth as a result of competition for growth factors its influence on site productivity estimates may be minor when dominant stand height is used as the criterion of site quality. Stocking experiments usually show that height growth of dominant trees is not affected greatly within desirable ranges of density. This response also has been verified by many pruning studies that reveal that the live crown must be reduced beyond any rational practical level before height increment is materially affected. On the other hand, "stagnation," or failure to express dominance, has been observed in severely overstocked stands of such species as slash, loblolly, red, lodgepole, and ponderosa pines on a variety of sites (Figure 13.1). Inadequate density of desirable growing stock is perhaps more limiting to site productivity than overstocking.

Cooper (1957) reported that stocking rates between 500 and 2500 slash pine stems per hectare had little influence on increment rate on poor sites, but it became increasingly important to maintain adequate density on the better sites. Barnes and Ralston (1955) showed a yield increase from 159 up to 175 cu m per hectare by increasing the density of slash pine from 1000 to 1500 stems per hectare on sites with a site index of 80 (feet) at 25 years. Crown thinning can produce a significant increase in basal area and volume production of overstocked stands, particularly on poor sites. Better sites offer a wider latitude of choice for stocking, as well as species and rotation length, depending on the relative value of the product to be grown (Ralston, 1965). Thinning may be practical for quality improvement in sawlog rotations, but it may not be for pulpwood production.

### Genetic Differences

Ralston (1964) pointed out that variations in genetic potential *between* species that occupy a site can be rather large and, thus, may have a substantial influence on estimates of productivity. However, this should cause little concern because individual species productivity can be readily evaluated by conventional mensurational methods. On the other hand, the magnitude of growth variations attributable to varietal differences *within* a species can seldom be identified in the field

**Figure 13.1** Overstocking (approximately 27,000 stems per hectare) severely limits productivity in this 11-year-old loblolly stand in Virginia.

and, consequently, must be regarded as an unknown error component in site studies. Carmean (1975) deemed such genetic variations in height growth to be relatively minor within small forest areas. Rapid height growth is an important survival factor for trees growing and competing in dense stands. Trees of uniform genetic potential for height growth would tend to make up a large percentage of wild populations, because the severe competition would soon eliminate the slow-growing genotypes.

Exceptions are species such as aspen, which regenerate mostly from clones that may occupy large areas. Crown competition within a clone is between individuals with the same genetic potential and competition between slow- and fast-growing clones only comes at their perimeters. Aspen clones growing on areas that appeared to be similar in site productivity had large differences in site index (Carmean, 1975). Selection of site trees from a number of different clones is suggested as a means of minimizing errors in site index estimation due to clonal differences.

The fact that rather large tree-to-tree variations exist in many forest species has permitted considerable yield improvements through selection and breeding. Genetic gains in volume production from one generation of breeding of southern pine, for example, approaches 20 percent (Kellison, 1975).

## Competition Control

Competing vegetation exerts a major effect on total forest productivity by preempting growing space that otherwise might be occupied by crop trees. Competition for available moisture, light, or nutrients can be critical to the survival and growth of planted forests. In the southeastern region of the United States, perpetuation of pine types normally hinges on control of competing vegetation during the regeneration period. As a consequence, considerable emphasis has been placed on the development of effective chemical and mechanical methods of weed control during plantation establishment and on the use of prescribed fire in older stands. Perennial grasses, bracken, and herbaceous broad-leaved weeds may have little effect on young tree growth unless they are so tall or dense that they provide excessive shade, and provided they do not collapse on trees at the end of the season causing death of the seedling by mechanical injury or as a result of disease problems.

Control of weedy vegetation in young slash pine plantations has resulted in significant initial increases in rates of tree growth under some conditions. For example, triazines applied at time of planting improved survival and resulted in up to 45 percent increase in slash pine height after one year on well-drained sands. Herbaceous vegetation is normally controlled for no more than two years by an application of herbicide and under some conditions this early control may not have a lasting effect on tree growth. Woods (1976) reported that bracken mowing in young plantations did not improve *Pinus radiata* productivity sufficiently to justify its expense on phosphorus deficient sands of South Australia. However, mowing combined with a phosphorus application was justified.

If survival is significantly decreased by excessive weed growth, competition control can have considerable influence on final yields. There is the possibility, however, that the presence of some herbaceous vegetation in young pine plantations may benefit wet sites as a result of increased transpirational water losses and by retention of mineralized nutrients during the early stages of stand development.

In uplands of the southern United States, soil moisture is seldom sufficient throughout the growing season for maximum pine growth, and understory hardwoods compete significantly for the limited amount available. In midsummer in southern Arkansas, water was lost about 25 percent faster from soils with hardwoods left in place than from soils with hardwoods removed (Zahner, 1968). Grano (1970) reported that while dense understories of hardwoods materially reduced the average annual growth of even-aged loblolly-shortleaf pine stands on upland sites there was no evidence that moisture was the limiting factor. Growth differences in favor of hardwood-free plots amounted to 14 to 25 cu m per hectare during periods of from 11 to 14 years following chemical treatment in 47- to 53-year-old pine stands. Regardless of moisture conditions during the growing season, pines on plots divested of hardwoods consistently outgrew those on control plots, and growth differences were not greater when moisture was deficient than under favorable moisture conditions as seen in Figure 13.2.

Richards (1967) found that the establishment of hoop pine *(Araucaria cunninghamii)* on the lateritic podzolic (Ultisol) soils of the coastal lowlands of south Queensland depended on a reduction in competition from blady grass *(Imperata cylindrica)* by preplant cultivation. He presented evidence to show that the grass competes primarily for nitrogen rather than water. Wilde et al. (1965) reported that the elimination of competing cover of heath plants by two spring cultivations tripled the volume of white pine stands within a period of 27 years. Exudates from the roots of heath plants may inhibit mycorrhizal development on roots of associated pine. Senna seymeria *(Seymeria cassioides),* an annual of the figwort family, is parasitic on roots of slash pine. This obligate parasite is found most often on moist, sandy sites of the lower coastal plain and can be a local problem in the establishment of pine.

Control of competing vegetation is especially important to the survival and early growth of hardwood plantations. In most instances, moisture appears to be the critical factor to the success of these plantings, although light and nutrients can become limiting under severe conditions. Broerman and Gatherum (1967) reported that a reduction in light from heavy competition resulted in good survival, but relatively poor growth of green ash planted on coal spoil banks.

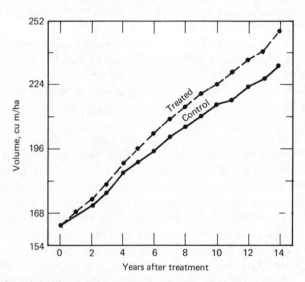

**Figure 13.2** Volume of a 53-year-old loblolly pine stand on a Hapludult and the volume increase for 14 years following removal of its hardwood understory (Grano, 1970).

## Disease and Insect Problems

Reductions in tree survival, growth, and yield as a result of attacks of insects and diseases have long been recognized. In most instances where substantial numbers of trees are killed or severely damaged, the effects of the pest on site productivity are easy to recognize and estimate. Such diseases as *Fomes annosus* and *Cronartium fusiforme* cause major damage to several species of pines. The symptoms of these diseases are well recognized but the damage to site productivity may be less well known. Rust may cause early abandonment of plantations, loss of growth of individual trees, and loss of saw-timber potential due to stem damage caused by the infections. Severe rust infections may cause losses in productivity of 40 percent or more. Annosus root rot appears in many temperate zone soils of the world and, apparently, the incidence of this disease is increasing in the eastern United States. Bjorkman (1967) reported that attack by *Fomes annosus* on pine was strongly favored in soils rich in lime. Soil nutrient levels have been implicated in increases in the incidence of fusiform rust, snow blight *(Phacidium infectans),* and other forest diseases.

Some insects, such as the southern pine beetle *(Dendroctonus frontalis),* can cause major tree damage and significant reductions in site productivity. Because of the suddeness with which the population builds up to outbreak proportions and the amount of timber killed, this is probably the most dreaded insect of southern pines. Insects such as the tip moths *(Rhyacionia* spp.) and pine sawfly *(Neodiprion)* may result in the loss of one or two years growth, but they have little lasting effect on productivity. Pales weevil *(Hylobius),* a serious pest to pine reproduction, and *Ips* beetles are sporadically devastating. In most instances the pest damages can be identified and assessed. Many of these insects are especially destructive in stands of poor vigor, where outbreaks are often triggered by soil moisture stresses and natural or human disturbances (Figure 13.3). Some insects, such as the white pine weevil *(Pissodes strobu)* preferentially attack white pine that have been fertilized or that exhibit rapid growth (Xydias and Leaf, 1964).

The attack of some pests is slow and insidious and may go unrecognized for years. *Phytophthora cinnamomi* (littleleaf disease) attacks the meristematic tips of fine roots. Reduction in the extent and amount of the effective root system interferes especially with the absorption of nitrogen, leading to a gradual cessation of growth and yellowing of the foliage followed by premature death. Some trees show top symptoms by age 20 while others on the same site may remain symptom free for 50 years or more. The threat of littleleaf disease has resulted in discrimination against shortleaf pine in reforestation programs. It is more severe on soils having poor aeration, low fertility, and periodic moisture stress. Hence it is most damaging on eroded and abandoned agricultural lands, especially where the subsoils are impervious clays.

The attack of parasitic nematodes is probably common on many tree

**Figure 13.3** Outbreaks of beetles in pines can be triggered by soil moisture stresses.

species, although the influence on tree growth and soil productivity is very difficult to measure. In a survey of young slash pine plantations in the coastal plain of the southeastern United States, plant parasitic nematodes were found to be associated with pine plantings in all 34 sites tested. Spiral *(Helicotylenchus)* and ring *(Criconemoides)* nematodes were found most frequently, although 11 other plant parasitic genera were also found. The actual effects of nematodes on tree survival and growth and on site productivity are not well known. Some striking increases in pine seedling growth have been obtained by field fumigation of well-drained sands prior to planting. A portion of the response may have resulted from a herbicidal effect, but the major benefit probably resulted from a reduction in numbers of parasitic nematodes (G. W. Bengtson, unpubl. data).

It has been noted that some tree species, such as *Pinus radiata,* perform better as exotics than in their natural habitat, and that their range of tolerances is considerably wider than the natural limits would suggest (Jackson, 1965). The reason for the greater than expected productivity in a completely new habitat may be the absence of competitors and natural pests and diseases that often inhabit soils of the ecosystem in which the species evolved.

There are indications of significant differences in mycorrhizae characteristics and that by inoculation with the proper fungi the normal characteristics of tree roots may be advantageously changed. Bowen (1965) observed that inoculation of radiata pine with various mycorrhizal fungi resulted in very large differences in the uptake of $P^{32}$ in short-term laboratory experiments. Marx (1977) reported an improvement in survival and early growth of pine seedlings planted on mine soils if they had been inoculated with the ectomycorrhizae *Pisolithus tinctorius* in the nursery.

## ABIOTIC FACTORS

Abiotic factors of the environment that affect tree growth can be broadly grouped into climatic, physiographic, and soil variables. The first two components do not lend themselves to appreciable manipulation by humans, but soil factors can often be altered for improved tree growth and will be discussed in a separate section. The influence of abiotic factors on tree growth are reasonably accurately measured by site index or by indirect methods, while biological factors are often difficult to identify and quantify.

### Climatic Components

Climatic variables, or changes in macroclimatic conditions, usually occur gradually over rather great distances and are associated with correspondingly slow changes in vegetation types. Thus, temperature, photoperiod, and other solar radiation growth factors change with latitude, but the effects are not apparent at a local level until one arrives in an area where the growth gradient due to a limiting climatic variable becomes rather steep. Thus, stunting of coniferous forests along temperature gradients is ordinarily observable only in northern areas such as Alaska, Canada, Scandinavia, and Russia, or in mountainous areas where changes in elevations cause great changes in temperature, or precipitation, over short distances.

Fourt et al. (1971) reported that climatic factors that most significantly influenced yields of *Pinus nigra* in south England were winter mean monthly temperature and growing-season sunshine. Frost heaving, in both spring and fall, has been observed as a cause of seedling mortality in the pumice soil region of Oregon and in the soils of southwestern United States. The heaving process is started when the temperature drops low enough to cause water to freeze in a thin layer. Water lower in the profile then begins to migrate upward in response to the potential gradient created by freezing. Ice damage may be a prime factor limiting the natural range of slash pine to the Gulf and southern Atlantic coastal states.

Compact foliage and brittle branches make this species particularly susceptible to canopy damage by ice glaze.

The effects of changes in precipitation may be readily illustrated if one compares widely separated geographic areas with large rainfall differentials. Simple correlations of growth and precipitation can be noted in local regions where large changes occur. In the southeastern United States significant decreases in site index of longleaf pine on upland soils have been observed along a decreasing rainfall gradient from Louisiana to East Texas, and a similar though smaller effect has been reported for loblolly pine in Mississippi. On poorly drained soils, a differential growth response has been noted in slash pine as growing season rainfall (March to August) and depth to slowly permeable horizons varies. Growing season rainfall in excess of 50 cm was detrimental to slash pine growth on shallow soils, but this effect diminished rapidly as permeable surface layers predominated.

*Pinus radiata* flourishes in Mediterranean and warm temperate climates. The range of latitude is reported to be between 32° and 46°, but day length rather than soil or air temperature is said to limit growth during winters in New Zealand (Raupach, 1967). Within the range of adaptation, rainfall is the factor that has the greatest influence on productivity. Annual rainfall of 76 cm or more is desirable for *P. radiata;* however, excellent growth has been obtained in areas of winter rainfall ranging from 56 to 64 cm. Hot, humid summer conditions may result in excessive damage from fungal attacks on many temperate region tree species.

In Hills' (1960) delineation of site regions in Ontario, climatic variation is used as a primary factor in stratifying the province into 12 subunits. The subunits represent zones of relatively uniform vegetational response to regional climate and, thus, eliminate the need for considering interactions of broad climatic effects with local site variables.

## Physiographic Factors

The influence of physiographic variables on forest productivity has been recognized longer than most other site components. For example, it is considered good practice to run cruise lines at right angles to topography in order to obtain a representative sample of timber types and volumes.

Topography exerts an effect on growth through local modification of climatic and edaphic variables, particularly moisture, light, and temperature regimes. Associations between site quality and physiography may be evaluated either in terms of measurable topographic characteristics such as altitude, aspect, and slope gradient, or as discrete topographic position categories, such as ridges, slopes, coves, and bottoms. The former treatment is most appropriate to areas of rugged mountainous terrain, while the latter method is the best system to use in regions of moderate relief.

Topographic factors are important features of many land classification systems (Hills, 1960; Steinbrenner, 1975). Steinbrenner found that elevation had a strong negative influence on Douglas-fir growth in Washington. Precipitation was significantly correlated to growth at low elevations but not at high elevations. Hills (1960) reported that a southern aspect may be favorable to tree growth in cold northern latitudes, but not in hot or dry climates. Aspect may have very little influence on growth in regions with relatively mild and humid climate. Because of the possibilities of appreciable modificiation of local climate and soil moisture regimes as a result of variation in physiographic features, Ralston (1964) stressed that excessive detail in topographic classification is self-defeating. Overrefinement leads to errors in classification and usually reflects minor physiographic variations that are meaningless in relation to site productivity.

## SOIL COMPONENTS

Soil properties comprise a third large group of abiotic factors that exert a significant influence on tree growth. Where climatic and physiographic factors can be held constant by appropriate stratification procedures, soil properties become the major factor of the physical environment that has an appreciable bearing on tree growth, and the one of greatest concern to the forest manager. These soil factors are often grouped into physical, chemical, and microbiological properties for convenience of discussion. However, it seems more appropriate when discussing site productivity to emphasize those feature which have the greatest impact on the soil's capacity to supply water, nutrients, and air in amounts required for optimum tree growth.

### Parent Material

Parent material is a major contributor to the process of soil development, and as such it has an indirect effect on tree growth. Because of the deep rooting habit of trees, soil parent material and condition of the geological substrata is a more important consideration to foresters than to agriculturists. The relationship between soil parent material and tree growth is most obvious in areas where the bedrock is sufficiently close to the surface to exert a continuing influence on soil properties, but the effect of parent material on tree growth is seen also in soils derived from transported materials such as marine sands and glacial till. Parent material influences productivity through its effect on the soil chemical, physical, and microbiological properties, but the extent of this influence can be modified by climate. For example, soils derived from similar parent material, but developed under different climatic conditions, may have vastly different properties and production potential due to variations in leaching of nutrients, accumulation of organic matter, and in soil acidity. Nevertheless, parent material has a greater effect on the mineral composition of soils than other soil-forming parameters and there is generally a good relationship between the mineralogical composition of soil parent material and the underlying rocks. Parent materials derived in large part from granite give rise to soil containing a larger proportion of quartz than do parent materials derived from diorite.

Within the same climatic zone in New York, soils derived from calcareous shales and igneous rock were noted (Lutz and Chandler, 1946) to support more exacting species, such as basswood, white ash, yellow-poplar, and hickory, while soils derived from acid sedimentary rock contained a higher percentage of beech, yellow birch, red maple, and certain oaks. It is generally accepted that the regeneration and growth of both northern white-cedar *(Thuja occidentalis)* and eastern redcedar *(Juniperus virginiana)* are better on calcareous soils than on adjacent acid soils. The presence of cabbage palm in the lower coastal plain of the southeastern United States is an indicator that limerock is sufficiently near the surface to influence soil reaction and base saturation.

Throughout most of the southeastern United States the effect of soil parent materials is expressed indirectly through derivation properties, such as soil texture or subsolum drainage characteristics. However, direct effects on soil fertility have been noted in certain areas. Parent material may have an overriding detrimental effect on the growth of certain species, such as on pine growth in the base-rich chalk soils of the Black Prairies on Alabama and Mississippi and in certain highly calcareous alluvial soils in east Texas. Parent material may be beneficial to tree growth, as noted in some acid soils of Florida and South Carolina with lower layers derived from phosphatic limestones. Ameliorating influences of old marine shell beds and marl deposits have also been observed on occasion around the margins and in the interiors of hardwood bays and river swamps.

In the event that parent material origins can be distinguished in the field and that they have some real or suspected connections with site productivity, the effect can be identified by stratification of site measurements by parent material categories. The influence of parent materials on site productivity was noted in Sweden, where parent rocks were grouped on the basis of calcium content. Soils derived from calcium-poor rock supported a poor forest of Scots pine; intermediate groups produced good soils for pine and mixed conifers; while basic igneous and calcareous sedimentary rocks resulted in productive soils with forests of Norway spruce and hardwoods in southern Sweden. Diagnostic features of petrographic materials were used by Hills (1961) to separate subordinate site districts within site regions in Ontario.

## Soil Depth

The volume of soil available to tree roots, as dictated by soil depth, influences tree growth to the extent that it affects nutrient and moisture supplies and root development and anchorage against windthrow. Trees growing on shallow soils are generally less well supplied with water and nutrients than trees on deep soils.

When soil depth is such that it defines the volume of growing space for tree roots above some restricting layer, such as a claypan, siltpan, bedrock, or other horizon of low permeability, depth measurements can be used with some precision to predict growth patterns in well drained soils. Growth normally follows a trend that can be expressed as a reciprocal function of soil depth with greatest decline in growth found on soils with less than 25 cm of effective depth (Figure 13.4).

The absolute and *effective* depths of a soil are not necessarily the same, because a high water table, toxic substances, or an impervious layer may completely restrict root penetration in a soil that would otherwise permit deep rooting. Some difficulty in using soil depth to estimate productivity also may be encountered where drought, erosion, or poor drainage are products of surface soil thickness or depth of soil above some restricting layer. In each of these instances, there may be soil fertility interactions important to tree growth that cannot be determined from depth measurement alone. For example, measurement of surface soil thickness as an indicator of erosion fails to discriminate between losses in fertility and reduction in effective rooting, according to Ralston (1964). Hanna (1968) found that depth of the surface soil (A horizon) was related to growth of white and black oaks in Indiana (Figure 13.5).

Barnes and Ralson (1955) reported that site productivity for slash pine on

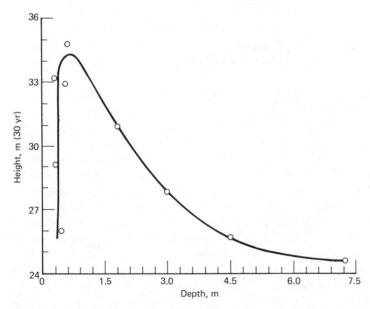

**Figure 13.4** Height of *Pinus radiata,* adjusted to 30 years, in relation to effective rooting depth in soils of South Australia (Raupach, 1967). Used with permission.

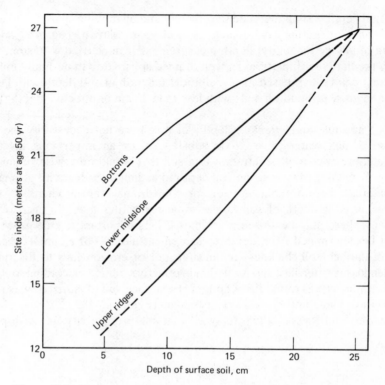

**Figure 13.5**  Site index of black oak as related to surface soil thickness and slope position in southern Indiana (Hanna, 1968).

sandy soils in Florida increased with depth to a fine-textured layer, with maximum growth obtained on soils with a fine-textured horizon at about 50 to 75 cm. They also found that depth to mottling was an index of productivity, with best growth found where mottling occured at 75 to 100 cm. Poor growth of *Pinus radiata* was obtained on soils with less than 46 cm of rooting depth in South Australia (Jackson, 1965). However, in California the same species grew to 20 m or more with as little as 15 cm of surface soil on weathred granite rubble, but roots apparently penetrated the parent material and tree growth rates were rather slow. Auten (1945) found that depth to a compacted subsoil was the most significant factor related to yellow-poplar growth. When such a restricting horizon was less than 60 cm below the surface, the site was poorer than average. Gilmore et al. (1968) reported that the height of 18-year-old poplar varied from 3 to 17 m in an old cultivated field where effective soil depth had been reduced in some parts by severe erosion and had been increased in other parts by deposition. Heights varied directly with depth above a fragipan and presumably resulted from an increase in available soil moisture. The influence of depth to a fragipan on the growth of sycamore and loblolly pine are shown in Figure 13.6. Zahner (1968) found the optimum effective depth for loblolly and shortleaf pines under a rainfall regime of 117 to 132 cm per annum to be about 46 to 50 cm.

The effective depth of soils can be increased under some conditions by drainage, deep tillage, or deep placement of lime or fertilizers. Breaking of cemented spodic horizons and mixing the organic materials with the A1 and A2 horizons have been proposed as a means of improving tree growth on Aquods but most attempts have not proven highly successful.

## Water Table

The interrelations between effective soil depth and available soil moisture and drainage have been reported by several researchers (Jackson, 1965; Coile, 1952; Gilmore et al., 1968). Drainage class is an inferred property, indicated by characteristic profile morphology and landscape position. It indicates the extent and duration of soil saturation in years of extreme wetness as well as in normal years (Dement and Stone, 1968). Controlled drainage for the removal of excess water may improve productivity for a particular species and, finally, affect profound changes on stand composition and regeneration. Douglas-fir cannot

**Figure 13.6** Effects of depth of a loess soil over a fragipan on the growth of 12-year-old sycamore (foreground) and 23-year-old loblolly pine on an Ochreptic Fraguidalf in southern Illinois. As depth of soil increases to the right, sycamore growth improves but pine growth decreases.

tolerate shallow water tables, but red alder, western redcedar, Sitka spruce, and western hemlock are all considered adaptable to shallow water tables. Where water tables were less than 15 cm below the surface, the "species responded differently to both depth and quality of the water-table" (Minore and Smith, 1971). Alder and redcedar were little affected by water table depth and grew reasonably well where stagnant water was near the surface during winter months. Spruce did not tolerate shallow stagnant water, but grew well where moving (aerated) water was near the surface.

Planted cottonwood grew best when the water table was about 60 cm deep, whether the trees were planted on soil with a high water table or the water table was raised a year after planting. Yields on soil with a 30-cm water table were about the same as where no attempt was made to maintain a water table in tanks (Broadfoot, 1973). Both slash and loblolly pines grew better where the water table in an Aquod was maintained at 46 cm than where it was maintained at 92 cm, during five years after planting (White et al., 1971).

## Soil Moisture

Tree growth is controlled more by water availability than by any other factor on most sites. The available water-holding capacity of a soil is influenced by a number of factors, but it is primarily determined by structure and texture. The use of soil texture as an estimator of water-holding capacity and, thus, of site productivity is complicated by the influence of texture on soil aeration, nutrient availability, and other soil fertility factors. If the growth potential of a species can be observed over a wide range of textures in a given area of well-drained soils, a curvilinear response can be expected. Growth improves with an increase in silt and clay content, as a result of more favorable moisture and nutrient supplies, to a point where further increases in the proportion of fine particles produces aeration difficulties (Ralston, 1964). Stone and gravel content can also modify soil moisture regimes. Moderate amounts of coarse fragments may favor deep penetration of light rains, thus reducing evaporation losses. However, large reductions in effective soil volume by stones decreases moisture retention storage.

Large root systems of loblolly pine and shortleaf pine developed when soil moisture was maintained close to field capacity. In contrast, very sparse root systems resulted from growing plants in soils that were allowed to dry down almost to permanent wilting percentage before rewatering to field capacity (Kozlowski, 1971). Mycorrhizal development was also much better in the soil maintained close to field capacity. The influence of mycorrhizae in increasing water and mineral uptake may have been important in improving root growth under high moisture levels. Loblolly growth has been noted to be greatly reduced when available moisture dropped below 40 percent. In comparison with controls, 40-year-old loblolly pine subjected to drought showed a marked increase in reducing sugars, nonreducing sugars, and total carbohydrates, and an approximate equivalent decrease in starch (Hodges and Lorio, 1969). The increase in sugars apparently resulted from a decrease in growth rate and not from the hydrolysis of starch.

Pearson (1931) wrote that "there appears to be but little difference in the ability of species to extract moisture from dry soil; that is, they all reduce it to the wilting point which in a given soil is about the same for all species. Probably the greatest difference between species lies in their ability to extend their roots and thus enlarge the sphere of their activity." The variation in drought tolerance among species is well recognized (Eyre, 1963), and soil moisture plays a prominent role in the adaption of species to site and their distribution among the climatic zones of the world. Shultz and Wilhite (1969) reported a significant difference in first-year survival and growth among families of commercial grade and genetically superior (in terms of growth) slash pine. Average survival of the commercial seedlings was 80 percent while that of the superior seedlings was only 60 percent. However, 4 of the 14 superior families (in terms of growth) survived better than did the commercial seedlings. Results suggested that it may be feasible to select tree lines for both drought resistance and superior growth rate.

## Soil Aeration

Aeration effects on site productivity are difficult to separate from those of soil depth, texture, and moisture. Lack of oxygen in a soil horizon prevents root penetration and exploitation of nutrients. Minore et al. (1969) related high soil density to poor root development of several western species. Forest site situations where such conditions may be observed in the southeastern United States are in upland soils with plastic claypans, siltpans, or occasionally other compacted or cemented "hardpan" layers at shallow depths. There are also upland and lowland soils whose medium- to fine-textured surface horizons have been compacted as a result of timber management operations or recreational use. In flatwoods areas additional damage can be incurred on fine-textured soils through impedance of surface drainage as a result of soil disturbance during timber harvesting. Restriction on the use of highspeed rubber-tired tractors may be advisable under some conditions.

Poor growth of pine is often found in shallow depressions in the lower coastal plain if surface drainage conditions are conducive to ponding and water stagnation. The amount of fluctuation in water table levels during the growing season also may be a critical growth factor in poorly drained areas. Large cyclical variation in water tables is undesirable, because roots permeating portions of the profile that become aerated during dry seasons are killed back by advancing water levels during wet periods. This periodic root pruning phenomenon tends to create an imbalance in the root:top ratio and reduce the tolerance of trees to drought conditions and increase their susceptibility to insect attack during stress periods.

## Soil Nutrients

Forest trees require the same elements for their growth and reproduction as other higher plants, but because of the conservative nature of nutrient cycling, deep-rooting habits of most trees, and the apparent capacity of mycorrhizal roots to extract some minimally available nutrients from soils, nutrient deficiencies are not common in undisturbed forests. Nutrient deficiency symptoms are rare in indigenous species. Ralston (1964) pointed out that the most important reason for lack of emphasis on fertility factors in studies of site productivity is the frequent correlation between variables used to describe other soil properties with nutrient supplies. He also mentioned the difficulty in diagnosing the fertility status of forest soils. The intensification of forest management and the increase in planta-tion forests, particularly plantations of exotic species, have served to focus attention on the importance of soil chemical properties to tree growth. A number of recent studies have shown that when other site factors are kept constant, soil nutrient levels are indeed related to site productivity (Möller, 1974; Pritchett and Gooding, 1975; Waring, 1973).

Nitrogen and phosphorus are the two elements most frequently deficient in forest soils, although deficiencies of potassium and the micronutrients have been reported for localized areas (Leaf, 1968; Stone, 1968). Nitrogen is applied operationally to mid-aged stands of Norway spruce *(Picea abies)* and Scots pine *(Pinus sylvestris)* in extensive areas in Scandinavia, to Douglas-fir in the Pacific Northwest, to pines in the southern United States, Japan, and a few other areas of the world (Hagner, 1971). Nitrogen is inherently deficient in some coastal sandy soils and abandoned agricultural lands with low organic matter contents. On these deficient soils, nitrogen may need to be applied to seedlings in order to insure early establishment and growth. Nitrogen deficiencies are most frequently found in cool climates where this element is immobilized in the accumulated forest floor organic layers of maturing stands (Rennie, 1971). In their examination of over 200 red pine plantations in Connecticut, Hicock et al. (1931) found that total nitrogen content of the A horizon showed a better correlation with site productivity than any other factor analyzed.

Phosphorus deficiencies have been reported in coniferous forest on siliceous sands and poorly drained soils of coastal areas (Pritchett and Smith, 1970) and in exotic pine plantations in the Southern Hemisphere (Ballard, 1971). Soils defi-cient in phosphorus are normally fertilized near time of planting to prevent stagnation of the young trees.

## INTERPRETING SOIL FACTORS IN SITE EVALUATIONS

In evaluating sites it is important to keep in mind that tree growth reflects the combined influence of many biotic and environmental factors. Some biological influences on growth, such as severe insect and disease attacks, are obvious and their effects can be minimized by careful site selection. Genetic and some other biotic variables may have to be considered as sampling error and ignored in site studies. On the other hand, variations in site productivity associated with physical factors are more easily identified and can often be resolved.

Climate can be treated as a continuous variable where obvious variations exist, but subdividing larger geographic areas into relatively homogeneous climatic provinces has proven a simpler way of resolving climatic differences (Hills, 1961). It is then generally possible to detect effects of physiographic variables when working within a uniform climatic region. Differences in microclimate and soil moisture regimes as influenced by altitude, aspect, and slope are usually sufficiently great to identify and stratify into physiographic regions.

Within uniform climatic zones and physiographic regions, one can usually separate site differences in productivity based on soil variables. Ralston (1964) and Carmean (1975) reviewed research in North America on soil factors affecting productivity. They found that most soil factors that correlated with site productivity were those attributes of soil profiles that reflect the status of soil moisture, nutrients, and aeration. Some of the soil properties related to productivity of several species are summarized in Table 13.1. Recent research has placed greater emphasis on soil fertility factors, but available water still appears to be the single most important determinant of productivity of many tree species. Broadfoot (1969) reported this to be true for southern hardwoods, but he pointed out that the use of multiple regression equations to predict site index of new populations has generally given poor results because of the inability to measure the true causes of productivity; that is soil moisture and nutrient availability during the

**Table 13.1**  Soil Properties Frequently Related to Site Productivity (from published North American reports[a])

| | |
|---|---|
| Southern pines | = subsoil depth and consistency (22); surface soil depth (21); surface and internal drainage (18); depth to least permeable horizon (14); depth to mottling (13); subsoil inhibitional water value (8); N, P, or K content (6); surface organic content (2). |
| Northern conifers | = surface N, P, or K content (17); surface soil texture (14); drainage class (11); depth of surface soil (8); A1 organic content (7); thickness of B horizon (5); stone content (5). |
| Eastern oaks | = surface soil depth (14); depth of A+B (14); subsoil texture (9); exchangeable base content (7); soil pH (6); surface soil texture (5); organic or N content (4). |
| Eastern hardwoods | = depth to pan or mottling (20); surface soil texture (19); soil drainage (11); nutrient content (8); depth to water table (7); depth of A horizon (7); subsoil texture (3); organic content (3). |
| Western conifers | = effective soil depth (18); available moisture (6); surface soil texture (6); soil fertility (4); subsoil texture (3); stone content (3). |

[a]Numbers in parenthesis indicates number of reports to 1978.

growing season, soil aeration, and physical condition including root growing space.

Medium-textured soils are usually the best forest sites and site quality decreases with increases in the percentage of either coarse textured or fine textured materials. Site quality may also decrease with increased amounts of stone. Both texture and stone contents are related to amounts of available moisture, nutrient levels, and soil aeration and drainage. Depth to mottling and subsoil color are often indicators of restricted internal aeration and drainage that limit root development.

Root growing space, measured as depth of soil above some root restricting layer or water table, has been used as an indicator of site productivity more than most physical factors because it can be measured with some accuracy in the field and it is an integrator of many other factors important to tree growth. Except in very shallow soils, depth has little direct influence on tree growth. It has considerable indirect effect as it influences available water storage capacity, nutrient availability, and aeration. Pragmatically, the goal of any evaluator of forest site productivity is to select a relatively few easily measured soil and site factors that integrate all other factors that are important to the growth of a particular species on a given site.

# Part 2

MANAGEMENT
AND
ITS
CONSEQUENCES

# 14

## CLASSIFICATION OF FOREST LANDS

Evaluation, classification, and mapping of forest lands are prerequisites for most any type of forest management. They are particularly important to intensive management programs and are essential first steps in the development of multiple land-use plans. Classification and mapping are two related but distinct steps in evaluating the soil inventory. Soil classification is analagous to taxonomic systems used for plants and animals, while soil mapping is an exercise in geography. Mapping is concerned with plotting on paper (aerial photographs) the boundaries of soil units that can be recognized in the field.

Soil and its associated environment can be thought of as the habitat or site of the forest trees and other organisms. Site includes position in space as well as the associated environment (Spurr and Barnes, 1973). Site quality, thus, is defined as the sum total of all factors influencing the capacity of the forest to produce trees or other vegetation. Most classification systems, therefore, attempt to integrate the various site factors so as to yield an estimate of site quality.

The primary goal of most classification systems in the past has been to accurately evaluate sites in terms of their potential volume production of wood. More recently, classification systems based on a variety of relevant site features, together with maps developed from those systems, are recognized as invaluable aids to a multitude of management purposes. As a consequence, classification of forest lands should be considered in the broad sense of rating sites in terms of their potential for withstanding recreational impacts, importance as wildlife habitats, yield of high quality water, hazards in road or engineering structures, and other uses, as well as their suitability for timber production.

## CLASSIFICATION SYSTEMS

Attempts at site classification can probably be traced back to the very beginning of silviculture as a science. A number of site factors have been used for this purpose. The most important of these are based on characteristics of the stand, such as growth rate, species composition, and nutrient content of the foliage; on the type of ground vegetation; or on physical and chemical properties of the soil. Historically, as forest management became more intensive, systems of site evaluation progressed from crude productivity ratings based on obvious characteristics of the forest and its physical environment to multifactor approaches. In the first quarter of the century, the question of site classification was subject to a great deal of research and discussion (Bates, 1918; Frothingham, 1918; Watson, 1917; Zon, 1913) with the general acceptance of site index as the primary means of evaluating site productivity. This empirical approach, based on height of dominant or codominant trees at a given age, as proposed by Frothingham (1918), is still widely used as a measure of forest productivity, but it has some major limitations as a method of site classification. Recent economic pressures for maximum efficiency in forest management have focused attention on attributes of forest sites other than their productivity in reference to one or two species. This has resulted in the development of a number of more advanced systems of evaluation, some of which approach total site classification. All of the systems have some usefulness under certain site conditions and management objectives. They can be conveniently classed into three general categories, listed in approximate order of complexity: (1) forest productivity, including site index and vegetation types, (2) soil properties, and (3) multifactor classifications.

## CLASSIFICATION ON THE BASIS OF FOREST PRODUCTIVITY

Methods of measuring productivity of forest sites are generally grouped into direct or indirect methods, depending on whether the estimate is based on some stand measurement or on some features of the local environment. However, short of harvesting a mature stand, there is really no *direct* method of measuring site productivity. What is usually measured is some quantitative aspect, or index, of forest productivity. Nonetheless, estimating productivity on a stand volume or weight basis is a reasonable approach, in spite of the serious shortcomings of these methods.

The main objection to the direct approach is that both volume and weight are affected greatly by variations in stand density. If attempts are made to hold stand density constant by measuring only "well-stocked" stands, one will be limited to a relatively small number of areas in developing site curves. On the other hand, if stocking is allowed to vary, forest productivity will have to be expressed as a function of both stand density and site quality. Because of the difficulties in locating sufficient stands with similiar densities and the fact that tree-height growth is not greatly affected by moderate changes in density of reasonably well-stocked stands, foresters have generally accepted height of the dominant stand at a specified age as an index of site productivity.

## Site Index

*Site index* is the term used to express the height of dominant and codominant trees of a stand projected to some particular standard age. In the eastern and southern United States, the base age is usually 50 years and the site index is the expected height growth at this age. Used in conjunction with volume tables, site index is the yardstick by which relative productivity of forest sites is measured. In stands younger or older than index age, a family of height/age curves is required for projecting measured height to height at index age (Beck, 1971). These curves are developed by measuring height and age of many stands at a single point in time, fitting an average curve of height-on-age to these data, and constructing a series of higher or lower curves with the same shape as the guide curve.

There is evidence that curves constructed in this manner often do not accurately represent actual stand growth conditions. For example, the guide curve is likely to be accurate only if the ranges of site indexes are equally represented at all ages (Spurr and Barnes, 1973). Unequal sampling may occur because of the timber harvesting and land abandonment trends in a particular region. It has been pointed out that trees reach merchantable size faster, and are often cut at a younger age, on high-quality sites. Consequently, a sample of stands selected at a particular time could result in a biased curve that would tend to underestimate site index of stands younger than index age and overestimate site index of older stands (Beck, 1971). It may also be important that the curves are applied in a manner consistent with their construction. That is, if a set of curves were developed on the basis of the 10 tallest trees per acre, they should be applied on other sites by using trees that have been selected in a similar manner.

Furthermore, the assumption that the shape of the curve does not vary from site to site and is, therefore, anamorphic is generally false (Beck and Trousdell, 1973; Spurr, 1952). The degree of diversity in curve shape seems to vary with species and location, but the pattern of growth with change in site quality may be similar for many species. Instead of the rates of growth being proportional at all ages for all qualities of sites, as generally depicted by conventional curves, the rate of height growth rises rapidly on the best quality sites and then become relatively slow. On the other hand, growth rates on poorer sites increase slowly during the early years but are maintained for a longer time, so that one might expect to find trees on sites with the poorest site indexes about equal to those on sites with the best indexes at some age older than index age, as illustrated from Beck and Trousdell (1973) in Figure 14.1*a*.

The bias introduced from the failure to use polymorphic curves is probably of little importance in the relatively short rotations of most intensively managed

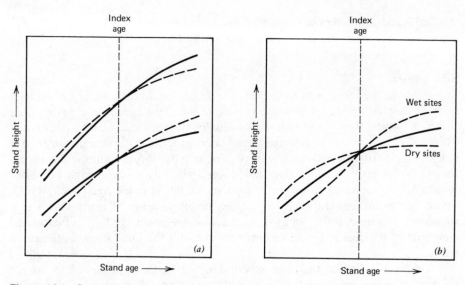

**Figure 14.1** Comparison of normal proportional curves (solid lines) with nonbiased, polymorphic curves (dashed lines).

forests. Nonetheless, variations in growth patterns among sites due to differences in certain soil conditions can be quite striking in the early years of stand development. For example, slash pine growth during the first 10 to 20 years is often quite slow on the wet savanna soils of the coastal flats of the southeastern United States. The developing stand gradually draws down the mean ground water table, thus, increasing the effective rooting volume of the soil. The increase in soil volume provides for a faster tree growth rate through improved nutrition and aeration during the latter stages of stand development. In contrast, slash pine planted on the well to excessively drained sand hills of the coastal plain make rather good growth during the first five to seven years after planting. However, growth tends to stagnate as moisture becomes a limiting factor for good stand development, as illustrated in Figure 14.1*b*. That different sites can have several shapes of the height-age curves, even though these sites may be of equal quality when measured at some common index age, has been reported for several species (Carmean, 1975).

It is apparent that the site index system of classification is highly empirical and it provides limited information except that concerning the immediate stand. It cannot be used for sites having no trees, for those lacking suitable trees, or for the conversion of species. Furthermore, it provides little understanding of the biological limitations of a site. That is, site index cannot be used to predict the potential productivity of a site subjected to intensive management by such practices as drainage, site preparation, fertilization, and use of superior planting stock—unless the curves were developed from sites managed in a similar manner.

Site index is, nonetheless, a convenient and useful guide to potential tree growth for a particular species under a given set of conditions and is widely used for this purpose (Beck, 1971; Carmean, 1975; Trousdell, Beck, and Lloyd, 1974). Site index curves and associated volume tables have been constructed for most of the commercial timber species in the United States and many studies have been conducted for the purpose of refining these tools. Height growth records from permanent growth and yield plots have been used in attempts to improve on harmonized site index curves. However, long-term records are not available for most species, and growth and yield plots are often not established in stands representing the full range of site, soil, topography, and climate where the species occurs. Stem analysis is the most reliable method for developing accurate site index curves and many new curves have been published in recent years based on this method (Carmean, 1975). Internode measurements have also been used as a means of constructing more refined site-index curves. This technique is similar to stem analysis and is most easily used with conifers having conspicuous limb whorls marking the course of annual height growth, such as in red pine, Douglas-fir, and eastern white pine. These new site index curves have confirmed that tree height growth is usually polymorphic.

In an effort to reduce the error relating to irregularities of the establishment period, the *growth intercept* index was first proposed by Wakeley and Marrero (1958). This index is based on height growth during the five-year period that begins with the year in which the breast height level (or some greater height) is attained. The technique is convenient for uninodal species such as red and white pines, but it is rather unreliable for the southern pines which may have several flushes of height growth during a single season. The relationship between height of red pine at age 25 and height predicted by the five-year growth intercept method is shown in Figure 14.2.

Site index comparison graphs and ratios have been developed for a number of forest species in various part of North America. They are useful for extending direct site index estimations, particularly in forest areas where soil and site vary greatly and where the forest manager has the problem of selecting the most desirable species for management from among several available species for a given site.

## Ground Vegetation Types

Of the indirect methods of classifying sites on the basis of productivity, systems employing ground vegetation types are probably the most popular. These early systems of classification by use of vegetational indicators are particularly useful where forest productivity can not be assessed directly by tree measurements. The site types, or ecological approaches, have gained considerable favor among

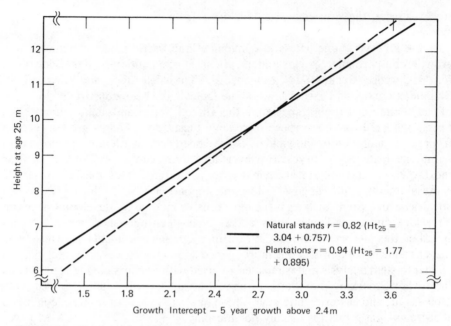

**Figure 14.2** Relationship between height at age 25 and 5-year growth intercept above 2.4 m (Alban, 1972).

many foresters in northern Europe and in Canada. Most of these classification schemes are based on the early work of Cajander (1926) and his Finnish coworkers. Cajander described five main vegetal classes, each divided into site types using characteristic understory plants. Finally, the various site types were ranked according to level of forest site quality.

Fundamental to the classification of site on the basis of vegetation is the presence of species with limited ecological adaptations; that is, species that tolerate only very narrow ranges of environmental conditions. These plants are usually ground vegetation species rather than trees, because the shallow root systems of ground vegetation make them more sensitive than trees to slight differences in site (Hodgkins, 1960). Indicator species can be used to characterize the site directly, or they can be used to characterize the plant community as part of the total site. Cajander (1926) reported that the community-characterizing species should be the dominant species in their particular strata of the forest. This system apparently works very well in Scandinavia because there are only two or three major tree species involved and they grow on almost all soils, while the more numerous ground cover species were rather sensitive to soil conditions (Figure 14.3). As one progresses southward from Scandinavia, however, the proportion of dominant ground vegetation species having narrow ecological

**Figure 14.3** Ground vegetation can be a good indicator of site productivity in some areas. An example is the use of *Vaccinium myrtillus* as a site indicator for spruce and pine in Sweden.

aptitudes seems to markedly decrease. It is apparently for this reason that the early systems of using indicator plants for site classification made little headway in central Europe. Instead, systems stressing fidelity and constancy of species presence have been more useful. In this latter system, refined by Braun-Blanquet (1932), analysis of the total plant life is stressed. Characterizing vegetation may be minor or obscure species and the communities, or forest site types, are identified through an inductive synthesis process applied to data from a region-wide survey (Hodgkins, 1960).

There have been a number of successful studies in site classification using Cajander's method in various parts of Canada (Ilvessalo, 1929; Rousseau, 1932). Heimburger (1941) studied ground vegetation-soil relationships in tolerant hardwood-spruce forests approaching the climax stage on shallow glacial till soils. He recognized four upland forest site types *(Cornus, Oxalis-Cornus, Viburnum-Oxalis,* and *Viburnum),* which were associated with four main tree covers (conifer, conifer-broadleaf, broadleaf-conifer, and broadleaf) and also with four main physiographic site types defined by depth of soil material and slope position. Hills (1961) classified forest sites by means of lesser vegetation, moisture regime, and soil permeability, with the last two factors linked to the physiognomy of the region. Vallée and Lowry (1970) reviewed other Canadian research that combined the floristic analysis phase of the Braun-Blanquet (1932) system with the forest-type system of the Finnish school. These systems generally took into account such ecological factors as soils, slope, and exposure in order to obtain a better definition of the plant community (Hills, 1961).

Research on the classification of forest lands on the basis of ground vegetation in the United States has been largely confined to the northeastern region (Heimburger, 1934; Sisam, 1938; Westveld, 1952). Spurr (1952) noted that single characterizing species were not very reliable and proposed an *indicator spectrum* of characterizing species ranging from the poorest site indicators to the best site indicators. However, the site-type systems have generally received little acceptance by American foresters because of the complexities of plant community relationships that change much more rapidly than soil and topographic conditions. While assays of ground vegetation, like site index, give a sort of integrated measure of environmental conditions, the systems have serious limitations. First, the indicator plants must be present to be of interpretative value. Fire and site disturbances may change the complex of index plants. Furthermore, lesser vegetation is often shallow rooted and its growth does not necessarily reflect subsoil conditions that may be important to tree growth. Stand density and the kind of overstory tree may affect the vigor and composition of understory plants. In spite of these limitations, indicator plants can be of considerable aid in land classification even in regions outside of the northern latitudes. For example, differences in lesser vegetation associated with variations in natural drainage patterns are often helpful in identifying site types in flatwood areas of the United States lower coastal plain. It would appear that when considered in conjunction with other site properties, vegetation types can be extremely helpful in classifying forest lands in most all regions.

## Habitat Types

Classification on the basis of habitat types is a somewhat more complex system of evaluation than systems based on ground vegetation alone. The system relies on the concept of the forest as an entity, composed of overstory and understory vegetation and certain features of its physical environment. One can consider the *major* forest ecosystems and their associated soils as the zonal concept of classification, in which climate has a predominant influence on their development, as discussed in an earlier chapter. In the same light one can also think of habitat types as an intrazonal concept in which some local feature of the environment, such as topography or drainage, has an overriding effect on soil and vegetation development. Remezov and Pogrebnyak (1969) gave a discussion of the idea of "forests as immense grouping of vegetation which indelibly marks the territory they occupy, including the soil and the climate, and that the vegetation is both a reflection and a product of that environment." The composition, structure, and productivity of natural forests, and to certain extent plantations, are reflections of differing soil and climatic conditions. Although the soil and atmosphere are major parts of the plant environment, neighboring plants, especially the trees, must be included in any classification of ecological factors. These features are divided into biotic, abiotic, and anthropogenous (cultural) factors, similar to the factors affecting site productivity, except that managment is included (Remezov and Pogrebnyak, 1969).

**Biotic Factors.** Biotic factors that influence forest productivity and species composition include all plants and animals that inhabit the forest ecosystem. Among plants, trees have the greatest influence on the forest environment, but mosses, lichen, and fungal and bacterial flora are also of great importance. Many members of the animal kingdom, from large wildlife to worms, thrips, nematodes, and rotifers are found almost exclusively in forests and play an important part in litter breakdown and in soil formation and productivity.

**Abiotic Factors.** Abiotic (inorganic) factors are grouped, as previously, into climatic, edaphic, and physiographic components. *Climatic* effects are those of the above-ground environment (light, temperature, precipitation, humidity, wind, evaporation, and carbon dioxide). Remezov and Pogrebnyak (1969) also include lightning and the resulting forest fires. *Edaphic* factors include soil moisture and substances dissolved therein; concentraton of salts; pH; toxic substances; physical properties (texture, porosity, aeration, hydrothermal properties); thickness of rhizosphere; and forest litter. *Physiographic* features of the terrain influence surface runoff, erosion, floods, and alluvial processes on slopes, valleys, and on flood plains, among others.

**Anthropogenous Factors.** Anthropogenous, or artificial, influences include felling, burning, cropping, pasturing, slash removal, site preparation, draining, fertilizing, and species introduction.

Habitat classifications, similiar to systems based on ground vegetation classifications, are predicated on the fact that plants have specific requirements regarding factors of the environment, including nutrients, light, moisture, and oxygen, but that some species are more tolerant than others to changes in these factors. This tolerance, largely as a result of anatomical and morphological features, places certain trees in a favorable competitive position for survival under adverse conditions. For example, pines are generally photophilic, drought resistant, and not very exacting as to soil conditions for survival. For this reason, they are generally one of the first species of higher plants to colonize an area after cutting or burning. They may, in turn, be replaced by more demanding shade tolerant species as the habitat is improved by the pioneer species.

**Examples of Habitat Types.**  Major forest habitat types of the world were discussed in considerable detail by Wilde (1958) and in various ecology texts. No attempt will be made to summarize this material, except to point out as examples the soil-forest types identified with the Atlantic and Gulf coastal plain physiographic region of the United States. Wilde listed them as (1) pine hills, (2) flatwoods, (3) sandy hammocks, (4) shell mounds, (5) table lands, (6) bluff soils, (7) lime prairies, (8) alluvial soils, (9) coastal peat, (10) and bayous. Each region has its own particular soil and assemblage of plants and colloquial terms to describe these local topographic conditions and habitat types. Most areas identified by these terms are only of local importance. Some examples of these colloquialisms used in the southeastern United States are described below:

1. *Sand hill, high pineland,* and *turkey oak ridge* all refer to sandy, dry, grassy, longleaf pineland that usually has a conspicuous understory of scrub oaks.
2. *Hammock,* an association of predominantly evergreen hardwoods on generally well-drained sites.
3. *Flatwood,* an ill-defined, nearly level zone of mostly imperfectly drained, acid sandy soils. It is underlain by limerock at varying depth, which accounts for the numerous lakes, sinkholes, and depressions. Spodosols are typical soils and wiregrass *(Aristida stricta)* and saw palmetto *(Serenoa repens)* make up a large percentage of ground vegetation. Gallberry *(Ilex glabra)* and waxmyrtle *(Myrica cerifera)* are common shrubs and longleaf and slash pines are predominant tree species.
4. *Savanna,* a more or less open, wet, grassy coastal lowland, usually with scattered trees or large scrubs, excluding palmetto. Savannas are sometimes referred to as *wet prairies, grass-sedge bogs,* and *pitcher-plant flats.*
5. *Swamps,* any low wet area, generally containing standing water and occupied by trees. A similar depression without trees is called a *bog.*
6. *Bay,* a swampy depression occupied by large shrubs and trees, many of which have broad, leathery, evergreen leaves (also collectively referred to as bays). *Titi swamps* are shallow bays occupied by shrubs and small trees called titi *(Cliftonia monophylla, Cryilla* spp.), with scattered slash and pond pines. In the Carolinas, a titi swamp with pond pine is called a *pocasin.* Seepage areas that form the headwaters of streams and are clothed by bays are called *bayheads.*

7. *Cypress domes,* or *cypress heads,* are cypress swamps that are more or less circular in shape. Trees in the center are taller than those along the edges, giving a rounded aspect to the canopy.

**Effects of Topography on Vegetation Types.**   An initial stratification into similiar zones, based on recognition of a few well-defined topographic categories, is the usual way of approaching forest site classification in many areas of the United States southeast. The separation of *upland* and *alluvial* sites, for example, immediately recognizes a substantial difference in site quality which ordinarily can be delineated on an aerial photograph. This initial stratification also separates soils of highly dissimilar moisture, nutrient, and aeration regimes. A few additional categories within each of these broad classes often will be helpful, but one should avoid refinement of topographic classifications to the point where photo or field identifications becomes doubtful. In flatwoods areas of the lower coastal plain, where rather small changes in surface elevation are important to soil drainage, recognition of a few vegetation types can be useful in site appraisal when used in conjunction with topographic categories. Such an association of soil-vegetation types was presented by Pritchett and Smith (1970), as shown in Figure 14.4.

On sand hill soils (Quartzipsamments) of southeastern United States, longleaf pine was the dominant species in the natural habitat, but planted sand often makes superior growth to longleaf pine on these dry sands (Figure 14.5).

Remezov and Pogrebnyak (1969) discussed the effects of relief on forest composition and production. They used an example of a "bor" complex, and ecological series of infertile sands and associated forest stands. In their example, the stand consisted of only two or three pine and birch species which are not very exacting in regard to soil. On top of a sand hill there was likely to be a dry (lichen) bor with pure pine, of moderately poor site class 3 or 4; further down the slope it gave way to a slightly moist association of pine and birch of class 1 or 2 (good). This in turn yielded to a moist site of class 3 birch, and the latter was succeeded by a very moist bor, and ends at the foot of the hill in "pine on bog" of class 5 (poor). Slope position in relation to productivity of *Pinus radiata* in New Zealand is dramatically shown in Figure 14.6.

The bor complex develops primarily as a result of varying levels of soil moisture. However, it is obvious that the diversity of species is not merely the result of varying moisture, but also of other factors related to relief: light, temperature, soil aeration, nutrients, and the position of the water table.

Similar series of habitat types associated with local variations in topography and moisture can be found in most forested areas of the world. For example, in the northern hardwoods of New England, one may find red pine growing on well-drained sandy soils with white spruce, balsam fir, sugar maple, white birch, and beech on the more favorable slopes, and gray birch, red maple, alder, and black

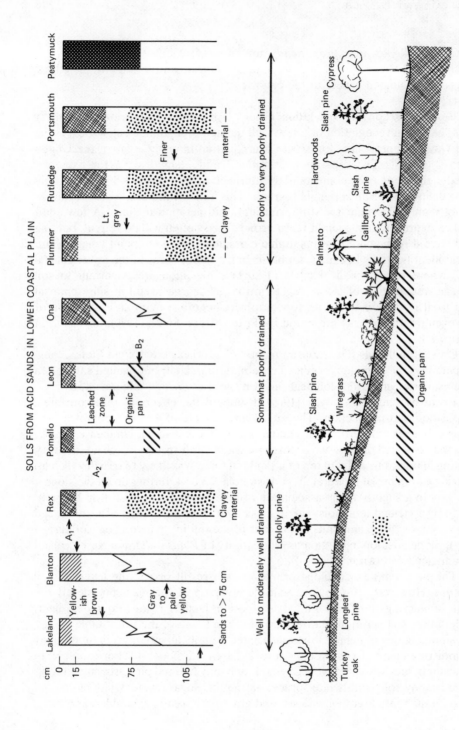

**Figure 14.4** Flatwood soils of the lower coastal plain with typical profile characteristics and drainage conditions and their associated vegetation (Pritchett and Smith, 1970). Reprinted from *Tree Growth and Forest Soils*, ed. by Youngberg and Davey. © Oregon State University Press.

spruce occupying the wet and swampy areas. Changes in vegetation and site productivity down a precipitous grade may also be the result of soil fertility differences. The gradual erosion of colloids and leaching of nutrients, usually at a time when the site was not under forest cover, results in more fertile soils on lower slopes and bottoms and in faster tree growth at the bottom than at the tops of the slope. In time, the tops of trees across shallow ravines may be essentially level due to this differential in growth rate (Figure 14.6).

## SOIL CLASSIFICATION AND SURVEY

Another general approach to site classification is one based specifically on observed or measured soil properties and topographic features. Several variations within the general concept have been used. Perhaps the best known are the uses made of the existing USDA Soil Conservation Service taxonomic system of classification and survey (Soil Survey Staff, 1975) for management planning and prediction of tree growth.

### USDA Soil Survey

This taxonomic method of soil classification has the important advantage of employing existing soil survey information based on a national system of classification. The system operates on the basic premise that soils can be identified as individual bodies and treated as integrated entities, so that knowledge of individual soil types can be interpreted in terms of their capacity to produce various crops. Several scientists have successfully used standard soil survey information and maps, by themselves or in combination with other observed properties such as drainage class, thickness of the surface horizon, and horizon depth, to predict site index (Haig, 1929; Coile, 1952; Stoeckeler, 1960; Broadfoot, 1969). Stephens (1965) reported that soil taxonomic unit at the series level provided an accurate prediction of Douglas-fir site index on zonal soils of the Oregon Cascades.

In spite of the considerable potential of detailed maps of the cooperative soil surveys, they have been viewed with mixed emotions by most foresters. Some forest scientists consider them a most valuable aid in management planning while others find them of little value, particularly for predicting productivity. Whether or not they are useful to an individual apparently depends to a large extent on the landform and species under consideration. Van Lear and Hosner (1967) reported evidence in their study that indicated "little, if any, usuable correlation between soil mapping units and the site index of yellow-poplar in Virginia." This conclusion was prompted by the wide variation of site indexes exhibited within each mapping unit. Other workers have encountered considerable frustration in attempting to group soils by series and type. Carmean (1961) pointed out the wide range of site values which may occur on single soil types. This same problem has been noted by several researchers in other areas (Broadfoot, 1969; Coile, 1952).

**Figure 14.5**  Sand pine (right) often outyields longleaf pine (left) and other southern pine in plantations on deep dry sands (Quartzipsamments) of the coastal plain.

**Figure 14.6**  Better soil moisture and nutrient conditions at the base of slope resulted in superior growth of *Pinus radiata* in New Zealand.

Many of the shortcomings of the soil survey maps probably derive from the fact that they were primarily developed for agricultural use. Some soil properties important to deep-rooted trees may not be considered in delineating taxonomic or mapping units, such as water tables at depths of 2 to 3 m and subsoil texture variations. At any rate, it is apparent that the standard USDA Soil Conservation Service soil survey, as presently conducted, is not as effective as it could be for forestry purposes. More attention needs to be given to phases within the series and types, as suggested by Carmean (1961), so that the survey more nearly reflects soil properties important to tree growth. The Soil Conservation Service has been collecting a wider range of information on soil series and types in recent years that should result in a more useful system of classification for foresters in the future.

## Special Soil Surveys
In the absence of standard soil surveys or in attempts to improve upon the USDA method of mapping soils for purposes of forest management, a number of special forest soil surveys have been initiated by contractors and large corporations. Excellent examples are the surveys conducted by Coile (1952) in the southeastern region of the United States and that by the Weyerhaeuser Company in the Pacific Northwest (Steinbrenner, 1975).

Coile found soil-site maps desirable for both moderately intensive and intensively managed forest lands. He stated that the maps should show the specific geographic extent of soil features and forest site classes useful in determining or making decisions on such things as (1) selection of species and spacing, (2) prediction of future yields, (3) drainage, (4) site preparation methods, (5) road construction, (6) definition of areas for seasonal logging, and (7) allowable costs for all phases of management based on expected returns.

Coile (1952) and his co-workers related tree growth to a limited number of easily measured or observed soil physical characteristics, such as texture of certain horizons, soil depth, consistency, and drainage characteristic, plus certain other selected features of topography, geology, and history of land use. The value of these surveys depends to a large extent on the development of working relationships through mathematical trial and error testing of many combinations of variables (Mader, 1964). However, relatively little attention has been given to explaining the basic biological-physiological relationships involved. They are based on the premise that a few factors will satisfactorily explain site differences over a wide range of conditions.

Coile (1952) stated that "the degree of success attained in demonstrating relationships between environmental factors and growth of trees is largely determined by the investigator's judgement in selecting the independent variables that are believed to be related to tree growth in various ways and in different combinations. How well the investigator samples the entire population of soil and other site factors determines the general applicability of the results." He limited his paired soil-forest stand observations to pure, even-aged pine stands over 20-year-old that were fairly well stocked.

In Weyerhaeuser's soil survey of mountainous terrain of the Pacific Northwest (Steinbrenner, 1975), topographic features are of paramount importance and the maps are based on a strong correlation between landform and soil series within a geologic unit, as seen in Figure 14.7. Topography as evidenced by landform is important to interpreting the survey for road construction, equipment use and, in some cases, soil productivity. Mapping for productivity is a primary objective and the interpretation for this purpose must be developed through research. In addition the maps are interpreted for land use, trafficability (for logging equipment), windthrow hazard, thinning potential, and engineering characteristics for road construction. Productivity interpretations are the basis for determining allowable cut and for intensive forest operations, such as regeneration methods, stocking control, and thinning.

Mapping units provide the logical basis for delineating cuts in the logging plan, according to Steinbrenner (1975). The interpretations also indicate the type

**Figure 14.7** Topographic features are especially important in classification of mountainous terrain (courtesy Weyerhaeuser).

of equipment required and the timing of the harvesting operation so that the impact on site quality can be minimized. The windthrow interpretation is utilized to minimize damage along cutlines. Thinning potential is used to assign a priority to all lands for intensive forest practice. Engineers use the map in determining the best location for roads, drainage problems, and location and size of culverts needed. The soil survey provides information that is basic to sound forest management and its usefulness increases as more interpretive detail is developed.

## MULTIFACTOR CLASSIFICATION SYSTEMS

The multifactor approaches to forest land classification differ from the special soil surveys (Coile, 1935; Steinbrenner, 1975) and some soil-vegetation surveys mainly in the direction of approach and degree or intensity of mapping. The number of land features, stand characteristics, and soil factors used in delineating mapping units are generally greater than those used in other systems, and as a consequence maps developed on these features provide the flexibility for classifying on the basis of goals in addition to wood production. They generally entail a combination of independent site variables, measured in the field or laboratory, superimposed as phasing elements on conventional soil series, to reflect variations within these units important to tree growth or other uses. Such soil factors as soil depth, available water capacity, texture, organic matter, chemical composition, and aeration, as well as radiation, ground vegetation, and landform have been studied individually and collectively.

Hills' (1952) "total site" classification of Ontario can be considered a multifactor approach to site evaluation. This *holistic* concept integrates the "complex of climate, relief, geological materials, soil profile, ground water, and communities of plant, animals, and man." Physiographic features were used as the framework for integrating and rating climate, moisture, and nutrients, and aerial photographs were widely used for classifying and mapping vegetation and physiography. The integration of the various factors of environment and vegetation at each level of this hierarchial scheme of classification involves much subjective judgment and intuition (Carmean, 1975). Nevertheless, the system provides a good framework for stratifying large inaccessible forest regions into broad subdivisions based on general features of vegetation, climate, landform, and soil associations.

One of the major problems in developing a multifactor system of classification is to determine which factors have significant influence on growth processes. Environmental measurements can either be selected on the basis of logic and experience or as many as possible can be measured and nonrelevant data discarded during statistical analyses (Mader, 1964). Certainly the factors that are important in mapping in one region are not necessarily important under other conditions. In the U.S. coastal plains, subsoil properties, including color, depth to fine textured layers, depth to mottling, or some subsoil property that influences drainage, aeration, or water retention are often important to tree growth. Coile (1935) reported that site index of shortleaf pine stands was influenced by the texture of the B horizon and its depth below the surface. He divided the percent silt plus clay in the B horizon by the depth of this horizon below the suface (inches) and found that there was an improvement in site quality with an increase in this texture-depth index up to about 5. Values above this index were usually associated with a decrease in site quality.

Gilmore, Geyer, and Boggess (1968) reported that the combination of depth of incorporated organic matter and depth to an impervious layer were the soil properties most closely related to microsite productivity for yellow-poplar in southern Illinois. Other investigators (Auten, 1945; Smalley, 1964) found depth to a tight subsoil or the amount of available water in the rooting zone were the most important features for evaluating sites for this species.

In a study of 124 sites in western Washington, Steinbrenner (1975) found that total depth, gravel content, effective depth, depth of A horizon, texture of B horizon, and a microscopic pore space in the B horizon had highly significant influences on site index of Douglas-fir. All properties had a positive effect on growth, except gravel content and microscopic pore space. He also found that increases in degree of slope reduced site index at high elevations, but not at low elevations. Many of these properties are those used in mapping soil series in standard surveys or else they can be used as phases in series and type designation.

Analyses of soils for pH, cation exchange capacity, organic matter content, and nutrient concentration have been tested as guides to classifying soils on the basis of productivity. However, the use of soil chemical analyses for this purpose has not always met with success. Gilmore et al. (1968) reported that pH was the only variable in the top 20 cm of a Gray-Brown podzolic soil that was statistically correlated with yellow-poplar height. They reported that the lack of correlation between site quality and concentration of elements in the top soil was due to the fact that trees obtain nutrients from the entire usable soil profile, which was as much as 150 cm in their area. In addition to the difficulties encountered in obtaining soil samples that adequately represent the area in which tree roots flourish, there are also problems in selecting solutions that will extract amounts of nutrients from the soil sample that can be correlated with nutrient availability to trees. Most standard extracting solutions were developed and evaluated for annual crops on agricultural soils and they have not been calibrated for use in forest soils. Alban (1972) extracted P from surface soils collected from red pine plantations in Minnesota by 10 different methods. He reported that methods that extracted small quantities of P (water, $0.002\ N\ H_2SO_4$, and $0.01\ N\ HCl$) gave better estimates of site index than strong extractants, or extractants of organic or total phosphorus. Nutrient quantification may be expressed in terms of concentration in the soil, total amounts per horizon or profile, or relative amounts in comparison to soil nutrient storage capacity or in comparison to other nutrients. Regardless of the method used in conjunction with proper sampling and analytical procedures, soil chemical analyses can be extremely useful in classifying sites on the basis of productivity, as long as the variable under test is the primary growth limiting factor (Ballard, 1971).

There are probably no systems of land classification that consider all factors of the environment in delineating mapping units. Because of the differences in management objectives and intensity of land use, the various multifactor systems rely on different features or combination of factors for characterizing site. Most systems involve physical features that can be measured in the field, with few of them requiring laboratory determinations. Exceptions to this latter point are the use of chemical analyses of soil and tissue samples, along with other site factors, for identifying areas for fertilization.

## Classification of National Forest Lands

An example of a multifactor classification system that provides the land manager with basic data about the soils and associated landscape features at the detail needed for intensive management is one undertaken by the National Forests for the Southern Region (Schlapfer, 1972). It is a resource inventory, in which soil is the basic resource, designed to provide information on the kind, location, size, and potential of lands available to the forest manager. Mapping is based on a hierarchical method in which landform, a natural manageable segment of the landscape, is the primary component. Other components complement the landform and are designed to reflect soil and management differences. In the coastal plain, the National Forest Soil Resource Inventory Gude uses five components in identifying ecological management units:

A.  Landform—an easily recognized natural landscape segment that characterizes different management ecosystems. They are (1) swamp, (2) flood plain, (3) stream terrace, (4) bay, (5) upland flat, (6) lower slope, (7) side slope, (8) steep side slope, and (9) ridge.
B.  Texture—average texture of the most significant horizon within a depth of 1.5 m. Classes are (1) coarse, (2) sandy, (3) loamy, (4) silty, (5) medium, (6) fine, (7) very fine, (8) organic (<50 cm), and (9) clay to surface.
C.  Water regime—the capacity of a soil to supply water to plant growth on the basis of soil thickness and texture. They are: (1) waterlogged, (2) wet, (3) moist, (4) dry, (5) droughty, and (6) very droughty.
D.  Accessory characteristics—those that reveal certain behavorial patterns of soils but are not in themselves considered critical. (1) none, (2) rhodic, (3) organic (60 to 120 cm), (4) organic (>120 cm), (5) pan (fragipan), (6) pan (plinthite), (7) organic pan (>50 cm deep), (8) pan (other), (9) sand cap (50 to 100 cm thick), and (10) sand cap (100 to 150 cm thick).
E.  Modifiers—conditions that affect normal management activities. (1) none,

(2) extremely acid (<pH 4.0), (3) extremely alkaline (>pH 8.0), (4) shallow soils (<50 cm to restrictive layer), (5) sodium accumulation (sufficient to retard growth), (6) water (on or near surface at all times), (7) high shrink-swell clays (> 40 percent of 2:1 lattice clays), and (8) gravel.

Scientists of the national forests have made considerable progress in classifying and mapping their forest lands for multiple-use management in which timber production is not necessarily the primary goal. Wherever possible, they have used the Soil Conservation Service taxonomic system in developing mapping units. Because of the wide range of uses of national forest lands, there will be a growing need for continual improvements in their system of classification and mapping.

## Soil-Site Evaluations

All classification systems designed for the purpose of site evaluation, directly or indirectly, involve the soil. Most indirect methods of site evaluation are based on soil or site factors that are easily recognized and measured in the field. The most popular of the latter systems is one making use of soil-site equations. This system is particularly useful for estimating site index for forest areas where trees are absent or where they are not suited for direct site index measurements. For this goal the correlated site features need not be causative factors, but they must be consistently correlated with site index.

Soil-site studies have received the greatest emphasis in the United States in areas where site quality, soil, and stand conditions are extremely variable. Carmean (1975) listed some 41 published reports on soil site studies for southern pines, 34 dealing with northern conifers, 23 reports on western conifers, 35 on eastern oaks, and 41 reports on other eastern hardwood. Features commonly correlated with site index are those site factors that significantly influence tree growth, as discussed in Chapter 13, and that are easily identified and mapped. In most cases the features of soil, topography, and climate that were found to be correlated with site index are indirect indices of some more basic growth controlling factors and conditions such as available moisture and nutrients, and microclimate factors that affect evapotranspiration and tree physiological processes. "Possibly the significant factors determined from soil-site regression studies should be viewed merely as links in the many chains connecting tree response (site index) to causitive factors such as moisture, nutrients, temperature, and light" (Carmean, 1975). Soil features most important in soil-site studies are usually those concerned with depth, texture, and drainage; that is, properties that determine the quality and quantity of growing space for tree roots (Coile, 1952).

The data-collecting phase of soil-site studies involves locating a large number of site plots, usually in older forest stands representing the range of soil, topography, and climate found within a designated forest area or region. Site

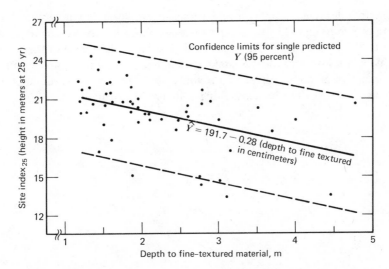

**Figure 14.8** Regression of site index for slash pine on depth to fine textured material in sand hills of west Florida (Hebb and Burns, 1975).

index is estimated from trees on the plots using height and age measurement or stem-analysis techniques. An effort should be made to see that there is a relatively wide range in site quality within the study area. The site index estimates are then correlated with associated features of soil and site using multiple regression methods. Sometimes a single soil property will have a predominant influence on site quality and a single-variable equation can be used for field purposes. Such an example was given by Hebb and Burns (1975) in which depth to fine-textured material was used as an indicator of site quality for slash pine in the coastal plain sand hills (Figure 14.8). Their equation showed that site index decreased approximately one meter for every 75 cm of depth to the layer.

The use of digital computers has resulted in more accurate and complete analysis of data and now transformations expressing curvilinear trends and interactions among independent variables can be tested with little effort. There are still shortcomings with the soil-site approach to site classification, mainly connected with the data collecting phase. For example, the determination of site index as the dependent variable by use of harmonized site index curves is subject to the errors previously discussed. The use of improved curves based on stem analysis or internode measurements can largely solve this problem. Carmean (1975) pointed out that quantitative values rather than qualitative rankings should be used for defining soil and topographic features as independent variables, but he acknowledged the difficulty of quantifying such features as slope shape, soil structure, and soil drainage. Regression equations have failed to accurately predict site index when applied to a very large and variable study area. Stratifying data or subdividing large areas into smaller more homogeneous units should help in those instances. Soil-site results should be applied only to the particular area studied and only to the soil and topographic conditions sampled within the study area. In spite of these limitations, the soil-site equation can be an extremely useful system for classifying soils for forest managers.

# 15

## SOILS AND SILVICULTURE

Throughout history the forest has provided the human race with many of the essentials for its survival. These vital commodities have included food, fuel, materials for shelter and tools and, more recently, the raw materials for many industrial products, such as paper and chemicals. Rapid population growth and industrial demands for wood products have resulted in a scarcity of forest resources in some parts of the world in modern times. Because the possibilities for expansion of the forest land base are quite limited in most countries, the increasing demands for timber generally have been met by some type of intensified forest land management. The primary objective of most management schemes has been increased production of lumber, pulp, and various wood products. In addition, forests are increasingly relied upon to provide other amenities, such as recreation and a quality environment, protection of watersheds, wildlife refuge, and grazing for domestic animals. Demands for these diverse forest benefits, along with great variations in climate, site, tree species, terrain and other natural conditions, and, more recently, economic factors and the development of efficient harvesting and transport systems have resulted in adoption of a multitude of silvicultural techniques.

### TYPES OF HARVESTS

After site classification and mapping, the primary consideration in developing a forest management program is choosing a system of harvest, if any, to use. The harvest system, therefore, sets the pattern for forest management. All forest cuttings can be considered as some version of the (1) selection, (2) shelterwood, (3) seed-tree, or (4) clear-cut system.

## Selection Cutting

The *selection* systems employ the removal of selected mature and immature trees, either singly or in groups, at specified intervals. Regeneration is almost continuous with trees of different ages or sizes intermingled singly or in groups. Selection on the basis of single trees in a mixed stand, leads to an increase in the proportion of shade-tolerant species, while group selection is used to maintain a higher percentage of less shade-tolerant species. Groups may vary from a small fraction of a hectare up to a hectare or more, in which latter case they resemble small clear-cut areas. The objective of the group section, however, is to create a balance of age or size classes in a mosaic of small contiguous groups. All other systems provide for some form of even-aged management.

## Systems for Even-aged Management

The *shelterwood* system aims for the removal of the mature stand in a series of partial cuts. A new stand is regenerated under a reduced canopy, which the final cut removes, thus permitting the stand to develop in the open as an even-aged stand. In the *seed-tree* system all the trees are removed in one cut except for a few of the better trees of the desired species which are left for reseeding purposes. The seed trees are then removed after regeneration is established. *Clear-cutting* provides for the removal of all trees in one cut from an area large enough to be a management unit, so that a new even-aged stand can be established by planting, seeding, or natural regeneration.

## Biological and Site Restraints

The choice of a silvicultural system is predicated on management objectives, biological factors, and site restraints. Reproductive habits and site requirements of the desired species have to be considered. Shade-tolerant species will dominate in forests regenerated under a canopy; while fast-growing, shade-intolerant trees usually dominate stands regenerated in full light. Furthermore, the age and vigor of the existing stand will be a factor. An even-aged stand of overmature trees should be harvested by the shelterwood, seed-tree, or clear-cut systems. The selection system is better suited to stands of trees with varying ages, sizes, and vigor. Wildlife browse is generally favored by clear-cut harvesting, but climbing animals, such as squirrels, are favored by cutting systems that result in abundant seed production and mature trees for nests.

Periodic use of prescribed fire reduces hazardous accumulations of flammable debris and undesirable undergrowth in some forest types, but it is normally only used in even-aged stands because of damage to young regeneration in all-aged stands. On the other hand, there is less debris for fuel resulting from any one cutting under the selection system. The maintenance of a canopy over reproduction may afford some protection against certain insects and against frost heaving in cold climates. However, when a stand is heavily infested by disease or insects it may be necessary to remove the affected trees or even the entire stand. Clear-cutting is also desirable on windy sites and in shallow rooted forests where the danger of windthrow of exposed trees is great.

The forest manager must consider the availability of labor, equipment, and capital, and the market for certain timber classes, when selecting a harvest system. He or she may wish to use genetically improved trees for the next crop, which are normally planted as even-aged stands in clear-cut areas. Perhaps even more important are the considerations of soil and terrain. Mountainous terrain may be too steep or unstable for the use of machinery in harvesting or road building. Soils that are too shallow or severe for good regeneration may be best left uncut. Flat areas with poor internal drainage may convert to swamps when the forest canopy is removed. Only partial cuts should be employed in these wet areas. On unproductive or erosive soils, the continual protection of a partial canopy may also be necessary.

Light cuttings or thinnings are required under the following conditions, according to Wilde (1958): (1) on rock outcrops, skeletal soils, and other shallow soils where there is danger of windfall; (2) on steep slopes, to prevent rapid runoff and erosion; (3) on hot and dry slopes or sites exposed to dry winds, to conserve moisture and retard decomposition of organic remains; (4) on highly fertile soils, to control the rank growth of competing vegetation; (5) on clay soils with unstable structure; and (6) on soils predisposed to a rise in the ground water table.

## TYPES OF MANAGED FORESTS

The way a forest is managed is determined essentially by the same factors as those that influence the selection of a harvest system. Stone (1975) stated that the classical notions of "intensive" and "extensive" management fail to describe the actual range and combinations of forest land management systems that exist in different parts of North America, or even in nearby tracts within the same forest type. He grouped systems into four types on the basis of management purposes and levels of skill and physical inputs. Excluding remote wildlands of Canada and Alaska, which receive little management inputs, groupings in order of intensity of management were labeled as *protected wild forests, exploited forests, regulated forests,* and *domesticated forests.*

### Protected Wild Forests

These protected forests are set aside in parks, preserves, and wilderness areas for nontimber uses. They make up a relative large and increasing percentage of the total forest landscape. They are managed to some degree "if only through benign neglect, and are now making small but acute demands for soil-related information" (Stone, 1975). They are often located on steep and fragile terrain where soils information is need in locating areas for roads, paths, or camp sites.

## Exploited Forests

Stone (1975) described expoited forests as those utilized by man with little or no investment of effort except in the extraction of wood or forage. They are exemplified by the vast areas of nonindustrial private lands in the United States and lands in boreal Canada. The areas generally receive fire protection but regeneration is left entirely to nature. A considerable part of the productive potential of these sites is never realized. Soil science plays a very minor role in the management of exploited forests because management input is at a minimum. To lift these forests out of the exploited class would require considerable knowledge of the soil resources, their capabilities and limitations, as a basis for selecting appropriate harvest systems.

## Regulated Forests

These forests are presented as the classical silviculture of most textbooks. Regulated forests are normally composed of native species handled in any of a number of harvest or silvicultural systems, in large or small units, and with a variety of objectives. The common denominators for these systems are assurances of adequate regeneration and controlled stand density and species compositions. Productivity is a fixed potential—an inherent property of the site that integrates soil and climatic effects. Silvicultural systems of regulated forests were developed as means of converting this potential into useful timber production, and at the same time giving some attention to demands for other goods and services provided by the forest. Successful management of these forests requires considerable knowledge of soil and site properties.

## Domesticated Forests

The systems of management that require the greatest amounts of energy and technological inputs and knowledge of soil and site factors are those of the domesticated or plantation forests. These intensively managed forests need not rely on nature for species or genotype, since they may be modified to obtain greater response to intensive culture and to obtain higher productivity. In these systems, site potential is not a fixed quantity, but it may be increased by soil modification or by the combination of soil treatments with the genotypes' response to them.

A large part of this chapter discusses techniques used in various areas of regulated and intensively managed modern forests because of the considerable potential for improving productivity through the proper use of soil and site information. Plantation management, which is considered the epitome of modern forestry, is emphasized.

## SOIL MANAGEMENT IN REGULATED FORESTS

The vast majority of the world's forests are naturally regenerated, and in spite of the considerable appeal of intensively managed plantation forestry, most of these forests should remain under some sort of minimal (extensive) management. This is because a large percentage of them are located in remote areas or on steep or rough terrain, which is not suited for mechanized equipment and intensive management. The absence of a market for the forest product, plus the fact that many of the forests are composed of low quality and nonuniform stands containing several species, many of which have low market value, make intensive management unattractive in such areas. In addition, land ownership patterns and history of land use are often unfavorable to investment in intensive management (Weetman, 1977). Of course, this is subject to change in some areas as market and population patterns shift.

Shelterwood cutting is a reliable method for establishing even-aged stands of most hardwoods. The system provides a partial canopy that encourages full stocking of desirable species, stimulates early growth, and restricts competition from grasses and herbaceous plants. It also affords continuous protection to the soil from sun and rain and results in a minimum disturbance of the forest floor. Consequently, the rate of decomposition of the floor material is only slightly increased as the result of harvest activities.

Cutting in such a fashion to ensure adequate and desirable regeneration is a major goal in managing all natural stands. Intolerant tree species are regenerated most effectively by either the clear-cut method or the seed-tree method and tolerant species by selection cuts or the shelterwood method. Forest stands that regenerate naturally following the harvest or natural destruction of forest vegetation are usually even-aged and a mixture of several species. The judicious application of thinning cuts provides the means of increasing yield and quality of merchantable timber and, where desired, for converting to unevenaged stand management (Filip, 1977).

Uneven-aged management requires considerable skill in selecting, marking, and harvesting trees to insure proper age and species mix. Furthermore, there is often difficulty in harvesting marked timber, as required in order to maintain a range of diameter classes, without damaging a high percentage of the residual growing stock. Logging damage may amount to 30 percent or more in northern hardwoods, if felling and skidding operations are not properly planned. Because of these difficulties, uneven-aged management is generally applied only on small timberland ownership of up to 100 ha or so. However, there are many portions of larger ownerships where uneven-aged management could well be used. These include areas of fragile soils, steep topography, or those that have high esthetic appeal. Uneven-aged management should also be considered for any area where public use or multiple-use management is important (Filip, 1977).

Present use of soils information in the extensive management of naturally regenerated stands is not very great. These stands generally are located on less productive soils that are not always responsive to management. Nevertheless, the prospects for improving management as a result of greater use of soils information are encouraging.

Land classification and mapping can serve many purposes in extensively managed forests, as in plantation forestry. They are invaluable as a basis for resource inventories and site selection. For example, southern hardwoods are found on both lowland and upland sites, but their soil requirements are rather exacting. Primary lowland sites are alluvial terraces and bottoms along streams. Localized areas of deep loamy sand with ample moisture and drainage often support mixtures of oak, hickory, ash, and sometimes yellow-poplar and magnolia. The best upland hardwood sites are deep loessal soils and the coves and ravines on slopes and bluffs above drainage courses. Maps based on appropriate classification schemes can also be used to identify habitat types and areas best suited for recreation, roads, and other nontimber uses. Improper location and design of roads and structures accentuate erosion problems and contribute to stream sediment in mountainous areas. When properly interpreted, maps may be used in determining site quality, the appropriate type of timber management and thinning regimes to follow, and soil areas that are particularly fragile, erosive, or deficient in nutrients.

## SOIL MANAGEMENT IN DOMESTICATED FORESTS

Intensive forest practices will replace extensive methods in parts of many countries as the demand for wood products increases. While intensive practices are generally more prevalent near areas of population, the trend is to bring ever greater areas of forest under intensive management for optimum sustained yield in more remote areas.

### Extent of Domesticated Forests

Chevasse (1974) pointed out that in mountainous areas in Russia and on other steep lands, stand establishment is mainly by planting rather than seeding. Of Japan's production forests that are less than 20 years old, the planted areas far exceed the areas of natural forests. The millions of hectares of world forests managed as plantations include a wide range of soil and climatic conditions and a variety of tree species.

In general, tree planting is used to affect (1) afforestation of open sites that have not supported forests for a long period, (2) reforestation of sites that have recently been clear-cut, and (3) conversion of one type of forest to a more desirable species. Examples of open lands that have been planted to forests are former agricultural lands in England, southwest France and parts of central Europe, and southern United States; sand dunes in New Zealand and Libya; arid zone sands in Israel and other Middle East countries; steep slopes in Europe and Japan; peatlands in the British Isles, Sweden, and Finland; former sea bottom in Holland; and spoil banks resulting from mining operations in many countries around the world.

Plantation establishment following clear-cutting of the same species is standard practice in many areas where intense silviculture is practiced. Examples of such reforestation efforts are in the Douglas-fir region of the Pacific Northwest, the southern pine forests of the United States, Australian *Eucalyptus* forests, and *Chamaecyparis* forests of Japan and Taiwan.

Plantation establishment on sites still supporting considerable forest cover often involves the conversion of native species to exotics, as in southeastern and southwestern Australia, replacement of maquis vegetation with pines or eucalypts in the Mediterranean areas, conversion of scrub hardwood and coppice in central Europe, and improvement of degraded tropical forests in several equatorial countries (Rennie, 1971).

Chevasse (1974) listed several countries where natural forests are being converted to plantations of widely different species, often from highly complex associations to exotic monocultures. These include plantations of *Pinus caribaea* and *Gmelina arborea* in Brazil; *Pinus radiata* in Australia, Chile, and New Zealand; *Pinus caribaea* in Fiji; *Tectona grandis, Eucalptus tereticornus, Cryptomeria japonica* in India; and *Triplochiton scleroxylon, Eucalyptus saligna,* and *Pinus* spp. in Africa. Whether regeneration is accomplished with native or exotic species, plantation forestry offers a unique opportunity to locate species and selections within species according to their soil-site requirements. It is not sufficient to plant hardwoods on "hardwood" sites and conifers on "conifer" sites, because of the tremendous differences in growth potential on different soils for different genera, species, and selections.

Some reasons given for the trend toward plantation forestry include (1) greater demands for forest products, (2) introduction of genetically improved stock, (3) easy manipulation of the stand for the most desired assortments, (4) higher yields, (5) full use of land by complete stocking, (6) improved conditions for the operation of machinery, (7) reduced cost of logging; (8) prompt restocking after harvesting; (9) more uniformity of tree size; and (10) improved log quality.

Plantation forestry generally involves an integrated system of cultural practices, including slash disposal, site preparation, and careful planting of cultivars, and may include such other silvicultural practices as soil drainage, weed and pest control, pruning, thinning, fertilization, and the use of genetically superior planting stock. They usually involve even-age management and clear-cut harvesting of monocultures.

Plantations are literally man-made forests, in the sense that they are established and maintained as the result of site manipulation. These efforts to improve the site and increase tree survival and growth may have profound influence on certain soil physical, chemical, and biological properties, particularly properties of the forest floor. Whenever heavy machinery is used in harvesting or site preparation there is an opportunity for undesirable soil disturbance. Soil disturbance is less for track than for wheel vehicles, less on flat than on steep terrain, less on sandy than on clay soils, and less on dry than on wet sites. However, long-term effects on soil properties and on long-term site productivity are mostly in the realm of conjecture because few well-controlled studies have been conducted for sufficient time to give a definitive answer (Evans, 1976).

## Land Clearing in Domesticated Forests

Development of detailed reforestation plans prior to logging can be invaluable in solving disposal problems. Such plans should take into consideration the soils, terrain, climate, and cover types and plans may include site and vegetation mapping. The type, amount, and distribution of logging waste or unwanted vegetation and terrain conditions will influence the choice of disposal method and machinery to use. Costly damage to the soil and site may be avoided by proper location of roads and selection of techniques and tools suited to each vegetation-terrain combination. On unstable locations such as moving sand dunes, erodable soil, or wetlands where the water table may approach the soil surface after removal of the transpiring canopy, it may be necessary to maintain a partial forest cover or to establish a vegetative cover in order to stabilize the land before a tree crop can be planted.

Logging debris and vegetation disposal may be accomplished by hand tools, heavy equipment, fire, or herbicides, depending on local conditions and available resources (Chevasse, 1974).

**Hand Tools for Land Clearing.**   Hand tools such as axes, slashers, machetes, and, more recently, lightweight power tools are widely used for land clearing and vegetation disposal in difficult terrain and areas where the labor supply is plentiful. They are generally used to sever small trees and shrubs so that they can dry sufficiently for burning. Hand tools are best suited for small areas of 10 ha or less. Because hand-clearing operations in large areas may require several months to complete, fuel curing occurs unevenly throughout the site and this may result in poor or irregular burning. Incomplete burns cause difficulty in planting and may result in heavy weed growth. Hand tools are also recommended for fragile soils where the use of heavy equipment could result in erosion, compaction, or excessive disturbance. As labor becomes more costly, machines will be used on the more accessible terrain, and hand methods confined to the steeper slopes. However, in densely populated countries, such as Japan, slopes as steep as 30° or more are used for plantation forestry in which hand tools play a major role.

**Use of Fire in Land Clearing.**   Fire has long been the principal tool in disposing of logging wastes and undesirable vegetation in many parts of the world. It is an inexpensive and versatile means of disposal of dry logging debris and unwanted vegetation. However, there has been a recent decline in burning in many highly populated, industrialized countries because of public objection to smoke pollution. This has led to restraints or prohibition against burning in some areas, even though wildfires may cause greater pollution and damage than controlled fires. There has also been a general reduction in the need for slash burning because of fuller tree utilization and the development of more powerful tractors and rugged plows to deal with slash. Nonetheless, fire is still a major tool in Australia and New Zealand and in the Pacific Northwest and the southern coastal plain of the United States.

Because controlled fires are relatively cool, they apparently do little damage to the mineral soil (Wells, 1971), but the intense heat of uncontrolled burning of slash heaps and windrows may destroy a significant portion of the soil organic matter. Destruction of organic matter often leads to degradation in soil structure, a reduction in macropores, and a decrease in water infiltration rates, as discussed in Chapter 22. However, Cromer (1967) reported an improvement in the growth of radiata pine planted in old ash beds. Decreased soil acidity and increased base supply in the ash bed were considered to have been partly responsible for the increased tree growth, but temporary alterations in populations of microflora and fauna in the surface soil may have played a large role. Slash burning may be either detrimental or beneficial to the soil, depending on conditions. In spite of the many objections to smoke, burning will continue to be used in less populated areas as an inexpensive means of disposal of logging wastes and unwanted vegetation.

**Heavy Equipment for Land Clearing.**  Machines are widely used for disposal of slash and excess vegetation in preparation for plantation establishment, except in steep areas not negotiable by crawler tractors or where there has been very complete tree utilization. Machines are extensively used, along with fire, in the conversion of unmerchantable eucalypt stands in Australia, in the disposal of low-quality hardwoods in preparation for pine plantations in southeastern United States, and in site clearing for Douglas-fir plantations in the Pacific Northwest. Some modification of a bulldozer blade is often used for the initial operation in site preparation (Figure 15-1).

In Australia, the trend to shorter rotations, and a need for reductions in site preparation costs have encouraged the use of bulldozers in conversions to exotic pines and faster growing native hardwoods. Standing residuals and logging wastes are first removed by burning, axing, herbicide injection, dozer pushing, or knocking down with anchor chains pulled between two tractors. Vegetation from 5 to 10 m high can be chained, but some species (*Leptospermum, Melaleuca, Acacia, Pomaderris*) must be roller crushed (Rennie, 1971). Nonstanding logging waste is heaped into windrows with large dozers equipped with root-rakes working in line abreast. Windrows are spaced 60 to 100 m apart to permit the longest possible runs by plows and planting machines, and for slopes over 15° they are aligned with the contour to reduce runoff and erosion. One burning of

**Figure 15.1**  Cutting woody vegetation with a shear blade as a first step in site preparation for plantation establishment in Oklahoma (courtesy Weyerhaeuser).

the windrows usually leaves 10 to 20 percent of the space unplantable by machine, but this can be reduced to 5 to 8 percent by hand planting in the burned windrows. Reheaping and reburning is done on lightly timbered lands.

Scalping is practiced where the residue is light and no plowing is needed. In this latter instance, a V-blade attached to the front of the planter-tractor provides a furrow about 5 cm deep and 1 m wide in which the seedlings are planted. Scalping displaces much of the organic-enriched top soil from immediate reach of the young plants and may result in prolonged slow growth on sandy soils. Scalping, as practiced on stony terrain in Scandinavia, consists of the removal of the surface debris and humus layer in regular small patches to permit hand planting in the mineral soil.

While tree species differ, techniques for the disposal of slash and woody vegetation in the United States are often similar to those used in parts of Australia and New Zealand. For example, successful attempts at conversion of wiregrass *(Aristida stricta)*-scrub oak *(Quercus laevis)* sandhills to southern pines requires temporary control of all native vegetation in order to conserve moisture and nutrients. This reduction in competition is apparently best accomplished by "chopping" with a drum-type cutter that severs and incorporates part of the existing vegetation into the soil without excessive disturbance of the topsoil. Because the silt plus clay fraction in these deep sands may be no more than 3 to 6 percent, conservation of the organic matter in the top soil is essential. Oaks are most effectively reduced when their root carbohydrate reserves are low; usually when leaves first reach full size in the spring. Two choppings spaced 6 to 18 weeks apart, beginning in early spring to summer are recommended (Burns and Hebb, 1972).

On hardwood hammocks, stream terraces, and areas of dense woody vegetation, bulldozers with V-shaped or serrated blades are often used to sever stems of residual hardwood. Root rakes may be used to uproot and windrow brush. The windrows are usually burned after a few months, but they still occupy a considerable portion of the land area that would otherwise be available for planting. Unfortunately, they often contain considerable top soil as well. Haines et al. (1975) reported that approximately 5 cm of the surface soil was scraped from the area and deposited in windrows during blading operations in the Carolina Piedmont. After 19 years, loblolly pine in an unbladed area contained 85 percent more volume than trees in a bladed area.

The removal of heavy vegetation from steep land is a major problem in some countries, although there have been some developments in terracing techniques and in the use of heavy tractors working along contours pulling over brush with anchor chains. Nevertheless, the use of heavy machinery may compact and disturb the soil and eventually lead to erosion, siltation, and water turbidity. Wet and fine-textured soils are most likely to present problems for heavy equipment. Sites where machines can be used to greatest advantage are large blocks of relatively level terrain with sandy soils and few obstructions. However, the advantages of machines in clearing large land areas of slash and unwanted vegetation in relatively short periods of time, and at lower costs than with hand tools, outweigh their disadvantages on most sites.

**Herbicides for Land Clearing.** Herbicides have been successfully used for controlling particularly persistant and inaccessable vegetation in many countries where intensive forestry is practiced. The use of herbicides expanded following an increase in the range of herbicides available and the development of specialized formulations. The following broad types were recognized by Chavasse (1974): (1) desiccants, or contact herbicides such as paraquat, often used to kill foliage in preparation for burning; (2) photosynthesis inhibitors such as aminotriazole; (3) germination inhibitors exemplified by the triazines; (4) translocated herbicides such as the phenoxyacetics, asulam, and picloram; and (5) root-absorbed herbicides such as bromacil.

Herbicides are available for most forest-vegetation situations, and because of the range of materials available, considerable care is required for their effective use. The forest manager must have a knowledge of their effects, the type of vegetation to which each should be applied, and the soil, weather, and plant conditions that give optimum results.

Chemicals are most often used to control shrubs and weedy vegetation in wet coastal areas, steep mountainous terrain, and other areas where heavy machinery or fire can not be effectively used. For example, herbicides are used in Japan on slopes too steep for machines, as a kind of "advance site preparation." Granular sodium chlorate has been successfully applied aerially 1 to 2 years before clear-cutting in order to obtain a clean forest floor for planting (Rennie, 1971). In Tennessee, injection of residual woody vegetation improved survival and growth of yellow-poplar, after 10 years, as compared to a planting where no residual woody vegetation was injected. Trees were twice as tall in plots where only competing stems greater than 5 cm in diameter were injected, but they were 2.6 times taller in areas where all residual woody vegetation was injected (McGee, 1977).

Chemical brush control has been used in Scandinavian countries for many years, but aerial spraying with herbicides has been virtually eliminated in those countries due to possible harmful environmental effects. The use of foliar sprays with ground equipment is only nominal in many countries because of difficulties in traversing rough terrain and environmental problems with spray drift. The use of granular materials and treatment of single trees by notching and injections are promising alternative methods. For example, *Betula* and other hardwoods have been notched and treated with amine salts of 2,4-D or 2,4,5-T esters with success in Finland (Rennie, 1971).

## Soil Preparation for Stand Establishment

Harvesting and the elimination of logging slash and unwanted vegetation may be considered as first steps in preparing a site for regeneration. However, the term *site preparation* is generally restricted to those soil manipulation techniques designed to improve conditions for seeding or planting that result in increased germination or seedling survival and tree growth. This may be accomplished by manual operations in some remote localities and on relatively clean sites. However, soil preparation is generally achieved by a variety of machines such as disc trenchers; disc, moldboard, or tine plows; choppers; and various others.

Ditching and water control in wet lands can be considered a kind of soil preparation, but these techniques are more appropriately discussed in a subsequent chapter on management of problem soils.

In cool to cold regions, particularly as found in boreal forests of Canada, Scandinavia, and Russia, exposure of mineral soil, elimination of competing vegetation, and improvement in water relations are primary goals in soil preparation operations. Because of the slow rate of litter decomposition, reactivation of the soil and release of nutrients for the young seedling are also important. This is generally affected by burning and by scarification with hand tools or with plows, discs, or anchor chains. Scarified areas of exposed and partially mixed mineral soil should be at least 0.4 m² per seedling and made in rows or patches. Screefing (pushing aside of humus layer to expose the mineral soil surface) is the most widely used method of soil preparation in Nordic countries. Harrowing has been used instead of screefing on soils with relatively thin humus layers. These harrows disturb the mineral soil to a small extent by creating a spoil bank composed of the humus layer. The seedlings are then planted in the bottom or on the side of the furrow. Improvements in water relations, increases in soil temperature, and better survival and early growth of seedlings have been attributed to these operations (Post, 1974).

Methods of soil preparation vary widely in central and western Europe, as might be expected from the great variations in terrain conditions of areas to be reforested. However, seeding is often accomplished following spot scarification or strip harrowing. Soil preparation is done manually where tractor use is impossible due to terrain configuration, or marshy conditions, or where the small size of plantation make such operations uneconomical.

Soil preparation for planting is most often accomplished by double moldboard plows. Trees are planted in the furrow, except in wet areas where the furrow may serve as a drainage ditch. In wet areas and peat soils, trees are generally planted on the tilt of the furrow slice, or else on beds or mounds created by throwing furrow slices together.

In Great Britain, sites are classified in advance to determine the nature of change to be achieved by the operation. Particular attention is given to depth of plowing in order to provide planted trees with conditions to develop deep roots so they can better adapt to prevailing wind conditions. Deep tine plowing is done on ironpan, peaty, and fragioglev soils with hard indurated layers. The plow tines to a depth of about 90 cm and has a moldboard that turns out a furrow up to 60 cm deep (Figure 15.2).

Soviet Union foresters have done extensive research and development on machinery for seedbed preparation, although site preparation is generally not carried out unless a humus layer is present. In cut-over areas on sandy soils and with little logging debris, preparation may consist of a double moldboard-plowed furrow or a rototilled or disced strip. Seeding is sometimes done in conjunction with plowing. In hilly terrain soil preparation is related to the degree of the slope, with total preparation on slopes up to 8°. On slopes of 8 to 12°, terracing and plowed strips are used. On steeper slopes bulldozers may be used to construct terraces if gullies are present, but in the absence of gullies, terracing is done with special plows or bulldozers (Post, 1974).

New Zealand and Japan also have special problems in preparing sites in

**Figure 15.2** Deep tine-plowed peaty-gley ironpan soils in a pastureland afforestation project in southeast Scotland. Sitka spruce is planted on resulting beds.

mountainous areas. In New Zealand the main preparation for planting bare-rooted exotic species is the removal of vegetation rather than soil cultivation. However, manipulating the soil by ripping, rotary hoeing, and discing in order to give a better rooting medium for trees and for better control of weeds after planting have been used.

For the regeneration of silver beech *(Nothofagus menziesii)* in New Zealand the ground is scarified before logging to provide a good seedbed to obtain some advance growth. However, in the podocarp forests logging operations are generally relied on to provide scarification. Radiata pine forests are clear-cut, the slash either burned or windrowed, and then the soil is ripped. Large gravity rollers are sometimes used in mountainous areas too steep for preparation by tractor equipment.

In the Pacific Northwest, the intermountain areas, and the great plains of United States and Canada, the variety of mechanical techniques and equipment for soil preparation covers the range available, according to Post (1974). On small areas, manual tools are still used to prepare for planting, but on larger areas a variety of plows and harrows are used. Environmental concerns, mainly aesthetics, have recently had a considerable influence on soil preparation techniques. Terracing and strip preparation on steep slopes have met with disfavor due to their appearance and possible adverse effects on overland flow of water to water-deficient areas. Spot preparation with bulldozers or backhoes may be acceptable alternatives.

The techniques for soil preparation for artificial regeneration of radiata pine in Australia, Scots and maritime pines of coastal England and southwest France, and the intensive culture of southern pines in the United States are quite similar. These relatively level and often sandy areas permit the use of heavy equipment for such site preparation operations as chopping, discing, bedding, and ditching. In Australia, Waring (1973) reported that early growth of radiata pine is very dependent on a range of site preparation techniques and degrees of weed control. He considered both the extent and persistence of a fertilizer response to be affected by these factors. It is the usual practice to cultivate thoroughly between windrows subsequent to burning in species conversion operations. This facilitates other operations, particularly planting, and improves the survival and growth of the planted stock.

**Chopping.**  This operation generally consists of passing a heavy (up to 20 tons) drum roller over the soil surface one or more times. Debris and vegetation, including small trees, are cut by sharp parallel blades attached along the length of the drum. Chopping concentrates the vegetation on the soil surface and facilitates burning as a disposal tool. Moreover, chopping may be the only soil preparation needed, particularly where slash and vegetation are not heavy and benefits can accrue from minimal soil disturbance. Because only a small part of the chopped vegetation and debris is incorporated into the mineral soil, this organic material acts as a mulch on the soil surface. It retards erosion, moderates soil temperature fluctuations, and reduces soil moisture losses. Chopping appears to be a practical means of conserving soil nutrients and is particularly useful on dry, sandy, and problem soils (Figure 15.3).

Double chopping with a well-timed interval was quite effective in reducing scrub oak competition in slash pine plantation establishment in the sandhills of southeastern United States (Burns and Hebb, 1972). After 15 years, survival of planted pine was 36 percent greater in chopped areas than in areas receiving no site preparation and tree heights and diameters were 133 and 137 percent greater. Treatments that removed vegetation and debris from the site (raking or blading) improved survival over that of the control (no site preparation or burned only) plots, but height of trees in those plots averaged only 58 to 82 percent of tree heights in double chopped plots.

While mound bedding is not generally recommended for dry sandy soils, because of a reduction in available moisture in the beds, some other type of intensive site preparation is often required in order to reduce competing vegetation. In South Australia, Woods (1976) reported that preparation of podzolized aeolian sands consisted of four cultivations over a two-year period prior to

**Figure 15.3** Site preparation with a rolling drum chopper on a dry site (courtesy Weyerhaeuser).

planting of radiata pine. The first two disc plowings were done 12 months apart and to a depth of 25 cm in order to uproot and expose the root system of the competing sclerophyllous vegetation. The third and fourth plowings were shallow cultivation with a twin disc harrow aimed at keeping the vegetation under control and destroying any invading surface weeds. Woods (1976) found that "multi-cultivation induced more rapid mineralization of the highly lignified sclerophyll-type organic matter which increased available nutrients and, thus, encouraged the invasion of annual or perennial herbaceous plants." This type of weed could then be controlled by amitrole-atrazine herbicides and resulted in better survival and early growth or "get-away" of the young pines.

**Harrowing and Plowing.** Soil cultivation achieved by harrowing differs from that of plowing mainly in depth and degree of disturbance of the soil. Harrowing is generally accomplished by multiple discs drawn through the surface soil that displace but do not necessarily invert the disc slice. Plowing can be done by either disc and moldboard plows. The latter is most often used on difficult sites.

They both can be effective in reducing competition from vegetation that develops a dense root mat just below the soil surface, such as saw palmetto *(Serenoa repens)*, wiregrass *(Aristida stricta)*, gallberry *(Ilex glabra)*, and runner oak *(Quercus pumila)*, according to Haines, Maki, and Sanderford (1975).

Blading or rootraking may have to be done prior to harrowing where particularly large quantities of logging wastes, stumps, or woody vegetation abound. Harrowing and plowing may result in a temporary immobilization of soil nutrients due to incorporation of large amounts of carbonaceous materials. However, upon decomposition of these materials nutrients are mobilized and some nutrients may be lost from the site, if not soon absorbed by vegetation.

Soil preparation is particularly beneficial in the establishment of hardwood plantations. Because these plantings are often made on rather productive soils, weed competition may be a problem. For example, sycamore height and survival were improved by disc harrowing prior to planting, but harrowing followed by two cultivations during the first year after planting improved growth by 190 percent over controls (Huppuch, 1960). Poplar plantation establishment in Yugoslavia and Italy begins with complete soil cultivation to a depth of about 60 cm (Post, 1974). A reduction in competition by weedy species for nutrients and moisture is not only essential for vigorous early growth of most hardwoods, but it also increases the efficiency of applied fertilizers.

**Bedding.** Bedding, or mound discing, is a standard soil preparation practice in many coastal and wet areas. Bedding concentrates surface soil, litter, and logging debris into a ridge 15 to 20 cm high and 1 to 2 m wide at the base. Planting on beds may improve early height growth of conifers and hardwoods on a variety of sites. The beneficial effects of bedding have been attributed to improved drainage; improved microsite environment for tree roots in regard to nutrients, aeration, temperature, and moisture; and reduced competition from weedy species (Haines et al., 1975).

On wet sites, growth and survival response from bedding derive primarily from removal of excess surface water. Surface drainage results in accelerated decomposition of organic matter and mineralization of nutrients. Where the A1 horizon is thin and the B horizon is either impervious to roots or nutrient poor, the concentration of topsoil, humus, and litter provides more favorable conditions for root development and, subsequently, top growth, according to Haines et al. (1975). They reported on the effects of chopped debris (0, 73, 145, and 290 tons per hectare placed in mounds of screened surface soil) on height growth of slash pine in the coastal flatwoods. Adding organic debris to the 15 cm high mounds significantly increased tree growth over that obtained where no debris was added. However, incorporation of debris in excess of 73 tons per hectare, the approximate amount left after logging, resulted in no additional improvement after four years.

Improvements in tree growth in response to different degrees of soil preparation depends largely on soil properties, site conditions, and the tree species involved. For example, a reduction in competing vegetation for soil moisture may be critical for survival and early growth on moderately dry sites, while improved surface drainage brought about by bedding may improve tree growth on wet sites. Scalping can be highly beneficial to germination and seedling survival on sites with thick humus layers, but it often results in poor seedling growth on dry and impoverished sites because the removal of organic matter from the root zone decreases available nutrients.

Worst (1964) reported an average of 36 percent greater height of slash pine on prepared plots than on nonprepared plots on coastal plain flatwood soils with varying degrees of natural drainage, after four years. Scalping the soil resulted in no better height growth than where the soils were not mechanically prepared, regardless of the natural soil drainage. Trees planted on low beds or in areas that were burned-harrowed were significantly higher than trees in scalped or nonprepared areas. After 17 years merchantable volumes of trees in burned-bedded and burned-harrowed plots were significantly greater than those of trees in control or burned-scalped plots on all soils. Flat harrowing resulted in greater volume than bedding on a well-drained soil, but bedding was the superior treatment for the poorly drained soil. There were only small volume differences between the flat harrowing and bedding treatments on the moderately well-drained site. Similar results were reported by Mann and McGilvray (1974) for site preparation techniques on Grossarenic Paleaquults varying in natural drainage (Figure 15.4).

There have been a number of other reports of improved tree growth from harrowing, bedding, and other methods of intensive site preparation in the southern pine region (Haines and Pritchett, 1964; Derr and Mann, 1970; Klawitter, 1970a), for maritime pine in southwest France (Sallenave, 1969), and for radiata pine in Australia (Woods, 1976). Benefits to early tree growth were generally attributed to improved soil moisture and aeration, deeper rooting, increased available nutrients, and a reduction in competing vegetation. Bedding on the contour has also been used to reduce erosion in hilly terrain (Haines et al., 1975). However, growth response to bedding and most other soil preparation techniques probably results from a combination of factors, including an increase in the rate of organic matter decomposition and supply of available nutrients, as evidenced by an experiment on a coastal wet savanna (Typic Albaquult) soil, described by Pritchett and Smith (1974).

This west Florida test combined three types of site preparation with varying rates of mineral fertilizers (Table 15.1). In this test, flat harrowing reduced ground vegetation, but it resulted in only slightly better tree growth than the control. Competition for moisture was not a critical factor on this wet site. Bedding, on the other hand, resulted in a 98 percent increase in tree heights and a 74 percent increase in total volume after 9 years of growth on unfertilized plots.

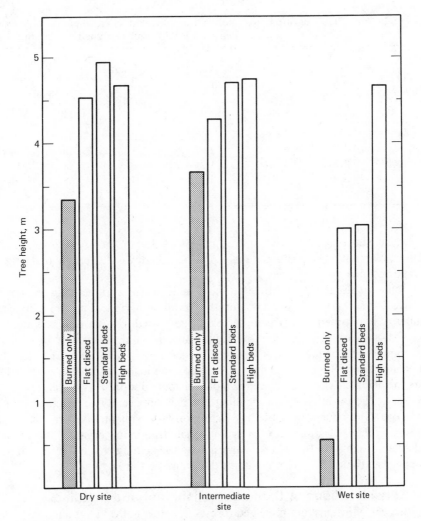

**Figure 15.4** Height of eight-year-old slash pines as influenced by seedbed treatment and soil moisture conditions (Mann and McGilvray, 1974).

The response to the fertilizer applications indicated that while part of the growth improvement from bedding derived from improved nutrition an equally important benefit came from a reduction in competition and improved surface drainage.

Wilde and Voigt (1967) reported that 14 years after white pine seedlings were planted in furrow bottoms in clay soils in Wisconsin survival was less than 10 percent and tree heights averaged only 1.4 m. Planting on scalped soil resulted in

**Table 15.1**  The Effects of Site Preparation and Fertilization on Height and Volume of Slash Pine after Nine Years on a Typic Albaquult (Pritchett and Smith, 1974)

| Nutrients Added | | Site Preparation Technique | | | | | |
|---|---|---|---|---|---|---|---|
| | | Nonprepared | | Harrowed | | Bedded | |
| N | P | Height | Volume | Height | Volume | Height | Volume |
| kg/ha | | m | m³/ha | m | m³/ha | m | m³/ha |
| 0 | 0 | 2.17 | 1.94 | 2.87 | 3.57 | 4.30 | 8.85 |
| 90 | 0 | 2.07 | 1.60 | 3.26 | 4.63 | 4.88 | 12.82 |
| 0 | 90 | 5.95 | 14.39 | 5.89 | 20.33 | 7.66 | 49.80 |
| 90 | 90 | 6.74 | 21.43 | 7.02 | 32.68 | 7.29 | 37.28 |

52 percent survival and average tree height of 3.2 m, but best results were obtained by planting on beds. On these plots survival was 80 percent and heights averaged 5.2 m.

Haines et al. (1975) reported that the beneficial effects of bedding, except for improved survival, may not last for the rotation period. They mentioned a number of experiments in which the early tree growth response to bedding had largely disappeared by the time trees were 10 years old. They found that slash pine planted on a bedded flatwood site averaged 1.83 m taller than trees on a flat disced area at age 8.5 years. Height differences diminished thereafter until at 14.5 years the height advantage of bedded plots was only one-third as large as it was at 8.5 years. The duration of the bedding effect is apparently determined by site conditions. The benefit may be long lasting on Histosols and on very wet sites, but where the advantage results from concentration of organic matter (and available nutrients) in the bed for early access to seedling roots, the benefit may disappear as the root system occupies the entire soil area.

## Soil Considerations in Stand Establishment and Tending

The success of intensively managed forests depends not only on adequate slash and vegetation disposal and soil preparation but also on vigorous planting stock of the correct tree species, provenance or clone, planted or seeded at the correct time and spacing, and weeded, thinned, and fertilized, if needed. The considerable investment involved in plantation establishment makes it imperative that all reasonable efforts be made to assure good tree survival and growth.

The selection of a species for plantation establishment should take into consideration its resistance to disease and insects and its commercial importance, but most importantly, the selection should be made on the basis of its adaptation to local soil and site conditions. There are usually several climatically adapted species to choose from for any given region and it is important for silviculturists to choose the species best suited to local conditions.

Problems have sometimes been encountered in adapting exotic species to local soil and site conditions. These problems have included various micronutrient deficiencies, lack of mycorrhizae and poor root development, and local soil moisture conditions. However, these problems are probably no greater than those encountered by shifting native species out of their ecological range or planting them "off site" within their range.

Nutrient-deficiency symptoms are seldom expressed in species growing in their native habitat, since natural selection has resulted in strains tolerant to local conditions. However, damage to the tree root systems by machines, diseases, or insects or some localized soil conditions, such as high pH or poor drainage, may interfere with nutrient absorption and result in symptoms in native species similar to those sometimes encountered in exotic species. Furthermore, species growing near the limit of their ecological range are more exacting in their soil requirements than when growing near the center of their range. This probably reflects their inability to compete successfully with local species under the former conditions. For example, black spruce grows well over a wide variety of soil conditions in the boreal forest of central Canada, but it is only found in wet swampy areas at the southern edge of its range in New England.

Selection of tree species tolerant to toxicants is essential in spoil bank afforestation projects. Often legumes or nonleguminous dinitrogen fixers have an advantage under these conditions. It also appears to be important in some areas to select species for afforestation that are tolerant to air pollutants.

*Genetically* improved stock are increasingly used for planting in short-rotation plantations in most major wood-producing countries. There are active genetic programs for hybridizing poplar in France, Italy, and the United States; for tree improvement through selection of superior strains for *Eucalyptus* spp. and *Pinus radiata* in Australia and New Zealand; for *Picea abies* and *Pinus sylvestris* in Scandinavian countries; for Douglas-fir in the Pacifc Nothwest; and for southern pine in southeastern United States. A large percentage of the approximately 300 thousand hectares of southern pine planted annually are of improved stock (Killison, 1975).

Trees are selected in genetic improvement programs on the basis of one or more desirable inherited characteristics. These include such features as straightness of stem, growth rate, specific gravity of wood, and resistance to disease or insects. These phenotypically superior trees are evaluated by planting their progeny in test sites covering a variety of soil and site characteristics. Genetic improvement, however, by no meaans provides a panacea for all soil-site problems. For example, Pritchett and Goddard (1967) reported progeny of selections that grow best on high fertility soils did not necessarily grow as rapidly on soils of low fertility as ordinary seedlings of the same species.

**Direct Seeding.** The term *plantation forest* is often restricted to even-aged forest established on prepared sites by planting nursery-grown seedlings. However, the term is more correctly applied to any artificially regenerated domesticated forest. This latter use would include direct seeding, as well as planting, on sites with varying degrees of seedbed preparation. The choice of operational method depends on terrain condition, labor costs, and machinery availability.

According to Appelroth (1974), approximately one-fifth of the plantation forests worldwide are established by direct seeding. This technique is used for regenerating both native and exotic species and it is particularly useful in wet and rugged terrain and other areas with such limited accessibility that ground machines can be used only with difficulty. Manual broadcast seeding is used in a number of countries, particularly for establishing hardwoods. Sites for seeding are often burned, harrowed, or strip plowed and the seed dispersed by hand or with a hand-operated cyclone seeder. Row and spot seeding are two modifications of the manual seeding techniques.

Terrain-crossing machines are also used for direct seeding in many countries practicing plantation forestry. Although broadcast seeding by surface machines is used mainly in Canada, Spain, and the United States, row and spot seeding by machines are quite widely used in these and many other countries. Probably the most widely used method of direct seeding is accomplished by helicopter or fixed-wing aircraft. Seeding from the air generally results in a broadcast pattern, but row seeding by aircraft has been tested (Appelroth, 1974).

Seeds are often given a protective coating to discourage consumption by rodents and birds and attack by pathogenic fungi. Pelletizing seeds with a coating of nutrient-enriched clay to facilitate sowing and to assist in early seedling establishment in infertile soils has been tried with such species as *Eucalyptus* and *Pinus*. However, success with pelletization of seeds has not been uniformly achieved. Salt concentration from small amounts of soluble fertilizers can interfere with germination.

Securing a uniform and acceptable stand density is one of the major problems associated with direct seeding. When germination and survival are good, the stand is likely to be so dense as to require early and extensive thinning; but, on the other hand, where survival is poor, the stand may be both thin and spotty, resulting in difficulties in tending and harvesting and in low yields. Spot scalping, strip harrowing, and bedding have been used with broadcast seeding as means of obtaining more uniform stands. Some results from seeding southern pines are shown in Table 15.2.

Direct seeding of exotic species in areas where they have not previously been grown sometimes results in poor inoculation with mycorrhizal fungi because of the absence of suitable sources of mycorrhizal fungi in the soil. Inoculation of nursery-grown seedlings is a relatively simple and straightforward process of treating the nursery soil with spores or mycelia of appropriate fungi prior to sowing, but direct seeding in new areas may require the introduction to those areas of small amounts of topsoil from a forest or coating the tree seeds with inoculum before seeding.

**Table 15.2**  Stocking, Height, and Diameters of Two Southern Pines, 10 Years After Seeding by Different Methods (Lohrey, 1974)

| Seedbed and Seeding Method | Loblolly Pine | | | Slash Pine | | |
|---|---|---|---|---|---|---|
| | Stems | Height | Diameter | Stems | Height | Diameter |
| | no./ha | m | cm | no./ha | m | cm |
| Broadcast sown, grass rough | 4744 | 7.9 | 12.2 | 2380 | 8.4 | 14.0 |
| Planted, grass rough | 2014 | 9.4 | 15.5 | 1470 | 8.6 | 14.5 |
| Strip sown, flat disced | 4826 | 8.8 | 13.2 | 2224 | 8.7 | 14.2 |
| Strip sown, bedded | 3657 | 9.2 | 14.7 | 1994 | 8.3 | 13.7 |

**Planting.**  Seedlings for plantation establishment can be broadly divided into (1) bare-root, nursery-grown stock, and (2) container-grown plants. Some advantages and disadvantages of each are discussed in Chapter 16. Methods of planting are somewhat similiar for both types of plants.

Manual planting methods are used extensively in most countries for establishing plantations (Figure 15.5). Of the manual planting tools, the mattock or hoe is used in the majority of countries. It is particularly useful for planting on steep terrain. Shovels or spades and dibbles, bars, and wedges are other tools for manual planting. Most are used for planting bare-root seedling but they can be adapted for planting container-grown stock. These tools generally open V-shaped slits about 20 cm deep in the soil into which the seedling roots are placed. The seedling roots are normally trimmed to correspond to the depth of the slit and are then placed in the slit slightly deeper than the root collar, but in a manner to insure a minimum of root folding at the bottom. The slit is closed by moving the wedge of soil back to its original position by a thrust of the tool and the planters heel. In gravelly or compacted soils that resist penetration of the planting bar, particular care must be taken to plant seedlings sufficiently deep to assure good survival. Furthermore, air pockets may be left at the bottoms of closed slits in fine-textured soils, which can result in excessive drying of the root system. Therefore, special care has to be taken when planting in difficult terrain to insure that slits are opened sufficiently deep and that they are well closed after planting.

Where only small numbers of trees are to be planted, or if the trees are particularly valuable as in a seed orchard or landscape planting, hole planting may be justified. This method may also be used on adverse or infertile sites where the seedlings may need additional nutrients or water during the early stages of growth. With this method, a hole is opened with a shovel or power equipment so that it is slightly larger than the extended roots of the seedling and compost or fertilizer materials can be mixed with the soil before it is packed around the roots.

Machines are the principal means of plantation establishment in Australia,

Canada, Russia, the United States, and other countries with sizable areas to plant. Manually carried (or pushed) powered augers are used as the major method of machine planting in Switzerland and are occasionally used in Czechoslovakia, Romania, and Yugoslavia (Appelroth, 1974). These machines are useful in hilly and stony terrain and where deep planting holes are needed. In accessible and less precipitous terrain, tractor-drawn tree planting machines of varying designs are extensively used. Planting machines usually consist of a plow for opening a planting slit, two discs for closing the slit, and two rubber-tired wheels for support of the operator and for packing the soil around the seedlings. Machines are constructed to plant one or two rows and they may have attachments to apply fertilizers or herbicides. A two-row planter for container-grown *Pinus caribaea* is used in Venezuela (Figure 15.6).

**Initial Spacings in Plantation.** The initial spacing used when planting forest crops have a considerable influence on various aspects of stand establishment as well as affecting later stages of crop development (Low and Tol, 1974). Use of wide initial spacings reduces the number of plants needed and planting costs per unit area and may reduce the need for soil preparation work. In addition, precommercial thinnings may be partially or completely avoided, and there is greater accessibility for machinery. On the other hand, Low and Tol (1974) pointed out some silvicultural advantages in using close initial spacings in plantations. These include early canopy closure, more rapid branch suppression, greater opportunity for crop improvement by selective cleaning, and less critical need for high initial survival.

Wilde et al. (1965) reported that optimum spacing depends largely on the nature of the soil, its productive capacity, and the actual or potential biomass of the weed cover. The poorer the soil, the wider should be tree spacing; the higher the soil fertility, the denser should be the initial stocking. Actual spacing of planted trees is also influenced by age of planting stock and method of site preparation. Mann and Dell (1971) estimated yields of unthinned 17-year-old loblolly pine planted at a variety of spacings over a range of site quality. Some of these estimates are given in Table 15.3.

There is a wide range in initial spacings used in plantation forestry throughout the world. However, these differences do not appear to be related to climatic zones or to tree species. There is as much variation in spacing among countries within a climatic region as among countries with very different climates. Spacings for conifers do not differ greatly from spacings used for broodleaf species. They vary from as close as 1 m by 1 m for oak and Scots pine in Austria and West Germany to as much as 3 m by 3 m, or wider, for teak and some eucalypts in India and radiata pine in New Zealand. An average spacing of about 2.5 m in each direction fits many situations. Rectangular spacing patterns, such as 1.5 m by 3 m, appears to be favored where there is a need for machinery access.

**Figure 15.5** Hand planting Douglas-fir seedling in the Pacific Northwest (courtesy Weyerhaeuser).

**Figure 15.6**  Two-row planter for container-grown *Pinus caribaea* in *llanos* of Venezuela.

Containerized seedlings are planted at wider spacing then bare-rooted seedlings because they generally have higher survival rates.

Although there is a general trend toward wider spacings in plantations as a result of increasing labor costs and lower profits for handling small timber sizes, this trend may be reversed with increased mechanization of tending operations. In the final analysis, spacing is primarily based on site quality with the lowest densities used on infertile and less productive sites, and on the ultimate use of the forest product. Trees grown for pulp are planted at much closer spacings than those grown for timber, unless thinning is to be part of the management scheme.

**Cleaning and Thinning.**   Cleaning refers to weeding the stand of undesirable vegetation, including a reduction in the number of unmerchantable stems where needed for better stand development. Thinning, while a type of cleaning, is more accurately applied to the removal of stems of marketable size.

**Stand Cleaning.**   Good site preparation is probably the most important factor in reducing weedy competition in the young stand, but even in well-prepared sites some cleaning and precommercial thinning may be required (Woods, 1976). The objectives of the cleaning operations are to reduce competition in the stand through which the development of the residual trees is encouraged; remove

**Table 15.3** Estimated Yields (outside bark) of 17-year-old Loblolly Pine Planted at Several Spacings on Soils of Varying Site Quality (Mann and Dell, 1971)

| Trees Planted per Hectare | Site Index (height in meters at 50 years) | | | |
|---|---|---|---|---|
| | 21.4 | 24.4 | 27.4 | 30.5 |
| no. | Volume, m³/ha | | | |
| 988 | 77.7 | 106.7 | 144.8 | 180.6 |
| 1483 | 97.6 | 135.2 | 172.9 | 211.6 |
| 1977 | 101.0 | 143.2 | 184.5 | 226.0 |
| 2471 | 95.3 | 142.2 | 187.0 | 231.5 |
| 2965 | 83.9 | 135.4 | 183.8 | 231.1 |

inferior trees; encourage desired species; produce suitable spacing; and increase stand resistance to storm damage. Cleaning in some form is used in most plantations, except in the cultivation of some fast-growing species on well-prepared sites that are used solely for pulpwood production.

Cleaning is widely practiced in many European countries and in Japan, Australia, and New Zealand by both mechanical and chemical means. For example, it is used in broad-leaved forests to reduce competition for nutrients and to allow quality development of the stands. Carter and White (1971) reported that cultivation increased nitrogen, phosphorus, and potassium concentrations in foliage of young cottonwood by 53 to 77 percent over that of noncultivated trees. Spruce stands are cleaned extensively to increase the girth and resistance to snow damage. The number of stems are reduced from a range of 4000 to 8000 per hectare to 2000 to 3000 per hectare by selective thinning or a combination of selective and row thinning. In Great Britain, the most important type of cleaning is annual weed control during two to three years after planting. The vegetation cleared is heather *(Calluna vulgaris)* on poor soils; grasses and bracken on more fertile soils; and grasses, bracken, and broad-leaved trees on lowland sites. The successful establishment of exotic forests in Australia and New Zealand is heavily dependent on the control of unwanted vegetation. Postplanting weed control is generally accomplished with herbicides applied by aircraft. The use of domestic livestock to improve stand access had also been tried in New Zealand (Hägglund, 1974), but animals can cause compaction of fine-textured soil.

Chemicals are more commonly used in countries where forest labor is scarce and expensive than in countries with low labor costs, but there are also a number of biological and environmental factors that influence their use. Herbicides are inactivated more rapidly in warm and moist climates than in cold climates. This tends to make them less effective, but by the same token the short inactivation time in warm climates makes it possible to use some chemicals there that could not be used in cold climates where they may remain in the environment for excessive periods. There are also indirect effects resulting from the use of herbicides that include the risk of insect attack when cleaning conifers in the selection-reduction stage. In a release stage when less valuable hardwood species are overtopping the crop species, a chemical treatment has the advantage of leaving the killed trees as a shelter and this also avoids blocking the ground with felled stems. Furthermore, the chemically killed, but standing, trees may have a stabilizing effect against snow and wind damage in dense stands (Hägglund, 1974). In mountainous terrain, chemical methods appear to be the only alternatives to manual cleaning.

Nominal competition from annual weeds is not necessarily damaging to the survival and growth of young pines planted on wet sites. A reduction in soil moisture that may result from transpirational losses by the ground vegetation can be beneficial to these poorly-drained sites. Annual vegetation may also help immobilize soluble nutrients, applied as fertilizers or those becoming available through mineralization, that might otherwise be lost before seedling root systems are sufficiently large to make effective use of them. However, competition from weeds can be disasterous to both hardwood and conifer planting on dry sites. Herbicides probably hold their greatest advantage under these conditions (Figure 15.7). For example, an application of simizine at about 5 kg per hectare (active ingredient) improved survival and early growth of radiata pine planted on well-drained sands in South Australia (Woods, 1976). Triazines are used to control weeds in young conifer plantations in Ontario and in afforestation of former agricultural lands to spruce and pine in Finland and Scotland. The reduction in heather vegetation by the use of herbicides or cultivation is required for satisfactory establishment of Sitka spruce on organic soils in Great Britain. The ericaceous vegetation may interfere with mycorrhizal root development on the spruce seedlings.

Chemicals are not widely used for plantation cleaning in North America because they have not always proven effective and because of the environmental consequences of their use. Herbicide use is likely to be confined to problem areas until more potent materials with sharper selectivity are developed and methods of application are devised to reduce handling and environmental hazards. In spite of these hazards, the use of chemicals may be the only feasible way of controlling vegetative competition under some conditions and the environmental risks appear to be acceptable for such chemicals as auxins, applied with due care. Examples of special application techniques include use of tree injectors for cleaning pine plantations of upland hardwoods, and the coating of simazine on fertilizer granules for localized control of weeds that might otherwise be stimulated by the fertilizer.

Fire is also used in some countries for the suppression of hardwoods in older

**Figure 15.7**   A localized application of herbicide to control grassy vegetation resulted in an early growth advantage for slash pine (right) on this moderately well-drained loamy sand.

pine plantations. Most notable, perhaps, is the use of prescription fire in southern pine plantations in the United States. Once pines have gained sufficient height so that crowns are relatively safe from scorch, controlled fires can be used periodically to suppress the more shade tolerant, but less fire resistant, hardwoods.

**Stand Thinning.**   Thinning is an intermediate cut in an immature forest stand between the time of its formation and the first regeneration cutting. It is made to enhance the value of the existing stand by controlling the composition, density, and growth through all stages of stand development.

Thinnings differ from regeneration cuttings in that no effort is made to secure regeneration and the creation of permanent openings in the crown is avoided. Commercial thinnings can be used in the management of uneven-aged stands as well as even-aged stands. However, the need for intermediate cuttings is more easily recognized and they are more economically applied in even-aged stands than in uneven-aged stands. The residual stocking to be retained after each cyclic thinning, in terms of volume or basal area, varies somewhat with production goals and with site quality. Thinning may produce a larger return on poor sites prone to stagnation than on better sites. However, cutting mature stands back to 50 to 60 percent of full stocking appears adequate for most stands.

Thinning has been used to enhance the response to fertilizers and other

**Table 15.4**   Average Basal Area Growth per Tree on Clay Loam Soil During 4 Years After Treatment of 30-year-old Site IV Douglas-Fir in Oregon (Miller and Williamson, 1974)

| Treatment | Unfertilized | Fertilized | Average |
|---|---|---|---|
| | $cm^2$ | $cm^2$ | $cm^2$ |
| Unthinned | 101.4 | 159.0 | 130.2 |
| Thinned | 155.3 | 196.2 | 175.8 |
| Average | 128.3 | 177.6 | 153.0 |

management practices applied to mid-aged stands (Table 15.4). The removal of intermediate and suppressed trees insures that any growth response is placed on prime trees. In short rotation plantations, economics will ultimately determine whether commercial thinning is a viable operational practice. Increased demands for lumber will make it more attractive to remove less desirable stems in pulpwood thinnings and grow the residual stems for timber, especially on the better sites and where fertilized. Both precommercial and commercial (pulpwood) thinnings are common in timber management of the Pacific Northwest.

From a biological standpoint, risk of windthrow can be reduced by wide spacings or heavy initial cleanings that encourage long and regular crowns; whereas during the second half of the rotation crown closure should be maintained by light low thinnings. Snow breakage in dense spruce stands can apparently be reduced by heavy thinnings, but strip or corridor thinnings may increase risk of windthrow. Heavy thinning can also change the micro-climatic conditions in the stand in that it reduces precipitation interception, results in greater soil moisture and higher temperatures with greater fluctuations. These conditions increase organic matter decomposition and nutrient mobilization with the nutrients shared by fewer trees. The improved nutrient conditions result in larger crowns and increased tree growth. The use of heavy machinery in strip thinning can also result in soil compaction and root and stem damage to the residual stand (Oswald, 1974).

**Other Management Practices.**   Irrigation, drainage, and fertilization are other soil management practices associated with domesticated forests. However, irrigation is largely confined to tree nurseries, seed orchards, and other high-value crops. Irrigation of large plantations is not economically feasible at the present time. Drainage and fertilization of planted forests, on the other hand, are viable options on wet and nutrient deficient sites. Because these operations are used only under special conditions, they are discussed in separate chapters. There are, no doubt, a variety of other special operations carried out in plantations to correct local problems.

It must be concluded that not all practices associated with intensive forest management are universally acceptable. In spite of the admitted appeal of domesticated forests as a means of satisfying the increasing demands for forest products, forest managers must not lose sight of the fact that their increased yields are often bought at the expense of soil physical, chemical, or biological properties, increased energy consumption, and a possible degrading of the environment. Some of these factors are discussed in Chapter 23.

# 16

# MANAGEMENT OF NURSERY AND SEED ORCHARD SOILS

The production of quality seed from genetically improved trees and the propagation in nurseries of sturdy planting stock is crucial to the successful operation of plantation forestry. Adequate supplies of seeds and seedlings are of prime concern in the regeneration of managed forests. While the total land area devoted to nurseries and seed orchards is very small compared to the millions of hectares of forest lands, the investment per unit area devoted to these operations is quite large. Furthermore, responses to improved soil management techniques in terms of yield and quality of seedlings can be particularly rewarding.

## NURSERY SOIL MANAGEMENT

The successful operation of a forest nursery involves large investments in labor, buildings, and equipment and only an operation sufficiently large to make effective use of the specialized machinery can normally be justified. Except in special situations where part of the equipment may be rented or else made to serve double duty, a nursery of 5 to 10 ha might be considered a minimum size for an economic operation. A nursery of this size can produce from 10 to 20 million 1-0 pine seedlings per year, providing seedlings for 5000 to 10,000 ha of plantations. A substantially larger area is needed if a rotation with a green manure or cover crop is followed, and in cool climates where two or more years are needed to produce seedlings of satisfactory size for field planting. Armson and Sadreika (1974) presented excellent instructions for nursery soil management for northern coniferous species.

Over 600 million seedlings were produced in state nurseries in 1976 while some 250 million seedlings were grown in privately owned nurseries in the

United States. Figure 16.1 shows a view of a pine nursery. The demand for both hardwood and coniferous seedlings will probably increase to meet the need for expanded areas of plantation forestry.

## Nursery Site Selection

Because nurseries are not easily moved from one site to another, it is especially important that much thought be given to soil and site factors that make for efficient operations in the production of quality seedlings. Soil depth and texture are of prime importance for nurseries. A soil depth of 1.0 to 1.5 m is normally desired, without a radical textural change between horizons, in order to insure adequate drainage and aeration and space for root development.

Areas exposed to extreme temperatures and strong winds should be avoided. In cold climates, location near large bodies of water will moderate temperature extremes. On the other hand, terrain should be selected that permits free drainage of cold air and thus reduces frost injury. The nursery site should be located as near as feasible to major areas of reforestation for convenience and to minimize differences in length of growing season and other climatic conditions between the two areas.

**Soil Texture.** Soil texture influences water intake, cation exchange capacity and nutrient retention, susceptibility to sand splash, surface washing, compaction, and ease of lifting. Loamy sands to sandy loams are generally preferred textures for nursery soils. Coarse-textured sands are acceptable for nurseries, but such textures place special demands on management of water, fertility, and organic matter. Soils at the other extreme of the textural range—those containing more than 30 to 40 percent silt plus clay fractions—are to be avoided. Soils containing a high percentage of colloidal materials are often difficult to cultivate and they require considerable effort to adjust moisture, pH, and nutrient levels. They are also slow to warm in the spring and, in cold climates, seedlings may suffer from frost heaving. Furthermore, seedling root systems are more likely to be injured during lifting from fine-textured soils than from sands.

**Leveling.** Level terrain is preferred for nurseries on sandy soils, but a slight slope may be necessary to provide surface drainage for finer-textured soils. Sufficient grading and leveling should be accomplished to minimize erosion of beds, translocation of fertilizer salts, and accumulation of water in depressions. On the other hand, the exposure of large areas of subsoil by grading should be avoided, especially where the movement of topsoil will result in shallow depths over clay or other impervious layers, or where the subsoil is calcareous. Where extensive areas of topsoil must be removed during leveling operations, care should be taken to stockpile the topsoil and redistribute it over areas that were exposed during grading. In the event such redistribution is not possible, a special

**Figure 16.1**  View of a pine nursery with a capacity of about 25 million seedlings per year.

effort should be made to increase the organic matter content of the disturbed soil before an attempt is made at seedling production.

A detailed soils map of the nursery area, including any reserve area for later expansion, should be made as soon as possible after selection. The map should indicate soil texture in both surface and subsurface horizons, depth to restricted drainage, pH, and possible nutrient levels and other properties critical to seedling growth. For example, hardwood seedlings generally require more fertile and less acid soils than conifer seedlings, so that if both hardwoods and conifers are grown in the same nursery, it may be possible to reserve the more fertile areas for hardwoods. Furthermore, a detailed map makes a convenient base for planning and record keeping—two essential steps in any nursery operation.

**Soil Fertility and Reaction (pH).**  A reasonable concentration of nutrients in the original soil is desired but, as they can be added as fertilizers, the soil reserve of nutrients is not of overriding importance. Of greater concern is that the soil not contain high concentrations of soluble salts, free carbonates or toxic materials, and that it contains a reasonable level of organic matter.

The optimum range for most tree species is between pH 5.2 and 6.2. The lower half of this range appears to be best for conifers, while many deciduous trees grow well in the range of pH 5.8 to 6.2. Although many species will thrive at a lower soil pH, the efficiency of fertilizer use is decreased in strongly acid soils. The amounts of native calcium, magnesium, and potassium are low in very acid soils, while added potassium is more easily leached and the availability of added phosphorus is reduced under these conditions. Aluminum, iron, and manganese tend to be much more soluble in acid soil. Normal high concentrations of these elements probably have little direct effect on conifer growth, but they may well affect some deciduous seedlings and the cover or green manure crops grown in rotation with the seedlings.

At the other end of the pH range, near neutrality, some species have difficulty obtaining sufficient iron and, sometimes, manganese for normal growth. Pine seedlings are particularly subject to iron deficiencies in soils above about pH 6.5. Most tree species, including pines, can be grown at reactions above pH 7.0 if their nutritional requirements for iron, manganese, and phosphorus can be met with chelates or foliar sprays. In fact, they can be successfully grown in some finer-textured soils near neutrality without micronutrient additions if the soils are well supplied with micronutrients and organic matter. Figure 16.2 illustrates the relationship of soil acidity to weight of red pine seedlings.

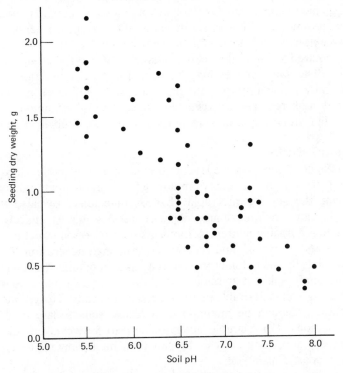

**Figure 16.2**  Relationship of soil acidity to mean total dry weight of 2+0 red pine seedlings (Armson and Sadreika, 1974). Courtesy, Ontario Ministry of Natural Resources.

Some root diseases are more prevalent in near-neutral and alkaline soils and such soils should be avoided for nursery sites, if possible. If irrigation is anticipated, the water should be checked for soluble salts. For example, 25 cm of water containing 500 ppm of soluble salts adds 1250 kg per hectare of these materials to the soil. This is sufficient salt to significantly affect the pH of weakly buffered sandy soils.

Simple test kits and pH meters are useful for checking soil reaction, but it is well to keep in mind that changes in soil acidity of as much as 0.5 to 1.0 pH units can take place due to seasonal variations. Furthermore, microsite differences due to leveling or organic matter applications often result in wide variation in pH among soil samples. Where adjustments in soil acidity are needed, additions of agricultural limestone or sulfur are usually made. However, small changes in soil acidity can be made by using fertilizer sources that have acid or basic residues. For example, the application of 200 kg nitrogen per hectare as ammonium sulfate can lower the pH of sandy soil as much as 0.5 units during a season.

Care should be exercised in using limestone or sulfur to alter the acidity of sandy soils to prevent excessive change in reaction. The kinds and amounts of clay and organic matter control the soil cation exchange capacity, which in turn determines the amount of lime or sulfur required to change soil acidity by a given amount. Generalized curves useful in determining the amount of agricultural limestone required to lower the acidity of some soil textural groups are given in Figure 16.3. The same curves may be used for predicting amounts of sulfur required to raise soil acidity from the present to a desired level. About one-third as much sulfur is required to lower soil reaction a given amount as that of agricultural limestone required to raise the reaction an equivalent amount.

## Soil Organic Matter

The maintenance of a reasonable level of organic matter is particularly important in sandy nursery soils as a means of retaining fertilizer elements against leaching and buffering the soil against rapid changes in acidity. Because these soils contain little clay, their exchange capacity resides almost entirely with the organic fraction. Organic matter additions also improve soil structure, friability, water intake and retention, and reduce soil crusting and erosion. The organic matter of a soil is the seat of microbial activity and it contains the reserve of most nutrients in sandy soils. A minimum of 2 to 4 percent organic matter in the surface 20 cm of soil is desirable in even the sandiest soils. The practical level of organic matter that can be maintained in sandy soils is largely dictated by climatic conditions. For example, maintaining 4 to 5 percent organic matter in Oregon nursery soils may be less difficult than maintaining 1 to 2 percent in southeastern coastal plain soils.

While soil organic matter can be increased by use of green manure crops and the additions of composts, sawdust, peat, and other organic material, such

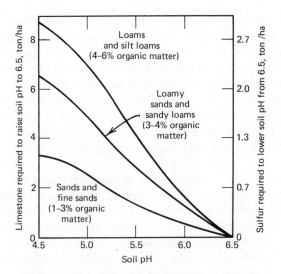

**Figure 16.3** Tons of agricultural limestone required to lower soil acidity or tons of sulfur required to increase soil acidity. Read differences in requirements corresponding to differences between present and desired pH.

increases are temporary due to the decomposition of these materials by soil organisms and they must be added on a regular schedule.

**Green Manure and Cover Crops.**   Nursery seedlings are often grown in rotation with green manure and cover crops in attempts to replace organic matter lost by the high biological activity associated with well aerated and fertilized nurseries. For example, each one percent of organic matter in the top 20 cm of a sandy soil is equivalent to 24 to 28 tons of dry matter per hectare. Therefore, a nursery soil may normally contain 50 or 60 tons of organic matter (dry). While a good green manure crop may add 20 to 30 tons of fresh material to the soil, only 5 to 6 tons is dry matter. Moreover, when mixed into the soil 50 to 75 percent of this material may be decomposed within a few months, leaving only 1 or 2 tons of relatively stable humus to add to the soil reserve. This is little more than that required to replace normal decomposition losses during a cropping season. While the benefits from green manure organic materials are rather shortlived, the use of green manure is, nonetheless, the most practical means of maintaining productivity of many nursery soils.

Rotations with green manure crops can also be an effective means of reducing infestations of nematodes and soil-borne diseases that often infest nursery soils, provided crops are selected that are resistant to the growth and development of these organisms. A cover crop may also be selected that will result in an increase in the population of certain endomycorrhizal fungi useful to the following hardwood seedlings. The selection of a suitable green manure crop should be further considered on the basis of soil and climatic conditions. For example, nursery soils are often too acid for good growth of many leguminous crops. Some species of vetch, lupine, soybeans, and field peas may be used, but in order to obtain rapid plant growth and abundant production of organic material, the nursery worker may have to resort to nonleguminous crops. Oats, rye, millet, sorghum, and corn have been used for this purpose in various regions.

Cover crop should be planted to protect the soil from wind and water erosion and to reduce nutrient leaching during any extended period that an area is not used for the production of nursery stock. A cover crop should be plowed under before it reaches maturity so as to avoid the nuisance of a volunteer crop in later nursery beds. However, the crop should be allowed to grow as long as feasible because of the increased dry matter production and because mature and lignified materials do not decompose as rapidly after turning.

**Composts and Peat.**   Composted and raw organic materials such as sawdust, straw, forest litter, and peat have long been important sources of materials for improving soil physical conditions and increasing soil organic matter. Since they are largely waste materials, they have been relatively inexpensive to obtain but costly to handle. More recently the supply of these materials has diminished to the point where they are used principally on nursery soils with critically low levels of organic matter or to correct problem areas, such as those resulting from leveling operations.

Composted materials consist of sawdust, straw and other materials of wide carbon-nitrogen ratio mixed with top soil and/or mineral fertilizers and allowed to partially decompose or ferment. During the fermentation process, $CO_2$ is evolved, the carbon-nitrogen ratio is decreased, and the mineral constituents are made more available. Such composted materials can be applied at fairly heavy rates of 40 to 60 tons per hectare and incorporated into the seedbed without danger of significant nitrogen immobilization. There is a danger of introduction of root disease pathogens with some composted materials. Fumigation of the materials prior to application, or of the nursery soils after application, will minimize this problem. Moreover, composted materials are quite expensive to use and raw organic materials and peats are the more common amendments for nursery soils. Care must be exercised in the use of raw organic materials to prevent excessive nutrient immobilization during critical periods.

Relatively inert materials, such as acid peats, fresh sawdust, and straw are slow to decompose and, thus, persist in the soil for longer periods. While they are less expensive to use than composted materials, they can result in nitrogen shortages at critical periods in seedling development, and more caution is required in their use. Allison (1965) reported that the nitrogen requirements of microorganisms that decomposed sawdust ranged from 0.3 percent of the dry weight of Douglas-fir wood to 1.4 percent for red oak. The corresponding value for wheat straw was 1.7 percent. These nitrogen requirement values serve as a reasonable quantitative estimate of the readiness with which the various woods are attacked biologically.

When raw sawdust, or other highly carbonaceous materials, are to be used directly in nursery soils, they should be applied sufficiently in advance of the seeding operation so that partial activation, or decomposition, takes place. These applications may be made prior to planting of a cover crop or green manure crop or up to a month prior to seeding the nursery beds. In any event, 6 to 12 kg of extra nitrogen may be needed per ton of dry organic materials added in order to hasten decomposition and minimize danger of a nitrogen shortage to the nursery crop. Since this extra nitrogen will become available to nursery stock after the added organic material has largely decomposed, care must be taken to prevent excessive levels of nitrogen in the plant bed at later stages of seedling development. This can be done by monitoring the concentration of nitrogen in seedling tissue, or available nitrogen in the soil, and making supplemental applications of nitrogen only when results indicate a need.

**Mulches.**   Mulches are applied on the surface of nursery beds to protect seeds or plants from erosive action of wind and water, to prevent soil puddling and crusting, to ameliorate frost heaving, and to reduce water evaporation from the soil (Armson and Sadreika, 1974). Gravel- sized (3 to 5 mm) grit mulches have been used in some northern nurseries, but organic materials such as straw, sawdust, and moss are more commonly used. Wood fiber mulches have been used with considerable success in many areas. They give good erosion protection and do not have to be removed at germination time.

## Fertilization Program
While the additions of composted materials, manures, and nonactivated organics are acceptable means of maintaining soil organic matter levels that contribute to good physical properties, they generally do not contain sufficient nutrients to replenish those lost during cropping. Consequently, the maintenance of nursery soil fertility depends largely on the use of commercial fertilizers, applied alone or in combination with organic materials and/or green manure crops.

**Crop Removal.**   Relatively heavy rates of fertilizers are often used in nurseries to replace those removed in the harvested crop plus those lost by leaching or fixed in the soil. Leaching losses can be rather high in irrigated sandy soils, even where organic matter content is maintained at a reasonable level. Crop removal may also be substantial because both roots and tops are removed. For example, slash pine seedlings in a Florida nursery, grown at the rate of 2.2 million plants per hectare, weighed 16,500 kg (green weight) when harvested at eight months of age. The moisture content of tops and roots averaged 71.6 and 70.5 percent, respectively. Oven dry weights averaged 3715 and 1012 kg per hectare of seedling tops and roots, respectively, giving a ratio of 3.7:1. The nutrient concentrations and total nutrient uptake by the seedlings are shown in Table 16.1. Armson and Sadreika (1974) reported larger removal of nutrients in two-year-old red pine and spruce than in slash pine (Table 16.2).

**Soil Testing.**   The additions of relatively large amounts of organic materials and mineral fertilizers, and the removal of nutrients in the harvest crop and as leaching losses, result in considerable fluctuation in nutrient levels in nursery soils during a cropping period. A soil test provides information on the fertility status of soils that is particularly useful prior to seedbed preparation. Soil acidity and nutrient status are normally adjusted prior to seeding the tree crop. Since plants are not available for tissue tests at this time, soil tests are the principal diagnostic aid. Soil samples should be collected from seed beds four to six weeks prior to seeding so as to allow adequate time for sample analyses and the purchase and application of any necessary fertilizers.

A composite soil sample should be taken from each nursery management unit or from areas within a management unit that differ in soil type, past management, or productivity. A composite sample is generally composed of 12 to 15 subsamples (cores) taken at random over a uniform area. (Samples should be taken within beds in a bedded area). Cores are normally taken to plow depth (approximately 20 cm) with a 2.5-cm-diameter soil tube and collected in a clean container. After air drying and mixing, about 250 ml are packaged, labeled, and submitted to a laboratory for testing. The location from which the sample was collected should be identified from a map of the nursery on which each manage-ment unit is designated. A detailed soils map can serve as a basis for record keeping as well as management planning. Chronological records of soil test results, fertilizer treatments, fumigation, seeding date and rate, irrigation, and other cropping practices are extremely useful to the manager in diagnosing problem areas and planning for future crops.

**Base Applications.**   Limestone or sulfur, when used for the purpose of adjusting soil acidity, and the base application of fertilizers are normally applied broadcast before fumigation (if any) and seeding or transplanting. In this way, the amend-ments can be well mixed throughout the rooting zone. If limestone or sulfur are needed, they should be mixed with the soil two or three months in advance of seeding, so as to allow time for these materials to react with soil components. On the other hand, most fertilizer materials are applied near time of seeding (or transplanting) so as to minimize the danger of leaching losses.

**Table 16.1** Nutrient Concentrations and Total Uptake per Hectare in Eight-month-old Slash Pine Seedlings

| Component | N | P | K | Ca | Mg | Al | Cu | Fe | Mn | Zn |
|---|---|---|---|---|---|---|---|---|---|---|
| | | | | | | Nutrient concentrations | | | | |
| | | | % | | | | | ppm | | |
| Tops | 1.20 | 0.15 | 0.60 | 0.43 | 0.09 | 408 | 4.8 | 118 | 161 | 48 |
| Roots | 0.86 | 0.23 | 0.59 | 0.36 | 0.08 | 1974 | 10.8 | 617 | 66 | 40 |
| | | | | | Nutrients Removed in Harvested Seedlings (kg/ha) | | | | | |
| Tops | 44.6 | 5.6 | 22.3 | 16.0 | 3.3 | 1.5 | 0.18 | 0.44 | 0.60 | 0.18 |
| Roots | 8.7 | 2.3 | 6.0 | 3.6 | 0.8 | 2.0 | 0.01 | 0.62 | 0.06 | 0.04 |
| Total | 53.3 | 7.9 | 28.3 | 19.6 | 4.1 | 3.5 | 0.19 | 1.06 | 0.66 | 0.22 |

**Table 16.2** Amounts of Nitrogen, Phosphorus, and Potassium in Kilograms per Hectare Contained in Two-year-old Red Pine and White Spruce at Four Seedbed Densities (Armson and Sadreika, 1974)[a]

| Nutrient | Seedbed Density (plants/m$^2$) | | | |
|---|---|---|---|---|
| | 108 | 215 | 430 | 861 |
| | Red Pine | | | |
| N | 113 | 119 | 164 | 220 |
| P | 16 | 22 | 27 | 35 |
| K | 56 | 80 | 109 | 133 |
| | White Spruce | | | |
| N | 88 | 107 | 150 | 167 |
| P | 19 | 24 | 31 | 41 |
| K | 44 | 60 | 77 | 114 |

[a]Courtesy, Ontario Ministry of Natural Resources.

Phosphorus deficiencies are generally corrected by the application of 40 to 100 kg phosphorus per hectare as ordinary or concentrated superphosphate. The actual amount needed depends on soil fixing or retention properties and the degree of deficiency, as indicated by soil test. In acid sandy soils, it is often more economical to apply larger amounts of less soluble forms of phosphorus, such as ground rock phosphate or basic slag. Phosphorus in these materials is less soluble and less likely to leach beyond the seedlings' shallow root system than that in soluble phosphates. Rates of 500 to 2000 kg per hectare are normally applied at one- to three-year intervals.

**Topdressings.** Nitrogen and potassium needs may be met by mixing part of the material into the seedbed prior to planting and applying the remainder in two or more topdress applications during the growing season. Split applications are particularly effective for nitrogen because of the ease with which it can be lost from the system and due to the high salt index of most nitrogen sources. Soluble fertilizer salts may result in damage to transplants when present in high concentrations. Usually no more than 30 to 50 kg of mineral nitrogen are applied to sandy soils prior to seeding. Higher rates may be used on fine-textured soils or when less soluble sources, such as ureaformaldehyde or isobutylidene diurea (IBDU), are used.

Most sources of nitrogen and potassium are completely water soluble and their solutions can be injected into the irrigation system for direct application, or solid materials can be applied as a topdress to the seedlings. The number of topdress applications needed depends on soil conditions and rainfall and irrigation patterns. However, three or four applications per year are not uncommon for sandy soils.

Potassium is sometimes topdressed on seedlings a few months prior to lifting from the nursery bed in an effort to harden and condition them to withstand cold temperatures and the transplanting shock.

## Tree Conditioning for Transplanting

Conditioning seedlings to withstand better the shock of transplanting and, thereby, improving survival and early growth is sometimes accomplished by slowing their growth rate in the months prior to lifting. Hardening through a reduction in growth rate may be accomplished by an induced drought following a decrease in irrigation. It has also been attempted by altering available nutrient levels or balance in the nursery bed. This generally involves a reduction in the level of available nitrogen and/or an increase in the potassium or micronutrient concentration in the soil solution. However, conditioning is more commonly achieved by root manipulations in the seed bed.

*Undercutting, wrenching,* and *root pruning* are widely used methods of conditioning seedlings of certain species. When properly scheduled and conducted, they can be means of controlling top-root ratio, carbohydrate reserve accumulation, and the onset of dormancy. A reduction in top-root ratio may improve success of plantings on droughty soils. Increased food reserves and early dormancy generally reduce cold damage and insure better survival on adverse sites, but, most importantly, an efficient root system is the key to success where survival is marginal due to moisture stress.

*Undercutting* consists of passing a thin, flat, sharp blade beneath seedlings at depths of 15 to 20 cm. Undercutting is usually performed only once in a bed of seedlings several months before lifting. The purpose of the operation is to sever seedling tap roots, reduce height growth, and promote development of fibrous root systems. It has been successfully used in nurseries for pines and other taprooted species.

*Wrenching* was described in detail by Dorsser and Rook (1972) as a method of conditioning and improving forest seedlings for outplanting. Their research on wrenching was largely confined to radiata pine, but the technique has been used on other species, including hardwood seedlings. Wrenching can be performed with a sharp spade inserted beneath the seedlings at an angle which severs their tap roots and then slightly lifting to aerate the root zone. However, it is usually accomplished by passing a sharp, thin, tilted blade beneath seedbeds at certain intervals following undercutting, to prevent renewal of deep rooting or height growth.

Timing of the conditioning sequence is important and it varies with tree species and soil and climatic factors. The sequence generally begins when seedlings are near the height desired for outplanting, but with at least two months of growing season remaining. With radiata pine in New Zealand, the operation begins when seedlings are about 20 cm in height. A thin, flat blade is passed beneath seedlings at a depth of about 8 to 10 cm with minimum disturbance to lateral roots. After undercutting, wrenching is performed at one- to four-week intervals for the remainder of the seedlings' growing season, using a thicker, broader blade, tilted 20° from horizontal, with front edge lower than the rear (Dorsser and Rook, 1972). The interval between wrenchings is dictated by the type of seedling desired. Frequent wrenching apparently produces seedlings with lower top-root ratios and slightly higher starch levels than seedlings not undercut or wrenched. Wrenching at monthly intervals causes a smaller decrease in top-root ratio and higher levels of starch and reducing sugars than wrenching every one or two weeks, according to the New Zealand research.

*Root pruning* is sometimes used to describe the undercutting operation outlined above. It has also been used to describe root trimming operations performed after seedlings have been removed from the ground. However, in New Zealand, and other countries where wrenching is commonly practiced, root pruning is accomplished by vertical cutting with a coulterlike blade between rows of seedlings at four- to six-week intervals. The vigorous lateral root growth initiated after the tap root is severed is partially controlled by cutting the roots 8 to 9 cm from the trees. Wrenching and vertical root pruning are often used with slow-growing conifer species, such as spruce and northern pines, in order to produce 2 + 0 seedlings with root area index equal to 1 + 1 or 2 + 1 transplants (Armson and Sadreika, 1974).

Conditioning by manipulations of the root system before outplanting can increase seedling tolerance to root exposure and extended cold storage. This will produce an efficient root system that supports high transpiration rates and reduces internal water deficits, thus increasing the probability of survival. While the degree of success from these operations apparently varies among species, as well as soil and climatic conditions, it is influenced by the method and timing of the conditioning operations and seedling density. Dorsser and Rook (1972) reported that to respond fully to the undercutting-wrenched treatment, radiata pine seedlings must be at least 5 cm apart in rows 17.5 cm apart. This is less than half the density at which many other conifer seedlings are grown in nursery beds.

## Biocides for Nursery Soils

Nursery crops are often grown as a single species of densely populated, rapidly growing succulent plants. Consequently, they are particularly vulnerable to the ravages of diseases, insects, and competing vegetation. Many nursery pests can be controlled, or their damage minimized, by manipulating the soil environment to insure sturdy, vigorous seedlings. For example, soil reaction should be kept below pH 6.0 so as to discourage the development of most pathogenic fungi.

Damping-off injury of very young seedlings, commonly associated with *Pythium, Rhizoctonia, Phytophthora, Fusarium,* and *Sclerotium* fungi is rarely serious if the nursery soil is maintained at pH 5.5 or lower. High soil nitrogen concentrations at time of germination should also be avoided. Low-to-moderate plant density and adequate moisture combined with good drainage of the surface soil generally result in optimum seedling growth without favoring disease and other pests. Rotating tree seedlings with green manure and cover crops that are less susceptible to attack of certain disease organisms, insects, and nematodes is an accepted method of reducing the pest problem. However in older nurseries, pests often cannot be controlled by soil management techniques alone and biocides must be employed.

Preplanting soil fumigation that controls most weeds, as well as soil-borne diseases, nematodes, and soil-inhabiting insects, is a common and economically feasible practice in tree nurseries of many parts of the world. In addition to controlling damping-off organisms, fumigation with such compounds as methyl bromide, ethylene dibromide, bichloropropene, and chloropicrin (250 to 500 kg per hectare) controls most root rots, probably the most destructive of all nursery soil pathogens. A variety of fungi are associated with root rots, but most can be controlled by methyl bromide, provided care is taken to insure that the fumigant is applied under soil conditions that maximize effectiveness.

Fumigants are applied as gases under plastic covers or injected into the soil as a solution. In either event, best results are obtained if the chemicals are applied when surface soil temperatures are between 10 and 30°C, when soil moisture is adequate for good seed germination, and when the soil has been recently cultivated to insure good penetration. Penetration is not as good in fine-textured or compacted soils as in sands and sandy loams, and organic matter reduces the effectiveness of fumigants by sorption and degradation. Therefore higher than normal rates may be needed on fine-textured soils with high levels of organic matter.

In addition to broad spectrum fumigants, herbicides such as triazine compounds, and fungicides may be applied against specific pests. Some nurseries may require multiple applications of these pesticides. All biocides alter the soil flora and/or fauna to a greater or lesser extent—sometimes rather dramatically and for an extended period. The temporary imbalance of the soil microbiology can result from the elimination of beneficial organisms as well as pests. Most desirable microorganisms reestablish themselves in the soil within a few weeks after fumigation. However, mycorrhizal fungi may not recolonize nursery soils in some regions for several months or even years. This is particularly true of endomycorrhizal fungi that produce soil-borne, rather than air-borne spores. Such conditions are more likely to be found in cold climates or where exotic species have been recently introduced. The time-honored methods for inoculating seed beds involve the mixing of a small amount of soil from an unfumigated nursery or the use of forest litter as a protective mulch. Because these methods may introduce pests into the nursery, the use of a commercial inoculum should be used when available.

Research in Australia (Bowen, 1965) and the United States (Marx, 1977) indicates that a more beneficial method of reinoculation of fumigated soils is by the use of inoculum of selected fungi known to have the potential for improving survival and growth of trees on a variety of adverse and routine reforestation sites. *Pisolithus tinctorius,* an ectomycorrhizal fungus of many conifers, oaks, and eucalyptus, shows considerable promise in nursery use for tailoring roots of tree seedlings for artificial regeneration programs, according to Marx (1977). The use of seedlings inoculated with selected fungi may be particularly beneficial for reforestation of coal mine spoils, as illustrated in Table 11.2.

## Container-grown Seedlings

Container-grown trees and woody ornamentals have been produced in modest quantities for many years in most countries throughout the world. However, only in the last decade have major research and development commitments and large investments in production facilities for mass production of container-grown seedlings been made in North America (Stein, Edwards, and Tinus, 1975). While only about 3 percent of the nearly one billion trees produced throughout the United States in 1973 for forest and windbarrier purposes were grown in containers, this method of seedling production is becoming widely accepted in certain areas. As might be expected, the popularity and use of container-grown stock varies among regions, because the relative advantage of this more labor-intensive system over the production of bare-root seedlings depends largely on shock tolerance of the species used, on climatic and soil conditions, and on planting methods.

**Areas of Greatest Use.**   Containers have their greatest appeal in areas where the growing seasons are relatively short or where species are grown that are particularly sensitive to bare-root handling. Container-grown Norway spruce and Scots pine seedlings comprise a relatively large percentage of trees planted in Scandinavian countries. In the Pacific Northwest of the United States and Canada, container production amounts to 15 to 20 percent of the total tree seedling production. In this region, Douglas-fir, western hemlock, spruces, some true firs, and a few pines are container-grown in large numbers, and some other conifers and hardwoods are produced in smaller quantities (Stein et al., 1975). Eucalypts and Caribbean pine seedlings are container-grown in most tropical and subtropical countries where these species are used in plantation forestry.

**Some Advantages of Containers.**   In cold climates where two or more years are required to produce plantable stock in nursery beds, the time interval may be shortened to a few months by the use of containers. The system also offers greater flexibility in the production and outplanting of seedlings in many areas. For example, container-grown stock may be planted at times when bare-rooted plants are not available or properly conditioned. Furthermore, lengthening the planting season permits the use of a smaller, more stable work force and better use of costly equipment (Stein et al., 1975).

The use of container-grown seedlings often results in better survival and early growth, particularly in regions where success with bare-root stock is marginal. Certain species are particularly difficult to grow in nursery beds or else they are sensitive to bare-root handling. Hemlocks, firs, spruces, oaks, and eucalypts are examples of forest trees that are best produced as container stock. The use of containers generally avoids the shock that bare-rooted stock experiences by losing part of its root system during lifting and the loss of vigor during protracted storage or desiccation during handling and planting.

Container-grown stock is often more uniform in size and quality than seedlings grown in nursery beds. It is also subject to better protection from pests and more controlled growing conditions. In some localities, container-grown seedlings provide a greater opportunity for mechanization (Figure 15.6) and for easier planting among stones and harvest residue.

**Other Considerations.** Successful use of containers in seedling production generally requires a higher level of technical knowledge and greater attention to daily conditions of the growing stock than that for bare-root nursery stock. Since roots are confined to a limited volume of substrate in containers, water and nutrient conditions are subject to more rapid fluctuations than are nursery soils. A greenhouse or shade may also be required for protection of container stock. Therefore, the total cost of acquiring the planting medium, seeding, and care of the seedlings is likely to be greater than for the production of a comparable number of trees in nursery beds. As a consequence, the system may continue to be limited in its appeal to those areas where it has special advantages.

Seeds may be germinated in flats or beds and then transplanted to containers when plants are a few weeks old; or seeds may be planted directly in the containers in which the seedling is to be grown for transplanting to the field. The latter system is less labor intensive and commonly used with most container-grown species. Some types of containers dictate this system of seeding.

**Types of Containers.** Containers may be grouped into three categories: tubes, blocks, and plugs. Tubes can be constructed of either biodegradable or nondegradable plastics, or of kraft or other paper. Tubes require filling with a soil mixture or other growth medium. Some control in the degradation rate of the material from which the tubes are constructed is important to the health of the seedling and for handling purposes.

Blocks are similar in shape and size to tubes, but they have no outer wall and require no filling. The block is both the container and the planting medium, and seeds are sown directly in the block and the entire package is later transplanted into the soil. They are molded from bonded softwood pulp, polyurethane foam, peat, peat-vermiculture mixtures, or similar materials in which nutrients may be incorporated. Blocks have given excellent results under many conditions, but, unless produced locally, freight cost can be prohibitive.

Plugs consist of seedlings grown in soil-filled molds, but, unlike tube- or block-grown seedlings, they must be removed from their containers before outplanting. Since the growth medium is bound only by the seedling roots, the plug can be rather fragile and not easily planted by machines.

In spite of the several disadvantages of the various systems of growing seedlings in containers, the technique has great appeal and should continue to grow in popularity in those areas where bare-root seedlings are difficult to manage. However, because of high labor costs, use of containers is not likely to replace conventional nurseries for pines and other species which lend themselves to bare-root planting.

**Greenhouses.**   Plastic greenhouse and slat shades are standard fixtures in the production of container-grown seedlings in many areas. These structures afford protection to seedlings from climatic extremes and they are used to extend the growing season. Growing trees in beds under plastic cover has also given good results in northern Europe, and especially in Finland where the practice is extensive (Phipps, 1973). Reported benefits are better seed germination and seedling survival, increased seedling growth rate, and fewer disease and weed problems. However, the use of greenhouses for nursery-grown seedlings appears to hold few advantages over standard nursery methods in temperate climates and even fewer in warm climates.

## SOIL MANAGEMENT FOR SEED PRODUCTION

Seeds are requisite to most reforestation projects. The exceptions are some hardwoods, such as *Populus* and *Platanus* species, which are more easily propagated by vegetative cuttings. Historically, seeds for direct seeding and nursery operations have been collected from conveniently located forest stands, as the seeds matured and dropped to the ground. Often these seeds were produced by less desirable trees that were residuals from previous harvesting operations. Seeds may be more conveniently collected from the branches of recently harvested trees than from a forest stand. Collecting from felled trees is almost essential for some conifers with serotinous cones. This system may also permit more selectivity as to source, but the opportunity for selecting superior trees as seed sources is often compromised for the sake of expediency. Consequently, seeds collected from the wild are often of questionable genetic quality.

## Seed Production Areas

The rapid expansion of plantation forestry has resulted in greatly increased demands for quality seed. In some areas, early efforts to meet this demand prompted the designation of selected stands as seed production areas. These areas were developed in attempts to secure rapid increases in seed production through management. Mature stands on high-quality sites were generally selected for management as seed production areas. The stands were then thinned to 60 to 75 trees per hectare, with phenotypically superior trees selected for seed trees. Trees on good sites generally produce more seeds than those on low-quality sites, but since site quality is not genetically controlled, progeny of trees on high-quality sites may be no better than those from low-quality sites. However, selection of uniformly straight, disease-free trees as seed sources is especially important.

Seed production areas are generally fertilized in accordance with soil conditions and species requirements, using soil or tissue tests as a guide. A complete (NPK) fertilizer is often used at rates of 5 to 10 kg per tree per year, broadcast beneath tree canopies. However, an application of nitrogen fertilizer made near the time of initiation of flower primordia is particularly important. This time varies among species, but for most species it falls between early and late spring. Management of seed production areas may also include programs for the control of understory vegetation and disease and insect pests. Proper management of these areas can significantly improve the yield and quality of seed and convenience of seed harvest, as compared to collecting from commercial forest stands, but seed production areas are largely used to bridge the time gap until seed orchards, composed of planted areas of superior trees, can be developed. Seed orchards are an essential part of successful programs in genetic improvement of forest trees.

## Seed Orchards

Both coniferous and hardwood seed orchards are established with seedlings or vegetatively propagated material obtained from parents that transmit desirable growth and wood quality characteristics to their progeny. The use of grafts conserves time and allows for a more efficient operation, because flowering and seed production on the physiologically mature portion (scion) of the graft commences within a few years after consumation. Similar production from trees raised from seedlings requires an additional 7 to 10 years for most species (Kellison, 1976).

Trees in an established seed orchard must be evaluated to determine their genetic worth. This is necessary because parent trees were originally selected on their phenotypic superiortiy (physical appearance), but the phenotype is influenced by both the environment and its genetic makeup. In order to separate and evaluate the genotypic influence, progeny of the trees are tested under selected environmental conditions. Progeny from known parental combinations are out-planted into monitored field trials so that the genotypic value of the parents can be determined under site conditions similar to those anticipated for future plantings. Parent trees found unacceptable by progeny tests are rogued from the orchard, leaving only the best to cross-pollinate for the production of genetically improved seed.

Seed orchards of some type have been established in almost all countries with substantial areas of plantation forestry. The objective of seed orchard programs is quantity production of superior seed, but few, if any, countries produce sufficient seed of improved genetic quality to supply their own needs. Hence it is important that the orchards be made as productive as possible. While the factors responsible for flower initiation are not fully understood, many techniques have been successfully used to increase flower production of various tree species. They include such treatments as thinning, pruning, banding or strangulation, girdling, root pruning, subsoiling, irrigation, and fertilization. Not all of these techniques are effective on all species and some of them, such as strangulation and girdling, may destroy the tree. Since seed orchards are expensive and laborious to establish and maintain, drastic measures for inducing flowering should be avoided in favor of the desirable soil management practices.

## Seed Orchard Soil Management

Seeds are the primary product of seed orchards and regardless of their outstanding volume, form, or wood quality, the selected trees must produce adequate seed for future generations. Cultural practices should be employed that ensure that the selected trees produce seeds abundantly.

**Site Selection.**    Seed orchards are costly to install and are rather permanent once installed. Therefore, considerable planning and care should be exercised in choosing a location. The orchard should be located on productive soils within the species natural range (or similar climatic zone in the case of exotics) and, if possible, convenient to an associated tree nursery or forest operation activity. Deep, well-drained loamy soils are generally selected for orchards, but in the final analysis the area selected should be a high-site type for the tree species involved. Steep slopes should be avoided because of the difficulty of maneuvering machinery and the danger of erosion.

The size of the area selected for seed orchard establishment is dictated largely by the anticipated volume of seed needed to achieve the silvicultural objectives within prescribed time limits. Calculations of annual seed needs must be tempered by the known periodicity or seeding habit of the species, plus an anticipated reduction in seed-producing trees due to roguing. For cross-pollinated species, isolation zones may be needed to restrict contamination. Chances of contamination from outside pollen are minimized by maintaining a buffer strip around the perimeter of the orchard. This zone of 150 to 200 m or more may consist of open area or it may be planted to tree species that will not pollinate the orchard trees.

**Ground Cover.** A grassy or leguminous cover is established in most orchard areas, even before trees are planted. This vegetative cover helps stabilize the soils, reduces erosion, and improves trafficability during wet periods. A low-growing pasture grass suited to local conditions is often used for cover, because such a grass can be maintained in a vigorous condition to resist encroachment of less desirable weed species. Mowing and occasional fertilization may be required to keep the grass in vigorous condition. If the cut grass is removed, additional fertilizer should be used to replace those nutrients taken off in the crop. Leguminous plants are sometimes used as an extra source of nitrogen, but the establishment of a good leguminous cover is sometimes difficult without liming the soil above the level needed for good tree growth.

Orchards on level, well-drained soils can be maintained without a ground cover. However, such systems are not widely used because they require that mechanical or chemical weed-control treatments be used periodically. Discing for weed control will prune tree roots and it may increase erosion hazards, result in damage from root rot, and reduce tree vigor. Root pruning has been noted to increase flowering in the years immediately following the operation, but it may reduce flowering in later years.

**Soil Fertility Maintenance.** Fertilizer programs for seed orchards may involve two separate operations to achieve two distinct objectives. First, a broadcast application of a complete (NPK) fertilizer is normally used to promote, or at least maintain, a good growth of ground cover. This application can be made in early spring, or else the fertilizer can be split and one half applied in the spring and the other in the summer. A soil or tissue test should be used as a guide as to rate and ratio of nutrients to use for the cover crop. In the absence of a test, an application of 400 to 600 kg per hectare per year of 20-10-10 ($N$-$P_2O_5$-$K_2O$) fertilizer (or equivalent) is often used for grass cover, while an equal amount of 0-10-20 fertilizer may be used for a leguminous crop.

Second, a separate application of fertilizer is often applied around individual trees to promote flowering. For most pine and some other conifers and hardwood species this application consists of nitrogen fertilizer. The rate of the application depends largely on the size or age of the trees, with about 0.5 to 2.0 kg nitrogen per tree per year applied around the periphery of young trees until they reach full flowering and up to twice this amount used for mature trees. Nitrogen is generally applied only once per year at a time to coincide with the initiation of flower primordia. Schmidtling (1975) reported that loblolly pine increased female flowering with progressively later nitrogen fertilizer applications until it peaked

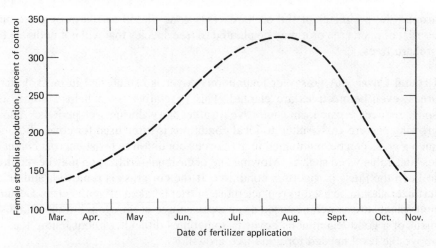

**Figure 16.4** Female strobilus production in young loblolly pine grafts in response to timing of fertilizer application. Dashed line represents the generalized response for three experiments (Schmidtling, 1975).

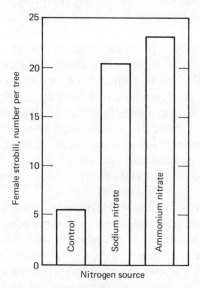

**Figure 16.5** Flowering response to different nitrogen fertilizers in young loblolly pine grafts (Schmidtling, 1975).

with the August treatment at about 300 percent over the untreated trees (Figure 16.4). He explained the results on the basis of growth and carbohydrate accumulation in relation to time of flower formation. Fertilizers should be applied early enough to increase flower initiation, or prevent abortion of already formed primordia, but late enough to minimize uptake by competitors or diversion of nutrients to growth rather than flowering.

In experiments dealing with the effects of nitrogen sources on female flowering of loblolly pine, Schmidtling (1975) found ammonium sources were as good as nitrate sources and that ammonium nitrate was as good or better than any other source tested (Figure 16.5). However, there are indications that nitrate sources may be more effective than ammonium sources for eliciting flowering response in Douglas-fir (Ebell, 1972).

Using irrigation in seed orchards to increase seed yields has met with mixed success. Irrigation used to supplement rainfall during drought periods generally does not stimulate flower production. In fact, a soil moisture deficit during the period prior to strobili initiation may favorably affect flowering. The moisture stress results in increased carbohydrate levels that appear to trigger differentiation of reproductive tissue. However, irrigation during the period of ovule fertilization may increase the number of sound seeds per cone (Schultz, 1971). It may also reduce conelet abortion and result in larger cones and heavier seeds if applied during the second and third years after flower initiation. Nevertheless, irrigation is used only on a limited scale in seed orchards because the increase in yield and quality of seed is often not sufficient to justify the added costs.

# 17

## MANAGEMENT OF PROBLEM SOILS

A relatively high percentage of the earth's forests are located on sites that have little value for other plant types. While the soils of most of these sites are unsuited for agricultural purposes, many are, or can be, highly productive forest lands. The vast majority of soils of better forest sites require no special management techniques for continued high productivity. But when disturbed or adverse sites are planted to trees in an effort to ameliorate soil conditions and improve aesthetic values special management techniques are usually required. The afforestation of spoil banks resulting from mining operations, seasonally flooded areas and peatlands, moving sand dunes, and old cultivated fields degraded by erosion can not be accomplished without some special effort. Some of the soils are nutrient deficient and can be brought into reasonably high productivity through the addition of fertilizers. The improvement of these latter soils will be discussed in some detail in chapters dealing with forest fertilization. Soils of other adverse sites may contain toxic substances or lack mycorrhizal fungi, while some are too dry or too wet for good tree growth. The one thing that all problem sites have in common is the need for additional management efforts in order to obtain acceptable stand establishment and tree growth.

### MANAGEMENT OF DRY SANDS
Rainfall exerts a major influence on the worldwide distribution of forests. But even in regions with ample rainfall there are areas where available soil moisture is too low for good tree growth. These areas are mostly deep sands with minimal water-holding capacity or else they are shallow soils with low moisture storage

capacity. Dry sands that are also exposed to strong winds offer a problem in dune stabilization.

## Dune Stabilization

The stabilization of moving sands is a special problem in certain coastal and inland areas when marine or fluvioglacial deposits of sand are disturbed. Blow sands that have required special afforestation effort are found in France, Holland, North Island of New Zealand, and Michigan, as well as some coastal and dry sandy areas on most continents. Dune fixation by tree planting was underway in Europe by the middle of the eighteenth century and in America by the early part of the nineteenth century (Lehotsky, 1972). One of the early attempts at dune stabilization in America was on Cape Cod (Baker, 1906), and other projects have been undertaken in Michigan (Lehotsky, 1972) and Cape Hatteras. Some of the sand areas, such as those in Michigan, became unstable and began to move following timber removal and farming.

Sand dune fixation requires the establishment of a permanent plant cover. The cover directs or lifts the prevailing wind currents away from the easily eroded sandy surface. This feat usually cannot be accomplished directly by planting trees. The young trees simply do not afford sufficient break to alter the wind currents and, furthermore, unprotected seedlings are often damaged by the abrasive sands. Consequently, blow sands are first temporarily stabilized by such devices as parallel rows of snow fence, closely spaced stakes driven into the sand, surface trash, plantings of drought-resistant grasses, or a combination of these techniques. Distance between rows of stakes or grass depends upon the intensity of erosion, but may be no more than 1 to 2 m.

Beachgrass *(Ammophila arenaria)* and other xerophytic clump grasses have been used successfully in stabilization efforts, but they normally cannot be established in infertile sands without the application of a mixed fertilizer. In New Zealand (Gadgil, 1971a), nitrogen fertilizer was used to effect establishment of a bunch grass (Marram grass) as a preliminary step in stabilizing blow sand. After about two years, lupine was broadcast seeded without additional fertilizer. In some instances the population of a leaf-eating caterpillar (Kowhai moth) increased in numbers to a sufficient level to defoliate the legume and permit planting of radiata pine after two or three years. In the absence of the caterpillar, the lupine was mechanically crushed or disced in order to reduce competition for planted pines. A planting operation on dunes in New Zealand is shown in Figure 17.1. It was estimated that the reseeding legume supplied the soil with from 200 to 300 kg nitrogen per hectare during the five to six years before it was overtopped by the pine.

As soon as the sand movement has been halted, the dunes are normally planted with a hardy tree species. Conifers have generally been superior to hardwoods on adverse sites, although both types have been used. Radiata pine is

**Figure 17.1**   Planting radiata pine on a stablized sand dune in New Zealand.

planted on dune sands in New Zealand and maritime pine is often used in southwest France. Jack pine, red pine, Scots pine, pitch pine, Virginia pine, and sand pine have been used on the less favorable sandy sites in other areas; while white pine, Norway spruce, locust, and sometimes alder have been planted on the more moist sands between dunes. Planting distances for the pine vary from 1.5 × 1.5 m on the eroding sites to 2.0 × 2.0 m on the submarginal lands (Lehotsky, 1972). A band application of a fertilizer equivalent to about 200 kg diammonium phosphate per hectare, or a spot application of the same material applied in an opening some 20 to 30 cm on either side of the seedling and 15 cm deep is generally made soon after planting. The amount of fertilizer at each spot should not normally exceed about 50 gm in order to reduce the danger of salt damage to the seedling root system. The nutrient ratio should be based on soil fertility conditions as determined by experience or by a soil test.

## Other Dry Sands

There are vast areas of deep dry sands in many countries that require some special management techniques for the establishment of desirable tree species. These marginal sites often support a woody scrub or grass vegetation and are, therefore, well stabilized. Droughty soils of the southeastern United States sandhills are examples of these adverse sites. These latter sands (mostly Entisols) support remnants of longleaf pine stands harvested around the turn of the century, plus scrub hardwoods and wiregrass *(Aristida stricta)*. Early harvesting methods made no provision for regeneration of the pine and subsequent efforts at reforestation, especially with slash pine, have not been highly successful. The understory scrub oaks and wiregrass prevent natural and artificial establishment of most pines (Burns and Hebb, 1972). A reduction in competition from this vegetation is apparently essential in order to lessen the drain on soil moisture and nutrients, but the small amount of surface litter and soil organic matter must not be destroyed. A heavy brush cutter or drum chopper, often weighing more than 20 tons, best eliminates standing woody and herbaceous vegetation with a minimum of disturbance to the topsoil. Chopping is most effective when done so as to interrupt growth processes and reduce the store of carbohydrates in woody roots of the scrub species. Late spring, after the leaves are fully developed, is probably the best time for chopping. Developing sprouts further reduce root reserves and a second chopping made 6 to 18 weeks later depletes stored food so that few new sprouts survive.

Soils disturbed by mechanical site preparation should be permitted to settle for at least three months before planting or seeding. Dormant seedlings should then be planted slightly deeper than they grew in the nursery. Choice of species is most important on the dry sandhills, for they must not only compete successfully for a low level of nutrients and moisture, but also be able to withstand certain needle and root diseases. Longleaf pine is an indigenous species of the sandhills, but planting difficulties, a prolonged grass stage, and brown spot needle blight have prevented its being widely planted. Slash and loblolly pines are better suited to more moist and fertile sites. It appears that sand pine *(Pinus clausa)* possesses the rapid growth characteristics and drought resistance needed for reforestation of these sandhills (Figure 14.5). Choctawhatchee sand pine has surpassed all other pines in survival and yield in regional trials (Burns and Hebb, 1972). This variety possesses considerable resistance to gall rust *(Cronartium quercuum)* and mushroom root rot *(Clitocybe tabescens)* that affect most other pines of the area. It appears likely that many of its less desirable characteristics, such as poor form and excessive branches, can be eliminated through genetic improvement.

Trees should be planted at comparatively wide spacings on dry sites. The actual density of planting depends on estimated mortality and whether the plantation is to be thinned, as well as on the species planted. Sand pine should be planted so as to have about 1000 stems per hectare at age 30 years. If the stand is to be thinned at about age 20, the planting should provide about 1500 stems per hectare at that time. A mechanical thinning can remove designated rows plus suppressed and diseased trees within remaining rows and leave some 750 to 1000 stems for the harvest cut at plantation age 30 years.

## Forest Irrigation

Irrigation is not a cultural practice that is expected to be extensively used in commercial forests in any area in the near future, although it is often used in seed orchards, nurseries, and other special conditions. There is little doubt that water stress slows tree growth during dry periods in many areas, but the increased growth resulting from supplementary water does not appear sufficiently great to warrant the added investment.

The few experiments that have been conducted with the use of supplementary water in commercial forests have generally been located on dry sands (Baker, 1973). In an experiment on unfertilized Astatula fine sand (a deep, dry Entisol of the southeastern United States coastal plain), above-ground dry matter productions of 7-year-old slash pine were 8198; 12,768; 13,978; and 16,173 kg per hectare in plots that were nonirrigated or irrigated at infrequent, intermediate, and frequent intervals, respectively. In a fertilized area, the same irrigation treatments resulted in 7325; 18,413; 19,824; and 25,469 kg per hectare (Personal Comm., G. W. Bengtson).

In a 37-year-old red pine plantation on potassium-deficient, deep outwash sand in upper New York, Leaf et al. (1970) reported a slight increase in tree radial growth associated with irrigation during a normal season, but not during a wet year. The combination of irrigation plus potassium fertilizer increased growth during both years.

## WETLAND MANAGEMENT

There are extensive areas of wetlands along coastal plains and in basins with impeded drainage in most countries with extensive forest areas. Lands with high natural or perched water tables consist of both seasonally flooded areas and those that are permanently wet, with the latter condition often resulting in peatlands.

### Seasonally Flooded Soils

These periodically wet soils occupy a rather large percentage of the tropical rain and monsoon rain forests, but because of the high evapotranspiration rate of these ecosystems, excess water is seldom a limiting factor for tree growth. Other extensive areas of seasonally wet forests are found in coastal flats and basins with underlying impervious layers. Examples of wetland forests are the nearly 8 million hectares along the coastal plain of southeastern United States, according to Klawitter (1970b). These complex wetlands are very broad and nearly level interstream areas of coastal flats. They are interspersed with *bays* and *pocosins,* which are characterized by high centers with natural drainage impeded because of elevated rims, sluggish outlets, and impermeable subsoils; and *swamps* and *ponds,* which occupy the lowest locations along streams, plains, and bottom lands. They support such hydrophytes as water tupelo, cypress, maple, and green ash. Nuttall oak, sycamore, and cottonwood can tolerate flooding during the dormant season and may be used for reforestation in areas where water recedes early in the year.

The bays and swamps are generally wet all year and are often covered with one to several feet of peat. However, an excessively high water table during only a few months of the year appears to be the primary limiting factor to intensive woodland management of most wet plains. During seasonally dry periods, the plains can be exceedingly dry, because their primary source of water is precipitation. Tree cover may consist of scattered suppressed stems of one of the southern pines or locally dense stands of pines and hardwoods with an understory of grasses (mostly *Aristida stricta*) and clumps of gallberry *(Ilex glabra)* and St. Johnswort *(Hypericum petiolatum)*. Occasional rises of a few centimeters in the land surface may retard surface water flow over large areas, except in the immediate vicinity of shallow streams. Water moves slowly through the soils because of slowly permeable subsoil layers and small hydraulic heads between the soil surface and stream flowline.

Some forms of water control, mostly through ditch drainage, have been applied to thousands of hectares of forests on wet flats and in coastal plains since 1960. The objectives have been to (1) improve trafficability of woods roads so as to provide ready access for harvesting, fire protection, and tending; (2) facilitate regeneration of pine on prepared sites and insure early survival of planted seedlings; and (3) increase growth or reduce rotation age (Hewlett, 1972). However, as these wetland forests often serve as the headwaters for countless streams and lakes, as well as recharge areas for underground water reserves and as a buffer against seawater intrusion, any change in their water relations may modify the whole wetland environment, including changes in recreation values and wildlife habitat. Such changes should, therefore, be made only after careful study of their impact on the total environment.

Surface drainage ditches may not remove water from soils of the wet flats very effectively, because the draw-down of the ground water-table along the ditch may extend only a few meters inland from the ditch. Young and Brendemuehl (1973) reported that a drainage system consisting of main drains 1.2 m deep and 800 m apart with lateral collecting ditches 0.5 m deep spaced about 200 m apart improved the site productivity of a Rains loamy sand (Typic Paleaquult), but that the system was not intensive enough to provide the pole-sized slash pine with continuous protection from excess water. Furthermore, natural aging of the system tended to return the area to its original state following several wet years. In drainage tests in northeastern France on "pseudogley soils with impermeable layers at 25 to 50 cm, drainage ditches at 20 m intervals improved the growth of Norway spruce, Douglas-fir, and larch almost as much as spacings of 10 m and significantly more than spacings of 40 m (Levy, 1972).

The benefits from draining wetlands generally derive from increased rooting depth, which in turn reduces windthrow, improves soil aeration, and increases the nutrient supply. The latter may come about through accelerated oxidation of soil organic matter, as well as through increases in the volume of root-exploitable soil. However, excessive drainage of permeable soils may actually reduce growth and affect regeneration of certain types of wet site trees, such as tupelo and cypress. In a drainage experiment on Leon fine sand, a coastal plain Aeric Haplaquod, seven-year-old slash pine growth was not improved by 0.5 or 1.5 m deep ditches. Poor growth near ditches apparently resulted from excessive removal of ground water, particularly during dry periods (Kaufman et al., 1977).

While some drainage ways are often necessary in order to provide area accessibility, drains should be kept to a minimum and other methods of water level control used where feasible. Stability of the water table may be more important than drainage to tree growth. In an experiment on Leon fine sand where the water table was controlled at 45 to 90 cm with a tile drain and irrigation system, both slash and loblolly pines grew better the first five years than where the water-table was allowed to fluctuate (White and Pritchett, 1970). Growth of both species was better with a 45 cm stable water table than with the water level maintained at 90 cm below the surface, and slash pine was superior to loblolly on this site, regardless of water table control. These results indicate that fairly high water table levels can be tolerated by southern pine as long as the water table is stable.

Bedding, or row mounding, may be an alternative to lateral ditching of wet flats. Klawitter (1970b) found that four-year-old slash pine planted on beds were 13 percent taller than trees on unbedded plots of a Typic Paleaquult, even though the experimental area was adjacent to a main drainage ditch. In another part of the coastal plain on an unditched soil with similar properties, slash pine averaged 67 percent taller on bedded plots than trees on unbedded plots after three years (Pritchett and Smith, 1974). Although it is not clear whether the early advantages from bedding will persist into the later stages of stand development, it appears that such may be the case in the wet flats where trees stagnate without some relief from excess water. Once vigorous growth is promoted and crown closure is achieved, evapotranspiration rates may be sufficient to keep the area from swamping during most of the growing season.

Vigorous tree growth may also be promoted through fertilization of most wet soils. They are frequently low in available phosphorus and because of the limited soil volume for rooting, this element often limits tree growth. In an experiment in western Florida, slash pines that received 90 kg phosphorus per hectare averaged 60 percent taller than trees that received no phosphate fertilizer (Pritchett and Smith, 1974). Tree growth on plots that received fertilizer without bedding was about the same as on bedded plots without fertilizer. The combined effect of bedding and fertilization resulted in a 10-fold increase in volume growth over that of the control plots after eight years.

There have been attempts to improve productivity of degraded forest Spodosols with ortstein layers. Deep plowing, which ruptures the humic ferrous hardpan, plus the addition of lime, organic materials, and mineral fertilizers improves the growth of pines and mixed stands. However, the economics of such treatments appears questionable because the amelioration of the sites apparently lasts for only 8 to 12 years. Attempts to improve growth of pine by deep placement of lime and phosphorus fertilizers during site preparation of an Aquod soil met with only limited success (Robertson et al., 1975).

An improvement in accessibility and increase in the growth rate of the stagnant stands of pine often found on the seasonally flooded flat lands appear to be desirable objectives. However, it is not at all clear whether these objectives can or should be obtained through major drainage projects. A combination of water control through perimeter ditching for the removal of excess surface water, and bedding and fertilization, where needed for improved tree survival and growth, appears to be a more appropriate management scheme. The forest manager must strive for multiple-use land productivity while maintaining a relatively stable ecosystem on these delicate sites.

## Peatlands

Peatlands occupy vast areas of permanently wet lands, especially in cool climate regions such as found in Scandinavia, Siberia, Canada, and parts of the British Isles. These organic soils occupy as much as 15 to 30 percent of the total land area of Sweden and Finland where they often support poor stands of spruce and pine. In Finland, swamp drainage for forestry purposes has been carried out on over 2.5 million hectares. Another 3 to 4 million hectares of peatlands are worthy of drainage in regions in which forest production is favorable, while the remaining 3 million hectares will be left to nature conservancy, according to Huikari (1973). Holmen (1971) reported that about 7 million hectares, 17 percent of the land area of Sweden, may be regarded as peatlands with little or no forest production because of adverse water and nutrient conditions. Another million hectares have been drained. Water is generally controlled at 20 to 40 cm below the soil surface with a series of interconnecting ditch drains and check dams (Figure 17.2).

Holmen (1969) pointed out that afforestation of drained peatlands is not generally successful without the use of fertilizers. Phosphorus and potassium are the elements most often deficient and applications of 45 kg phosphorus and 120 kg potassium per hectare should suffice for 15 to 20 years. On poorly decomposed peat, it is also advisable to apply about 50 kg nitrogen per hectare, but micronutrients are not generally deficient.

Water control in peatland can do more than increase forest production. It may improve the habitat and supply of food for wildlife and enhance recreation possibilities and other forms of multiple use. However, overdrainage in some areas may be a problem that will have to be faced in the future.

## STRIP MINE SPOILS

Revegetation of land that has been drastically disturbed by strip mining for coal, iron ore, sand, phosphates, and other resources is a slow process at best. These

**Figure 17.2**  Main lateral ditch in water control system in Finnish peatland planted to Scots pine.

wastelands must often be graded, partially resurfaced with fertile soil, fertilized and limed, and stabilized before afforestation is attempted. To be sure, the kind and degree of amelioration depends on the physical, chemical, and biological properties of the spoils. These properties vary depending on the lithology, spoil age, degree of weathering, and erosion processes. The bare spoils cause serious environmental problems in many places by drastically affecting the adjacent land, water, forests, or aesthetic values. Because of the widespread concern for their degrading effect on the environment, reclamation of spoil banks has received considerable research attention in recent years (Czapowskyj, 1973). There are a number of tree species adapted to most classes of spoils, but there are few shrub and herbaceous species adapted to the most severe sites. In the north-central coal region, crownvetch provides a low-maintenance leguminous ground cover. Birdsfoot trefoil and service lespedeza are more acid tolerant than crownvetch but they generally require higher fertility levels.

Grasses, such as tall fescue, redtop, switchgrass, or oats are often sown with the legumes. Following the seeding the slopes are generally mulched with straw (4 to 5 ton per hectare) and the mulch anchored with asphalt emulsion, prior to tree planting.

Some spoils will support excellent tree growth without any site preparation or soil amendments, while others require grading, terracing, or harrowing. Grading of spoils high in clay content may cause compaction, and thereby reduce air and water infiltration, root penetration, and tree growth. Czapowskyj (1973) reported on Pennsylvania studies that indicated coarse-textured anthracite spoils were improved by compaction resulting from grading operations. He also found evidence that calcium, magnesium, and potassium concentrations were higher in certain graded spoil types than in their ungraded counterparts.

All essential elements are found in most overburden, but concentrations vary from very low to toxic. Some investigators (Schramm, 1966) have concluded that most coal-mine spoils are deficient in nitrogen and that vegetative growth on these spoils results from nitrogen added by rainfall or by biological fixation. In Alabama, loblolly pine on coal mine spoils are generally fertilized with 100 kg of nitrogen and phosphorus. However, Cornwell and Stone (1968) found that certain anthracite and black shale spoils supplied sufficient nitrogen for good tree growth as a result of weathering of these materials. Spoils of other mining operations are generally devoid of organic material and nitrogen. May et al. (1973) reported that while kaolin-mining spoils were extremely variable, most of them were deficient in nitrogen as well as phosphorus, potassium, and calcium. Bengtson et al. (1969) found that loblolly pine seedlings responded to both nitrogen and phosphorus fertilizers in pot studies with spoil material. They also reported that the response was enhanced when microbial inoculum of fresh pine duff was added. Marx (1977) obtained improved survival and growth of pine seedlings on spoil banks when the seedlings had been inoculated in the nursery with select strains of ectomycorrhizal fungi.

Levels of available iron, manganese, and zinc may be sufficiently high in very acid soil materials to pose a serious problem in the establishment of vegetation. Cornwell and Stone (1968) reported that gray birch was one species that was exceptionally tolerant to acidity and to high concentrations of heavy metals. Fortunately, toxic strip mine spoils are not widespread, according to Czapowskyj (1973), and most toxic conditions can be eliminated with applications of limestone. Although lime is a relatively inexpensive material, the extremely high application rates sometimes needed on the more acid mine waste can be quite costly. From 11 to 90 tons per hectare were required to neutralize the acids contained in some spoil materials before legumes and grasses could be established as stabilizing agents on Pennsylvania mine spoils (Czapowskyj, 1973).

Because of the deficiency of organic matter and nitrogen in most spoils, wide use should be made of trees and shrubs such as black locust and European alder, which have the capacity to fix atmospheric nitrogen (Figure 17.3). Other pioneer species that are used in afforestation of mine spoils include jack pine, pitch pine, Virginia pine, Austrian pine, eastern redcedar, juniper, and larch. Three species of birch and three *Eleagnus* species survived and grew well on a range of surface

**Figure 17.3** Reclaimed coal mine spoils in northern Bavaria. After leveling, a mixture of alder, birch, oak, and larch was planted.

mine spoil sites in eastern Kentucky (Plass, 1975). Acid-tolerant hardwood, such as northern red oak, sweetgum, and sycamore are also used. Performance of several species on anthracite strip-mine spoils is shown in Table 17.1.

## NUTRIENT-DEPLETED SITES
Nutrient-depleted sites include abandoned agricultural fields, severely disturbed areas, and inherently infertile soils to which forestry is often relegated. Many of these sites can be successfully reforested through the judicious use of commercial fertilizers as discussed in following chapters. In addition to these infertile areas, there are soils with low nitrogen reserves and other problems which may be improved by interplanting with legumes or other nitrogen-fixing and soil-conserving plants.

### Underplanting Conifers with Hardwoods
There have been attempts (Hesselman, 1926; Lutz and Chandler, 1946) to improve forest soils through the introduction of hardwood species into conifer stands in Europe and the United States. In some instances, an understory of shrubs and trees was created that accumulated high concentrations of bases in their leaves. The resulting forest floor favored the decomposition of organic debris, reduced acidity, and stimulated nitrification and mineral cycling. Alway et al. (1933) reported that the addition of maple and basswood leaves increased

**Table 17.1** Average Survival and Height of Several Tree Species Planted on Graded Anthracite Strip-mine Spoils After Five Years (Czapowskyj, 1970)

| Species | Survival | | | | Tree Height | | | |
|---|---|---|---|---|---|---|---|---|
| | I[a] | II | III | IV | I | II | III | IV |
| | % | | | | m | | | |
| Hybrid poplar | 90 | 78 | 97 | 42 | 4.5 | 2.2 | 5.9 | 1.5 |
| European alder | 88 | 69 | 85 | 31 | 2.0 | 1.6 | 2.3 | 1.6 |
| Black locust | 82 | 26 | 80 | 66 | 3.8 | 2.0 | 3.3 | 1.9 |
| Jack pine | 75 | 96 | 97 | 72 | 0.7 | 1.5 | 1.1 | 0.7 |
| Red pine | 77 | 89 | 87 | 65 | 0.4 | 0.6 | 0.4 | 0.3 |
| Scots pine | 70 | 86 | 82 | 56 | 0.8 | 1.1 | 0.8 | 0.8 |
| Virginia pine | 62 | 64 | 59 | 38 | 0.9 | 1.6 | 1.2 | 0.6 |
| European larch | 49 | 65 | 51 | 34 | 1.1 | 0.9 | 0.7 | 0.5 |
| White spruce | 54 | 27 | 51 | 10 | 0.3 | 0.3 | 0.3 | 0.2 |

[a]Denotes spoil type:
I = Black Carbonaceous shale
II = Gray to yellow shale
III = Sandstones and conglomerates
IV = Glacial till and surface deposits

calcium and nitrogen content and reduced acidity of the surface soil. White birch is reported to exert an influence on exchangeable potassium in the surface of potassium-deficient sandy soils. As a result, young white pines grew well under birch crowns, while those in adjacent openings showed symptoms of potassium deficiency (Walker, 1955). Eastern redcedar has also been found to influence chemical and physical properties of soils when compared to soils under adjacent pine stands.

However, interplanting has not always improved site productivity. Underplanting spruce and beech in pine stands improved certain physical and chemical properties of sandy soil but reduced overstory growth due to moisture depletion. The decomposition of beech leaves appeared to be too slow on sandy soils to be of much value. Some hardwoods, such as oaks, may not be sufficiently shade tolerant to thrive under conifers, and others are too site demanding to grow on the sandy soils on which conifers are often found. Consequently, soil improvement by the technique of underplanting conifers with deep-rooted hardwoods has met with varying degrees of success. Considering the high costs involved, it is not generally considered to be a practical procedure.

## Dinitrogen Fixation in Forests
Planting or encouraging the growth of legumes and other nitrogen-fixing plants in forests is an intriguing means of increasing soil nitrogen levels without the

environmental risk inherent in the use of mineral fertilizers. In addition to free-living microorganisms in the soil—mainly blue-green algae and the bacteria *Azotobacter* and *Clostridum*—many higher plants have the capacity for biological fixation of nitrogen from the atmosphere. In fact, free-living microbes are not believed to be very important nitrogen-fixers in forest ecosystems. The blue-green algae are mainly active in areas of higher light intensity than found in most forest floors. Denison (1973) estimated that blue-green algae associated with the lichen *Lobaria* may fix from 2 to 12 kg nitrogen hectare per year in Douglas-fir forests in Oregon. However, green algae appear to be algae most frequently associated with lichens in other forests. The C:N ratio of organic materials may be wide enough in most soils to sustain high nitrogen-fixing activity of the saprophytic bacteria. Furthermore, most forest soils are too acid for any appreciable nitrogen fixation by *Azotobacter*, according to Waksman (1952).

**Leguminous Plants.**   Symbiotic nitrogen fixation may be of considerable importance in maintaining the productivity of forest lands. In a review of the subject, Youngberg and Wollum (1970) pointed out there were many leguminous trees in tropical forests, but that they were not so common in temperate region forests. Only six species of leguminous trees are reported to occur in commercial forests in North America. A large number of herbaceous and shrubby legumes grow in wild land ecosystems and some of them may be important in forested areas. As a rule, they are shade intolerant and require higher pH and nutrient levels than found in most forest soils. Nevertheless, the introduction of legumes into forest ecosystems holds some promise as a means of maintaining soil nitrogen without the use of nitrogenous fertilizers.

Lupines have been planted in young radiata pine plantations on sandy soils in New Zealand. They reportedly (Gadgil, 1971b) increased the soil nitrogen to a sufficient level before crown closure to maintain good tree growth for a decade or so thereafter. This reseeding legume later reestablished itself upon opening of the stand. Lupines *(Lupines polyphyllus)* have also been planted in mature spruce stands in Germany (Rehfuess and Schmidt, 1971) in an effort to restore the productivity of soils depleted of nitrogen through generations of litter gathering (Kreutzer, 1972). Discing of the forest floor is often required in established stands in order to obtain adequate germination and survival of the seeded legume (Figure 17.4). Discing is generally a difficult task under such conditions and may result in increased *Fomes annosus* fungus rot from root injury of some tree species. The correction of soil pH and phosphorus or potassium deficiencies is necessary for successful establishment and growth of legumes on most sites. Infertile acid conditions of the soils appear to be the major deterrent to the use of legumes in young pine plantations in most coastal areas. However, with adequate seedbed preparation and additions of phosphates for good tree growth, it may be possible to establish legumes in plantations with little additional expense.

**Figure 17.4** *Lupinus polyphyllus* seeded in mature stand of Scots pine in an effort to ameliorate soil impoverished by earlier litter harvesting in Bavaria (courtesy K. Rehfuess).

**Nonleguminous Plants.** Symbiotic nitrogen fixation involving nonleguminous plants may be significant in forests of both hemispheres. Nodules have been observed on at least 76 species of gymnosperms, the majority of which occur in the Southern Hemisphere (Allen and Allen, 1965). More than 100 species of nodulated angiosperms have been identified in the Northern Hemisphere. Ten of the 15 genera represented in this group of nitrogen fixers are components of climax vegetation or pioneer species that usually appear after disturbances of North American forests and range ecosystems. For example, *Ceanothus, Shepherdia,* and *Alnus* are genera of understory plants in climax forests in the Pacific Northwest. *Myrica, Ceanothus, Dryas,* and *Alnus* contain species of pioneer plants that invade or colonize disturbed areas in many parts of North America.

The microbial endophytes involved in the symbiotic nitrogen fixation in nonlegumes have not been well identified. Youngberg and Wollum (1970) reported that there was strong evidence that *Streptomyces* is involved in nodule formation in angiosperms. Whatever the microorganisms involved, there appears to be considerable variation in nodulation. The level of the endophyte population, as well as environmental factors, undoubtedly influences nodulation. Some host plants are intolerant to shade and in their absence the endophyte population decreases. On the other hand, high light intensity and moderately high temperature and soil moisture generally favor nodulation of both leguminous and nonleguminous plants.

The presence of nodules on the host roots does not in itself prove that nitrogen fixation is taking or has taken place. However, some species of all but three of the naturally occurring genera of nodulated angiosperms in North America have been shown to fix nitrogen (Youngberg and Wollum, 1970). The amount of nitrogen fixed by a stand of nonlegumes may be considerable. Crocker and Major (1955) estimated that more than 60 kg nitrogen per hectare per year were fixed by species of *Dryas, Sheperdia,* and *Alnus* colonizing recessional moraines in Alaska. In other studies, *Alnus* spp. have been shown to fix from 40 to 84 kg nitrogen per hectare per year by Voigt and Steucek (1969) to as much as 160 kg per hectare per year in Ontario (Daly, 1966). Tarrant et al. (1969) found that the nitrogen accumulated in the top 92 cm of a forest soil under conifers totaled 13,126 kg per hectare while in soil beneath a mixed stand of pine-red alder it totaled 14,157 and beneath a pure alder stand it totaled 18,682 kg per hectare. Table 17.2 shows the amount of nitrogen returned to the forest floor in red alder litter fall. European alder has been interplanted with spruce, on a limited scale, on nitrogen-deficient sandy soils in Germany. However, it does not appear that this rather expensive practice is feasible for large forest holdings, except in some very intensively managed forests.

Whether the planting of either leguminous or nonleguminous plants that fix nitrogen is a practical method of supplying nitrogen to forests is an open question that needs further research. However, there appears to be little question that efforts should be made in site preparation and management of forests to encourage the proliferation of understory vegetation that will help maintain or improve the site nutrient status.

## OTHER PROBLEM SITES

There are many localized areas where special measures must be taken to assure successful afforestations. It would be difficult to mention all such problem areas, but a few examples will suffice to illustrate the difficulties.

### Compacted Soils

As a result of the increasing use of heavy logging and wheeled equipment, soil compaction is a factor of growing concern in forestry. Foil and Ralston (1967) found that root weights of loblolly pine seedlings grown in the greenhouse were negatively correlated with bulk density over a range in density from 0.8 to 1.4 g per cubic centimeter. They concluded, however, that no single factor was entirely responsible for the retarded growth at high soil densities. For example, root growth may be restricted by mechanical resistance, low oxygen supply, and possibly from high concentrations of carbon dioxide or other toxic gases resulting from poor aeration. In a greenhouse experiment, establishment of loblolly

**Table 17.2**  Annual Litterfall, Concentration of Nitrogen and Total Nitrogen in Litterfall, and C:N Ratio in Pure Conifer, Pure Alder, and Mixed Stands in Oregon (Tarrant et al., 1969)

|                              | Forest Type |               |           |
| ---------------------------- | ----------- | ------------- | --------- |
|                              | Conifer     | Conifer-Alder | Red Alder |
| Litterfall, kg/ha            | 4740        | 6642          | 5029      |
| Litterfall N, %              | 0.79        | 1.82          | 2.17      |
| Total N in litterfall, kg/ha | 36          | 116           | 112       |
| C:N ratio in litterfall      | 69          | 32            | 27        |

pine seedlings was lower on compacted soils than on noncompacted soils, and loosening the compacted surface soil greatly increased establishment on four of the five textural classes (Hatchell, 1970). Losses during germination were primarily due to restricted penetration of the radicle. There were highly significant reductions in dry shoot weight on compacted soils as compared to noncompacted soils.

Soil compaction, resulting from loaded pulpwood trucks, reduced shoot weight significantly more on fine textured soils than on loamy sands. Compaction also reduced the root dry weight but not the shoot/root ratio. Soil compaction caused decreased infiltration rates and subsequent increases in surface runoff and erosion.

In some instances compacted soils on skid trails, roads, and loading yards can be alleviated by discing or tine plowing, but as far as possible one should avoid situations that result in the problem. These are generally associated with multiple trips with heavily loaded wheeled vehicles over a given area. Soil moisture content and soil texture are two important factors affecting compaction. Medium textured soils (loam and silt loams) appear to compact to greater densities than fine or coarse textured soils. Soil moisture resulting in the greatest compaction is about midway between field capacity and the permanent wilting point (Swanston and Dyrness, 1973).

The mechanical shattering of dense spodic horizons in Spodosols may also facilitate root penetration and tree growth, although more research is needed before this technique can be widely recommended.

## Eroded Soils

Erosion results from accelerated removal of soil materials by running water, wind, or gravitational creep. Erosion is a natural process that has existed throughout geological time, but some human activities have accelerated the process far beyond the normal rate. Normal surface erosion usually proceeds at a very slow pace in forested areas and is not generally a problem even in most mountainous regions. Landslides, creeps, slips, and flows are the most conspicuous forms of natural erosion, but fortunately only small areas of timberland are vulnerable, and these are generally confined to the steepest slopes. Surface soil losses under native forests and pine plantations are minimal, seldom exceeding 100 kg per hectare per year (Megahan, 1972). However, soil disturbance by logging, road building, and site preparation may cause soil movement and subsequent downstream sedimentation except on relatively flat terrain. The amount of movement is influenced by the duration and intensity of rainfall, slope length and gradient, and inherent soil properties affecting stability and infiltration rate, as well as ground cover and degree of disturbance.

Planning is important in harvest operations because it is much easier to prevent erosion than to control it once it has started. Plans normally include (1) stratification of the land according to erosion hazards and proper development of road access and timber-sale areas, (2) minimizing roads by proper selection of the logging system, and (3) employing followup procedures to assure erosion control. The inherent erodability of a site is influenced by length and degree of slope, meteorological factors, and water infiltration rates.

**Surface Erosion.**  Plant and litter cover is the greatest deterrent to surface erosion. The tremendous amounts of kinetic energy expended by falling rain are mostly absorbed by vegetation and litter in undisturbed forests. Disturbances caused by logging or other activities reduce infiltration rates and increase surface runoff and erosion. Dominant mass erosion processes on forested slopes are slow, downslope movements involving subtle deformation of the soil mantle and discrete failures, both of the slump-earthflow type or the rapid, shallow soil and organic debris movement down hill slopes (Swanston and Dyrness, 1973).

The land manager should choose a silvicultural system to minimize surface erosion on steep lands. If the erosion hazard is high, he or she may decide to forego logging altogether and practice protective management. Certainly the harvest intensity should be guided by the erosion potential. For example, clear-cutting may be used on stable sites, but shelterwood or selection cutting may be necessary to protect the site in erodable areas. Logging methods can also be selected to minimize surface erosion. Swanston and Dyrness (1973) determined the average disturbance caused by four yarding methods used in clear-cut operations in northwestern United States, as shown in Table 17.3.

Tractor logging is generally the most efficient and least expensive on moderately sloping lands. However, tractor logging must be confined to periods when the surface soils are dry enough to minimize compaction, and it should be avoided completely on the steeper slopes. Highlead logging generally results in considerable soil disturbance and creates water paths that lead to gullying. Duff and brush may be removed from as much as 50 percent of the soil surface, which may improve regeneration but increase soil erodability. In the running skyline method of logging, the logs are carried partially in the air with the front end lifted

**Table 17.3**  Average Site Disturbance by Four Methods of Clear-cut Harvesting in Mountainous Terrain (Swanston and Dyrness, 1973)[a]

|  | Percent Bare Soil | Percent Compacted Soil |
|---|---|---|
| Tractor | 35.1 | 26.4 |
| Highlead | 14.8 | 9.1 |
| Skyline | 12.1 | 3.4 |
| Balloon | 6.0 | 1.7 |

[a]Used with permission by the Journal of Forestry.

clear of the forest floor. This method produces little disturbance of ground vegetation on the forest floor, and some supplementary scarification may be needed to expose sufficient soil for proper seeding or planting. Balloon and helicopter logging have many advantages from the standpoint of soil stabilization, but they are expensive and, where they are used, the area is often left without sufficient roads for future forest operations.

Site preparation activities that include full tree utilization, scarification, and burning affect soil properties and erodibility of some soils. Fires increase surface erosion through the removal of protective vegetation and litter. Sufficiently hot fires can cause a breakdown of water-soluble aggregates, decrease organic matter content, decrease water infiltration rate, and form hydrophobic layers in sandy soils. A special effort should be made to revegetate bare and compacted areas as soon as possible. Natural revegetation may occur so slowly that seeding, fertilizing, and mulching may be needed to hasten this process.

Mass movement or slumping of the soil mantle on forest lands is normally confined to slopes steeper than 60 percent, but it is influenced by soil moisture, soil physical properties, bedrock type, and the rooting characteristics of the trees and understory vegetation. The resistance of a soil to sliding involves a complex interrelationship between soil and slope characteristics. Basaltic soils have more cohesiveness and more fine-textured material than granitic soils. The former have an angle of repose of about 72 percent, while that of the granitic soil is about 68 percent.

Once unstable areas are identified and characterized, the forest land manager has several techniques available for minimizing erosion damage. These include various kinds of stand manipulations ranging from no logging to clear-cutting and replanting, selection of logging methods that can help to minimize surface disturbance and destruction of stabilizing vegetation, and judicious location and design of forest roads. Road cut and fill slopes can often be stabilized by shrub plantings. Blue elderberry *(Sambucus cerulea)* and bush penstemon *(Penstemon fruticosus)* appear to be suited to a wide variety of disturbed sites in the western United States. *Ceanothus* spp. have also been successfully planted on a number of road cuts and disturbed areas (Tiedemann et al., 1976). Grass-legume seeds, fertilizers, and straw mulch have been applied to road backslopes to check erosion in a manner similar to their use on mine spoilbanks.

**Gully Control.**   Regardless of the care taken to prevent erosion, some erosion does occur in many areas and foresters are often requested to rehabilitate soils that have already been devastated. Some of the most difficult reforestation problems are the result of accelerated erosion on mismanaged agricultural lands. Millions of hectares are in need of reforestation for erosion control in the United States alone (Lutz and Chandler, 1946). Trees are particularly valuable on areas where erosion is extreme and where permanent maximum protection is needed. The trees best adapted for planting in gullies and severely eroded sites usually are native species that are shade intolerant but have a low requirement for moisture and nutrients. These might include black locust and other drought-resistant trees with fibrous root systems for the slopes and drier bottoms of gullies, and willows or cottonwood for the moister areas.

Gully control is largely a matter of the development of an erosion-resistant cover of vegetation that will stabilize the soil. However, before planting is attempted, some mechanical measures may need to be taken. For example, it is sometimes necessary to prepare for planting by breaking down and sloping the gully walls and planting grasses and leguminous plants. It may also be necessary to divert runoff water before it enters the gully. If water diversion is not feasible, it may be desirable to install low check dams or other structures to halt further erosion until vegetation can become established. These temporary structures may be made from brush, poles, rocks, heavy woven wire, or debris that will help accumulate soil material in which trees or other vegetation can be planted. Multiple low dams less than 0.5 m high are preferred, but they should be securely anchored at the bottom and sides to prevent undercutting.

Large gullies, particularly those with a vertical drop at the head, may require permanent structures to halt erosion (Lutz and Chandler, 1946). These may be expensive affairs constructed of masonry or logs and earth and would be used only in extreme situations.

## Steep Oxisols

There are many other difficult sites that require some special attention for reforestation. The list could include sites for shelter-belt and urban plantings, alkaline and toxic soils, and soils over shallow bedrock. It seems logical that many Oxisols on steep terrain should be placed on the critical list. Because these old red and yellow tropical soils have relatively rapid permeability and are generally found on gentle slopes, they are not highly erodible. However, care will have to be exercised in the harvest and reforestation of steeply sloping Oxisols to preserve the surface soil layer. Almost the entire nutrient reserve resides in the top few centimeters of the soil-root zone, and if this top soil is destroyed nutrients will have to be restored by fertilization. Unfortunately, little research has been conducted on reforestation of these fragile soils but it appears that many of these soils should remain in forests.

# 18

---

# DIAGNOSIS AND CORRECTION
# OF NUTRIENT DEFICIENCIES

Worldwide increases in areas of intensively managed forests have been accompanied by parallel increases in nutritional problems. Part of the difficulty results from the relegation of large tracts of infertile soils to forest plantations, but the problem is compounded by the greater demands made on soil nutrient reserves by the faster growing man-made forests. Whatever the cause of nutritional disturbances, the growing demand for increased wood production on an ever diminishing land base emphasizes the need for reliable methods of identifying and correcting deficient sites.

## HISTORICAL PERSPECTIVE

The use of fertilizers to correct nutrient deficiencies in forest lands and to improve timber yields is not a new concept. Probably the first experiments with inorganic fertilizers applied to forest soils were initiated in France as early as 1847 (Baule and Fricker, 1970). In these tests, yields were increased from 17 to 26 percent by applications of wood ashes, ammonium salts, or basic slag. The same authors reported that the use of phosphorus and potassium fertilizers in the afforestation of heath lands of Belgium and France was thoroughly discussed at the 1910 meeting of the International Union of Forestry Research Organizations. A number of empirical tests with fertilizers were established in Germany, Belgium, and Denmark soon after 1900 (Tamm, 1968), and a firm scientific foundation on which to base forest fertilization had been established in northern Europe by midcentury. Furthermore, nutritional problems encountered following the introduction of exotic conifers to parts of Australia and New Zealand resulted in successful trials with phosphate and certain micronutrients during the first quarter of this century (Stoate, 1950), and led to operational use of fertilizers in those areas.

Fertilization of forest stands received little attention from American foresters during the first half of this century. They were concerned by the more pressing problems of protecting their vast timberlands from fire and bringing them under some sort of management. The seemingly endless supply of timber, plus the prevailing attitude that soil physical properties were the only really significant factors influencing forest site productivity, dampened interest in soil chemical parameters. The generally efficient and conservative nature of nutrient cycling in natural forest ecosystems precluded widespread deficiencies in natural stands. This, plus the general lack of technology for conducting field tests and measuring response, probably explains why many of the early forest fertilization trials met with mixed success (White and Leaf, 1956; Arneman, 1960; Mustanoja and Leaf, 1965). Even the more significant later successes such as the growth increases of planted red pine from applications of potassium on glacial outwash sands in New York (Heiberg and White, 1951); slash pine responses to phosphorus in plantations in the southeastern United States flatwoods (Barnes and Ralston, 1953; Pritchett and Swinford, 1961); and the substantial volume increases resulting from nitrogen fertilization of Douglas-fir in the Pacific Northwest (Gessel and Walker, 1956), created little immediate interest in operational fertilization in the United States. Fertilizer use continued to be largely confined to tree nurseries, Christmas tree farms, and seed orchards, until about the middle 1960s. Progress in the development of reliable diagnostic techniques for delineating deficient areas was largely responsible for a major increase in fertilization activity at that time.

## DIAGNOSIS OF NUTRIENT DEFICIENCIES

Diagnostic techniques are particularly useful in a forest fertilization program as a means of assuring the most effective use of the fertilizer material. The most commonly used methods of diagnosis are (1) visual symptoms, (2) plant tissue analysis, (3) soil analysis, (4) pot cultures, (5) bioassays, (6) field trials, and (7) indicator plants.

### Visual Symptoms

A deficiency of a particular nutrient may be considered to exist if the addition of that element to the soil-plant system in a suitable form and rate results in an increase in growth. However, a tree growth response to an added nutrient may be obtained long before a visual symptom of a deficiency is evident. Therefore, if one relies solely on visual symptoms of deficiencies much growth potential may be lost in the interim before corrective action is taken.

Nonetheless, visual symptoms have a useful place in diagnosing instances of severe nutrient deficiencies, if care is taken in interpretation. Leaf (1968) pointed out that most of the essential nutrients perform such functions in the tree that a characteristic symptom generally develops if one of the elements is missing. A great deal of observation (and photography), description, experimentation, and chemical analyses are needed to associate specific symptoms for a given species with a known nutrient deficiency. Once these peculiarities are documented, visual symptoms can be a valuable tool in identifying deficient areas.

Much research has been conducted in culture studies; some tests have been carried out in nurseries, and a few in plantations to relate visual symptoms to a known deficiency of an element. However, tree deficiency symptoms developed in pot studies do not always have application under field conditions. Furthermore, visual symptoms of deficiencies developed in the field may be difficult to interpret because environmental factors such as climatic and soil conditions and disease and insect problems that affect tree growth are highly variable and subject to change. On some particularly poor sites, multiple nutrient deficiencies may occur and complicate symptom diagnosis.

An important aspect of visual deficiency symptom use is species specificity. A particular nutrient deficiency will manifest itself in several ways depending on the severity of the deficiency and on the tree species involved. This is readily apparent when comparing conifers with hardwoods, but even racial and provenance variations within species may alter the outward appearance of a nutrient deficiency.

On a particular site, Leaf (1968) pointed out that visual deficiency symptoms produced by a stand of trees will vary markedly from species to species; from nil for low nutrient-requiring pioneer species to pronounced specific symptoms for high nutrient-demanding species. He stressed the need to modify the term "deficient" to embrace not only the nutrient but also the tree species involved. To be sure, deficiency symptoms may take many forms that make them difficult to recognize in the field. For example, a small but continuous supply of boron is required by plants for the formation of new tissue. Boron availability fluctuates under the influence of changes in soil environment, and acute deficiencies may develop under unfavorable conditions, resulting in the death of the apical meristems. When dieback recurs frequently the result is a characteristic bushlike tree (Figure 18.1). In less severe conditions, symptoms resemble those of drought-induced dieback (Snowdon, 1973). Copper deficiency in conifers often results in long, twisted apical leaders that have difficulty remaining upright (Figure 18.2). Low soil copper levels that are sufficient for normal tree growth may become inadequate when growth rates are accelerated by nitrogen fertilization or other cultural practices. Severe phosphorus deficiency is most often associated with slow growth, short needles and early needle abscission resulting in a very sparse crown in pines (Figure 18.3).

Although visual deficiency symptoms range from general abnormalities in tree growth and development to specific foliage characteristics, many symptoms

1. Nitrogen-deficient second-rotation *Pinus radiata* in South Island of New Zealand. Note yellow-green color and nonthrifty appearance.

2. Stunted Corsican pines in foreground are phosphorus deficient. Those in background were fertilized with 40 kg phosphorus per hectare at planting, some 40 years previously, on wet coastal soils in south England.

3. Potassium deficiency symptoms (*Picea abies* on deep peat in Sweden) are often expressed as necrotic tips of older leaves.

4. Magnesium deficiency in *Quercus borealis* in Germany.

5. Zinc deficiency in *Pinus radiata* in Australia.

6

7

8

9

10

6. Boron deficiency in *Pinus elliottii*, induced by liming sandy soils in the coastal plain of the United States.

7. Manganese deficiency symptom exhibited on *Picea abies* on high-lime soils of Götland, Sweden.

8. Iron deficient *Pinus elliottii* seedling (left) induced by liming Aquod surface soil.

9. Sulfur deficiency in Coconut palm.

10. The major symptom of copper deficiency in conifers is enlongated, twisted terminal and lateral branches.

**Figure 18.1** Boron deficiency in *Pinus pinaster* near Nelson, New Zealand. Note multiple dead leaders resulting in bushlike appearance.

are related to foliage color patterns. Care must be taken to avoid confusing seasonal and racial variations in foliage color with those denoting nutrient deficiency. Color photographs can provide useful and long-lasting records, although true colors are not always easy to achieve. The use of standard primary color chips in each photograph helps in evaluating "trueness" of the foliage color. The use of remote sensing techniques to detect the occurrence and distribution of deficient areas appears to have some value for elements whose deficiency produces a characteristic color change.

As noted, deficiency symptoms in trees may not develop until considerable loss in volume growth has occurred. Furthermore, visual symptoms of incipient deficiencies may be of such a general nature (mild chlorosis) that they are of little or no value for detailed diagnostic purposes (Figure 18.4). Other techniques such

**Figure 18.2** The most obvious symptom of copper deficiency in *Pinus radiata* is long, twisted leaders. Deficiencies of most micronutrients are compounded by increases in soil reaction.

as tissue analysis, though laborious, permit possible detection of nutrient deficiencies in incipient stages before serious losses in production occur.

## Plant Tissue Analysis

The concept of plant tissue analysis is based on the recognized relationship between the nutrient concentration in certain tissues of a tree and the growth rate of that tree.

**Tissue Sampling.**   Various tissues are used for determining nutrient uptake and distribution in trees. These include such tree parts as petioles, bark, latex, phloem, fruits, or rootlets, but for most species tissue analysis is almost synonymous with *foliage* analysis.

The optimum sampling time is often a compromise between selecting the period of maximum sensitivity to site difference, the time of year with a maximum between-tree variability, and the need for a stable period in which to complete a sampling program. For diagnostic purposes, the current year's near-terminal foliage of conifers and uppermost sun leaves of hardwoods, collected at the end of the active growth period and before hardwood autumn foliage coloration, has generally been used. Wells and Metz (1963) have suggested that for some tree species and nutrients, foliage from other portions of the tree crown

**Figure 18.3** Phosphorus deficiency in pine results in sparse crowns and slow growth, as illustrated by this 20-year-old slash pine plantation on a coastal Typic Ochraquult.

and, possibly, other than current year's production, and collected at other seasons of the year may be better for diagnostic sampling. Magwick (1964) noted that growth of red pine correlated better with potassium levels in the lowest portion of the live crown than in other portions of the crown. It has been suggested that since tissue levels of the more mobile elements are influenced by environmental stress to a greater extent during the growing season than in the dormant period, samples for diagnostic purposes should be collected during the former period for such elements as nitrogen, phosphorus, magnesium, and zinc (Mead and Pritchett, 1974). These authors noted that potassium does not have sufficient stability until autumn and sampling for this element, as well as the less mobile calcium, manganese, and aluminum, should be made during the dormant period.

Trees with extremely heavy seed crops, parts of the branches beyond the point of cone attachment, or trees that suffer severe disease or insect attack should not be sampled. Due to biological and microsite variability among trees, aliquots from several trees should be collected to obtain a reliable sample. The number of sample trees needed to represent stand conditions varies with the elements tested. Mead and Pritchett (1974) reported that only 2 to 7 trees needed to be sampled in order to detect 10 percent difference between means for nitrogen, phosphorus, and zinc, but for potassium, calcium, magnesium, man-

**Figure 18.4** Potassium deficiency in Sitka spruce on deep peat in Scotland was induced by phosphorus fertilization. Because potassium is mobile, symptoms are first noted on older needles.

ganese, and aluminum determination approximately 20 trees had to be sampled to obtain the same precision. If greater accuracy is desired, more trees should be sampled instead of more branches from each tree, because there is greater variation between trees than within trees (Tamm, 1964). It seems clear that standardization is needed in sampling in order to minimize the effects of season of the year, age of tissue, position within the crown of the tree, and position of the tree in the stand on nutrient levels of the sample.

**Critical Concentrations.**   If plant tissue analysis is to have value as a technique for diagnosing deficient conditions, threshold concentrations must be identified below which a significant yield response to the application of that nutrient can be expected. Richards and Bevege (1972) have defined the critical level as the concentration which is associated with 90 percent of the maximum yield. But they point out that the critical level for a particular nutrient is not independent of the level of other nutrients. In fact, the critical level is an ill-defined value, because other nutrient or environmental conditions may limit the response that would otherwise be obtained from the application of the deficient element. In the example of boron deficiency mentioned above, Snowdon (1973) reported that *Pinus radiata* will apparently remain in a healthy condition with foliar boron levels as low as 4 ppm during years of favorable rainfall. However, trees with 8 ppm boron can be considered deficient and dieback can be attributed to a lack of boron during dry years. The critical level, therefore, should be considered as a range of concentrations rather than a well-defined point.

Critical concentrations are generally determined by analyzing plant tissue from unfertilized trees in experimental areas exhibiting varying degrees of response to the addition of the element under study, as illustrated in Figure 18.5. Another approach is to relate tissue nutrient concentration to response in plots fertilized by varying rates, as with red oak in Figure 18.6.

The basis for the use of nutrient concentrations for diagnostic purposes is Leibig's 1843 concept than when all but one nutrient are in adequate supply, a plant grows in proportion to the supply of the limiting element and it tends to keep a fairly constant concentration of that nutrient for a given tissue and condition. In spite of the complicating tendency of certain nutrients, notably potassium, toward "luxury consumption" when present in excessive amounts

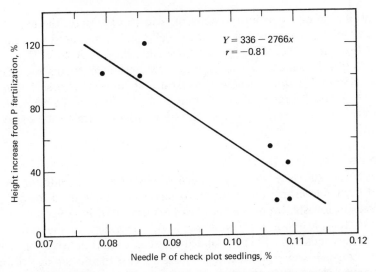

**Figure 18.5** First-year height growth response of loblolly pine seedlings from applications of phosphate fertilizer in relation to percentage phosphorus in needles of check plot (no P) seedlings (Wells and Crutchfield, 1969).

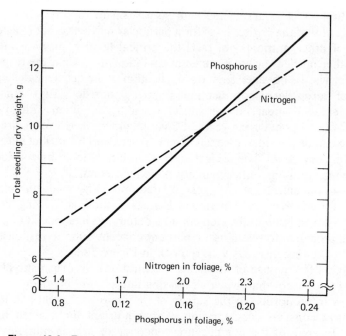

**Figure 18.6**   Total dry weight of red oak seedlings as related to foliar concentrations of nitrogen and phosphorus (Phares, 1971).

and by a dilution effect when the nutrient is in short supply, the use of tissue analyses to predict nutrient deficiencies is valid and, when used in conjunction with soil tests and other aids, it is an invaluable diagnostic tool. Examples of possible "critical" ranges in concentration for certain nutrients in foliage are listed in Table 18.1. *Critical* is defined as a range in nutrient concentration above which trees would not respond significantly to applications of the element, but at which trees with a lower tissue concentration would normally be expected to respond.

## Soil Tests

*Soil analyses,* like tissue tests, have been widely used as a guide for fertilizing field and vegetable crops, but neither test has been widely used for diagnosing deficiencies in forests. Some success has been reported in developing soil test procedures for phosphorus in New Zealand, Australia, southeastern United States, Netherlands, and Finland, and the test is used extensively in some of these countries. However, there has been less success in developing soil analysis for diagnosing other nutrient deficiencies. Some of the difficulties encountered with soil tests are (1) lack of basic information on nutrient requirements of our

**Table 18.1**  Approximate Critical Ranges for Foliar Concentration in Two Pine Species

| Species | Macronutrient Concentrations, % | | | | | |
|---|---|---|---|---|---|---|
| | N | P | K | Ca | Mg | S |
| Slash pine | 1.0–1.2 | .085–0.09 | 0.25–0.30 | 0.13–0.16 | 0.04–0.08 | 0.08–0.10 |
| Radiata pine | 1.2–1.4 | 0.09–0.10 | 0.30–0.35 | 0.13–0.14 | 0.06–0.08 | — |

| Species | Micronutrient Concentrations, ppm | | | | | |
|---|---|---|---|---|---|---|
| | Fe | Mn | Zn | B | Cu | Mo |
| Slash pine | 15–35 | 50–100 | 20–50 | 4–8 | 2–4 | — |
| Radiata pine | 40–60 | 140–800 | 30–50 | 6–10 | 3–7 | 0.03–0.05 |

principal forest species, (2) lack of correlation data useful for interpreting test results in terms of tree response to fertilizers, (3) difficulty in securing a representative sample in forested areas, (4) lack of information on what soil layer (depth) to sample, and (5) uncertainty as to what nutrient form or fraction to extract. For example, total soil phosphorus could be a meaningful test for trees in view of the possible influence of ectomycorrhizae on the availability of less soluble forms of phosphorus. Therefore analytical procedures primarily designed for cultivated soils may not be adequate for forest soils.

**Soil Sampling.**  Sampling is complicated by soil heterogeneity and variations in root feeding volumes. Variability in soil properties may occur in the horizontal, vertical, and time dimensions, or in a combination of these. The vertical dimension is especially important in forest soil sampling because of the deep rooting habit of most trees. However, since most tree feeder roots are located in the surface 0 to 20 cm, soil samples are normally collected from this zone.

The sampling intensity required is dependent on soil heterogeneity and the intensity of management intended. The magnitude and pattern of variability is not the same for all nutrient elements and the number of samples (cores) needed to adequately represent an area is generally a compromise between those elements whose true levels can only be determined by intensive sampling and those elemental concentrations adequately detected with a relatively few samples. A composite of 12 to 15 cores (subsamples taken at random from a "uniform" area) usually forms an adequate sample for operational purposes. A uniform area is delineated on the basis of similar tree species, age, and stand condition or on understory vegetation, soil type (slope, color, texture), and past use. All areas of a management unit that differ significantly from each other should be sampled separately (except small distinct areas may be ignored). The use of a soil or stand map is invaluable in delineating uniform areas and recording sampling locations.

A field description of soil and/or stand conditions should be noted at time of sampling to identify unusual drainage or profile properties or disease or insect problems that might influence the availability and use of nutrients by trees.

The size of the area represented by a composite soil sample is not critical, as long as the area is essentially uniform. In actual practice, the intensity with which an area is managed influences the size of the sampling area. In intensively managed nurseries, one composite sample per hectare is not unusual, while an area two or three times this size would be adequate for a seed orchard. In field plantations, a composite sample may represent 10 to 20 hectares or more, depending on uniformity.

**Analytical Procedures.** All analytical procedures are generally quite similar, except for the solution used to extract the "available" fraction of nutrients from soils. Extracting solutions vary from distilled water or weak bases and salt solutions to strong acids. They have been developed primarily by agricultural chemists for use with annual crops, but some of them appear to have application for forestry use if properly calibrated. While there is some justification for the larger variety of extracting solutions based on differences in crop needs and soil properties, theoretically almost any solution can be used for diagnostic purposes, if the extractable nutrient level has been adequately calibrated against crop response to the added nutrient.

Because of the long-term nature of the forest crop, solutions that extract the "readily available" (generally quite soluble) nutrient available at one point in time are of less importance to the overall growth and development of trees than to annual crops. The latter require relatively high levels of readily available nutrients in a more restricted soil volume; while the *nutrient supplying power* of a soil may be more important for tree crops. This supplying power is dependent on the total soil environment, the inorganic and organic components of the soil, and activity of soil microorganisms. For example, it is more important to have a knowledge of the phosphorus sorption capacity of acid forest soils than for agricultural soils, because the former soils vary more widely in their capacity to retain phosphates over long periods of time. Acid sands, such as found in some coastal plains, often contain insufficient aluminum or iron to effectively retain soluble phosphates for more than a year or so. On the other hand, the aluminum and iron activity of acid clays, particularly wet clays, may be so great as to rapidly fix added phosphates into plant unavailable forms. In either area, the extracting solution should be one that gives a measure of the aluminum and iron activity in the rooting zone.

Soil tests are particularly useful in reforestation in cut-over areas, but even with adequate samples and a well-calibrated extracting solution, soil test results should be considered only as guides for delineating nutrient deficient areas. When possible, they should be used in conjunction with tissue analyses and a knowledge of the soil and plant cover of the test area. Analytical procedures for determining nutrient concentration in extracting solutions and in tissue are detailed in a number of books (Jackson, 1958; Black, 1965).

## Pot Cultures

Pot cultures have been long used as a tool for diagnosing nutrient deficiencies and to give rough quantitative evaluations of fertilizer requirements and nutrient interactions. They allow use of more complicated designs under less variations in soil and climatic conditions. Results can also be obtained in a shorter time period and are generally less difficult to evaluate than field tests. Will and Knight (1968) used an intensive repetitive cropping system in potted soil to assess quantities of major elements available to trees in various layers of a pumice soil. Others (Bengtson et al., 1974) have used greenhouse experiments to evaluate fertilizer sources or to study soil moisture-plant relationships. Many nutrient-physiological studies have employed sand or solution culture techniques to determine relationships between nutrient supply, uptake, growth, and symptoms of deficiencies or toxicities (Ingestad, 1963; Swan, 1972; Will, 1961).

The main disadvantage of using pot trials for diagnosing nutritional problems in trees is the difficulty of extrapolating results to field conditions. In pot trials, trees are grown only to seedling size; they are in an artificial climate, in a disturbed soil; and often the experimenter uses only topsoil from a limited area without regard to variations in the field. The nutrient requirements of seedlings are different from those of adult trees. Tamm (1964) indicated that results of pot trials may give excellent information on nutrition of nursery seedlings, but these results are not necessarily applicable to more mature trees. Richards (1968), while recognizing that these criticisms may be valid, stated that they do not detract from the usefulness of pot trials in determining limiting nutrients.

Mead and Pritchett (1971) reported on a test to compare tree responses to fertilizers in the field with those in pot trials. Eight uniform fertilizer experiments with *Pinus elliottii* were established between 1959 and 1962 on six soil series ranging from excessively drained fine sands to poorly drained sandy loams. Soils of the A1 horizon from control plots in these experiments were used in pots to which the same fertilizer treatments were applied as used in the field. Seedlings were grown in pots for eight months and in the field for seven years. No single measure of seedling response to treatments (height, diameter at soil surface, dry weight of tops, total dry weight) consistently correlated with field responses, expressed in terms of average tree height. There were large differences between field and greenhouse experiments in both type and degree of response obtained from the various soils. The best correlation was between seedling total dry weight and tree height at seven years. Height of greenhouse seedlings was the poorest method of those tested for predicting field response. All results from pot experiments were more poorly correlated with tree heights in the field at three years than they were seven years. Tree growth and the correlation between results obtained in the field with those in potted soils will depend upon the amount of stress imposed on the soil by the trees. In a pot trial, stress is related to seedling density and the length of time seedlings are grown. It would appear that a deficiency could be induced in almost any potted soil, if the seedlings are planted dense enough and grown for sufficient time. Maximum stress might normally be expected within one or two years in pots, but not until canopy closure, or later, in the field (Mead and Pritchett, 1971).

The control of soil moisture conditions and nutrient leaching in pots to simulate field conditions (in the field) is very difficult due to impeding horizons, fluctuating water tables, and seasonal droughts in the field. A commonly used watering procedure is to keep soil moisture near field capacity by periodic watering of weighed pots. This method produces near optimum moisture conditions for tree growth. Although not well suited to greenhouse experiments in which field conditions are simulated, it is excellent for experiments designed to eliminate all limiting factors except the one under test.

To be sure, many factors prevent the direct projection of greenhouse results to field conditions. This is particularly true where the objective is to predict the magnitude of fertilizer response and/or to determine optimum fertilizer rates to use in the field. Greenhouse studies are better suited for use as a diagnostic tool, but even here results should be treated with caution and used mainly as a supplement for field tests.

## Bioassays

Bioassays are methods of diagnosing nutrient deficiencies that combine techniques of tissue analysis and pot tests. One of the early methods was that of Neubauer, in which large numbers (up to 100) of seedlings are grown for two to three weeks in a small amount of soil (about 100 g). The plants are then separated from the soil, dried, and carefully analyzed. It is presumed that the total amount of a nutrient in the plants minus that present in the seed, reflects the amount of that element available to plants in the soil. The growth rate of fungi has also been used to measure available nutrients in small samples of incubated soil. Stanford and Hanway (1955) reported on a simplified technique for determining relative nitrate production in soils during a two-week incubation period.

A number of other modifications of the Neubauer method, or of bioassays, have been attempted. Most of them involve plant extraction of native and/or added nutrients from small amounts of soil over relatively short periods of time. Although these methods generally have not been used for diagnosing deficiencies of deep-rooted and long-lived tree crops, they can be used for testing solubilities of fertilizer materials and relative availability of nutrients in soils.

## Field Tests

Field testing is perhaps the oldest and most reliable method of diagnosing nutrient deficiencies, but it is a costly and time-consuming procedure, and unless the test sites are carefully characterized, the results do not have application over wide areas. Field tests with forest trees can be grouped into two broad categories: (1) those installed at, or near, time of planting; and (2) those installed on established stands. With either method, response to fertilizer treatment can be measured by comparing fertilized with unfertilized trees, or plots of trees; and in older stands by comparing tree growth rate before and after fertilization. The former method gives the most accurate measure of response, because it automatically makes allowance for the effects on growth rate of other factors, such as rainfall and temperature, which vary from year to year.

**Tests on Young Stands.**   Field tests should normally be used in *young stands* in areas of suspected severe deficiency, or if the element(s) under test has a long-lasting effect on growth, such as phosphorus. Net plot size is generally a compromise between one considered large enough to include ample trees to adequately represent plot conditions, after genetic and microsite variations are taken into consideration, and plots small enough to fit into the available area of uniform soil. Soils in relatively flat areas generally have smaller microsite differences and plot sizes can be larger than is possible in sloping and rough terrain. In the coastal plain, net plots of about 1/20 hectare are considered satisfactory for most experiments. If there is not adequate uniform area to include sufficient replication of treatments, it is sometimes necessary to place one replication on one soil condition and another on a distinctly different soil condition. This is an acceptable solution to the space problem as long as due caution is taken to see that there is minimum variation within blocks, and that the soil on which each block is located is well characterized as to chemical and physical properties.

Borders generally consist of two to three rows on all sides of each plot, or a total of four to six rows between net plots. Even this may not be sufficient on some sites, as shown by wide border effect in a slash pine experiment after 20 years (Figure 18.7). There is little evidence of significant cross feeding until trees close canopy, or about 6 to 8 years of age for southern pines. Experiments that are to be terminated at an early age may, therefore, need little or no borders, unless there is physical movement of the fertilizer from plot to plot. Furthermore, if it is suspected that cross feeding is developing into a problem, a "coulter" can be run along the edge of the plots at yearly intervals to sever feeder roots, or the width of the border may be increased by reducing the net plot size.

Height is probably the easiest and most meaningful single measurement of response in young trees, although diameter measurements are also used. Measurements are generally made during the dormant period and diameters can be taken at a standard height or at half height until half height of trees equals breast height (DBH) measurements. In measuring response to fertilizers in either young

**Figure 18.7**  The border effect in this 20-year-old slash pine fertilizer experiment extends to the center of the 10-row-wide check plot.

or old stands one should also obtain an estimate of the effects of treatments on (1) disease and insect susceptibility of trees, (2) secondary vegetation which may offer competition to the trees or provide browse for wildlife, (3) tree survival, and (4) changes in tree form, growth pattern, and wood properties.

**Tests in Established Stands.**  *Established stands* often suffer from varying degrees of nutrient deficiencies. These range from stands that have stagnated due to severe deficiencies to others where only small response can be expected from fertilization. Nitrogen appears to be the element most often deficient in older stands and comparatively small responses may be economical because of the relatively short investment period. Most of the forest fertilization in Scandinavia involves nitrogen applications to spruce stands 50 to 70 years of age, where much of the soil nitrogen has been immobilized in the biomass.

    *Single tree* plots (Viro, 1967) have been used primarily to screen for nutrient deficiencies in older stands. Typically this technique involves a series of circular 1/100 hectare treated plots, each around a selected "measurement" tree. This central tree is marked (at breast height), numbered and measured for height and diameter. All other trees within the circle are measured for diameter and the area volume is calculated on the basis of the basal area-volume ratio of the "measurement" tree. Single-tree plots are not widely used in nutrition studies because of the danger that cross feeding by trees outside the small plots might invalidate results.

Some form of *randomized complete blocks* is often used in established stands, as in young stands. However, extraordinary control is sometimes needed in order to detect relatively small difference in growth increment obtained during short test periods in established stands. This problem is often complicated by increased mortality of intermediate and suppressed trees as a result of fertilizer treatments. Increased mortality of Douglas-fir following nitrogen fertilization is shown in Table 18.2. Using only dominant and codominant trees as measurement trees or thinning from below prior to fertilization are approaches to the problem. Figure 18.8 shows relationship of initial diameter to nitrogen response. A more promising solution is to determine the basal area in all completely randomized plots. The plots are then stratified according to basal area into groups equaling the desired number of replications. That is, one group of low basal area plots will receive a replication of treatments, a group of intermediate basal area plots will receive another replication of treatments, and so on. The net plots in older established stands do not differ greatly in size from those in young stands, although a wider border is needed in older stands to reduce cross feeding.

Growth response to fertilization in older stands is usually expressed as periodic volume or basal area increment. Ideally the former should be obtained by measuring height and diameter of every tree in the net plot at time of treatment and at specified intervals thereafter (usually two to five years). If time does not permit measuring heights of all trees, subsampling must be used. Some random method of selecting a specified number of trees can be used. The number of "measurement" trees needed to adequately represent the plot depends on stand variability and the degree of accuracy desired, but probably should not be less than six. One method involves stratifying diameters of all trees in the plot into 2.5-cm-diameter classes. Next, a sample tree is selected at random from

Table 18.2   Winter Damage (ice, snow, and wind) and Stocking 7 Years After Treatment of 35-year-old Douglas-Fir with Nitrogen Fertilizer (Miller and Pienaar, 1973)

| Nitrogen Added | Initial Stocking | Winter Breakage | Stocking 7 Years Later |
|---|---|---|---|
| kg/ha | | Stems/ha | |
| 0 | 1483 | 279 | 230 |
| 157 | 1416 | 264 | 166 |
| 314 | 1549 | 544 | 477 |
| 470 | 1532 | 642 | 561 |

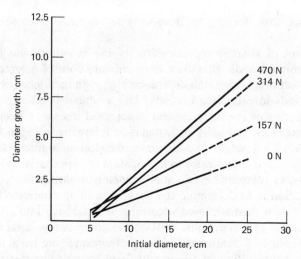

**Figure 18.8** Relation of initial tree diameter to 7-year-diameter (DBH) growth response of Douglas-fir to nitrogen fertilizer. Trees were 35 years old at time of application of 157, 314, or 470 kg nitrogen per hectare (Miller and Pienaar, 1973).

among the first six trees in each diameter class, and every sixth tree thereafter in each class is measured. Plot volumes can be estimated by using the "measurement" trees to establish a regression relationship between volume and diameter. A separate regression is fitted for each plot and the regression used to establish predicted volumes for all trees in the plot.

Unless stand density varies more than 50 percent from plot to plot, there appears to be little need to use this factor as a covariant. On the other hand, the relative size of the trees in a given stand influences their capacity to compete and respond to fertilization. Consequently, initial basal area should be used in a covariant analyses, unless basal area was used in stratifying the plots. There are indications of changes in tree form following treatments and, when possible, volume equations should be used that take this factor into consideration.

Basal area increment is easier to obtain and may be a more sensitive measure of response in certain species of older trees than volume measurements. Basal area response can be obtained by comparing periodic increments of treated versus untreated trees, or by comparing growth before and after treatments by measuring widths of annual rings.

## Indicator Plants

*Ground vegetation* has been used with varying degrees of success in site classification, but it has not often been useful an an indicator for the deficiency of a particular nutrient. Undoubtedly certain understory species have a higher requirement for some nutrients than other species and it is not unusual to see a change in the spectrum of understory species following the application of fertilizers to a deficient site. However, competition and other site factors often exert an overriding influence on the establishment and growth of most species so that subtle changes in soil nutrient levels can not be detected. There are a few instances of species associated exclusively with nutrient-rich or nutrient-deficient sites, such as certain plants growing on zinc– or copper–enriched areas and *Epilobium angustifolium,* an indicator for high nitrate sites (Zöttl, 1973).

Zöttl (1973) proposed that the analysis of understory vegetation might be a sensitive index to suitability of areas for reforestation where tree species are absent. Other diagnostic techniques have been used under special conditions, particularly for the diagnosis of micronutrient deficiencies. These include the application of small amounts of readily soluble fertilizer around the tree, injecting a micronutrient solution into the stem or dipping chlorotic needles into a dilute solution of the nutrient suspected of being deficient. The treated trees are checked from time to time for a growth response or disappearance of the deficiency symptom. Most of the diagnostic methods described herein are useful when properly applied, but few, if any, are infallible. A combination of two or more of the techniques will give a more reliable answer than any one test.

## CORRECTION OF NUTRIENT DEFICIENCIES

Although nutrient deficiencies in forests can sometimes be avoided by proper matching of site and species or, perhaps, corrected by cultural practices, the remedy for most nutrient deficiencies is the application of the proper kind and amount of chemical fertilizers.

### Operational Fertilization of Forests

Results from a large number of well-designed fertilizer trials, established within the past two decades and indicating favorable economic returns on the fertilizer investment, have helped establish fertilization as a viable management technique for intensively cultured forests. It is only with the development of sound scientific principles that the use of fertilizers in forest management can be closely integrated with other cultural practices, compatible with the establishment and development of forest stands on given sites to meet management objectives efficiently and effectively (Hagner and Leaf, 1973). Baule (1973) reported that 4 million hectares of forests have been fertilized worldwide and that by 1980 the total area would probably reach 15 or 16 million hectares. Since some of this area represents repeat applications, he estimates that this represents 0.20 to 0.25 percent of the world's forest lands.

Fertilizers are generally applied to forests (1) at or near time of stand establishment, or (2) to established stands well after crown closure, although a combination of the two is sometimes used. Hagner and Leaf (1973) stated that the choice of application time in relation to stand development depends largely on the existence of nutrient deficiencies at time of planting, longevity of the response to the added nutrient, site conditions, tree species, stocking and length of rotation, compatibility with other forest cultural practices, use and value of the wood produced, and other economic considerations.

## Fertilizing Young Stands

Fertilizing near time of planting may be essential to assure stand establishment and early growth where lack of available nutrients is the major limiting factor. It may also hasten crown closure, reduce costs associated with weeding, and reduce the time that young trees are subjected to intensive animal browsing. Fertilization at time of stand establishment may involve a localized application made concurrently with site preparation; it can be applied in a band down the row of seedlings, or made in a broadcast application with either air or ground equipment. It may also be used in conjunction with herbicides to minimize fertilizer stimulation of competitive ground cover vegetation (Hagner and Leaf, 1973). On sites diagnosed as deficient in phosphorus, potassium, or micronutrients, single applications of fertilizers are applied early in the rotation because of the longevity of the response. In the lower coastal plain of the southeastern United States, southwest France, England, parts of Australia, New Zealand and a few other countries of the Southern Hemisphere, phosphorus appears to be the nutrient most often deficient for good growth of conifers (Gentle et al., 1965; Pritchett and Smith, 1970; Waring, 1973).

**Phosphorus Deficiencies in Young Stands.**   Forest soils generally contain lower levels of phosphorus than do regularly fertilized cultivated soils. Nevertheless, phosphorus deficiencies are not widespread in forests, and many species apparently make good growth at relatively low levels of available soil phosphorus. This ability to survive at low phosphorus levels may be due to the trees' capacity to exploit larger volumes of soil and use less available forms of soil phosphorus than annual crops. At any rate, trees have covered much of the earth's land surface for millennia without yield declines due to phosphorus exhaustion. Crucial to this continued productivity has been the establishment in mature forests of near equilibrium in which nutrients, taken up in the forest vegetation, are released for reuse after the litter falls to the forest floor and slowly decomposes. It is usually following management practices associated with efforts to increase forest yields by more intensive management that phosphorus deficiencies are noted.

The relatively small demands for phosphorus made on the soil by forest species indicates that most world soils have sufficient reserves to replace phosphorus removed in tree harvests. However, some clay and clay loam soils with high phosphorus-fixing capacities and coastal sands and peatlands with very low phosphorus reserves cannot meet the requirements for sustained high tree yields under short-rotation management; particularly if management includes such practices as whole-tree harvests, intensive site preparations, and the use of more demanding, genetically improved planting stock. Nutrient reserves in sandy soils reside almost entirely in the surface organic layers, and the total content of phosphorus may not exceed 40 to 80 kg per hectare. Much of the phosphorus in the humus layers is unavailable, but the debris and surface organic materials are subject to rapid decompsition upon removal of the tree crop and the disturbance resulting from site preparation for reforestation. In some coastal areas, acid quartz sands containing low iron and aluminum concentrations and low organic matter contents have a small capacity to fix phosphorus and, consequently, soluble phosphate compounds may be easily leached from the zone of feeder roots into lower horizons. In contrast, many clay soils, particularly wet clays, contain such high concentrations of active iron and aluminum that any soluble phosphates entering the soil solution as a result of mineralization are soon converted to insoluble forms which may not be readily available to trees (Humphreys and Pritchett, 1971). Phosphorus deficiencies are likely to develop in intensively managed forests under either of these soil conditions, thereby requiring the use of phosphate fertilizers if adequate survival and good tree growth rates are to be maintained.

**Nitrogen Deficiencies in Young Stands.**   Nitrogen is normally not applied to young conifers, except for plantings on abandoned agricultural lands and other areas where the nitrogen and organic matter contents are extremely low. Even in these situations and for hardwood plantings, nitrogen applications are often delayed for two or three years after planting in an effort to (1) minimize mortality resulting from excessive soluble fertilizer salts, particularly on droughty soils; (2) reduce competition from excessive weed growth; and (3) prevent significant nitrogen leaching losses. Root systems of seedlings are generally not sufficiently well developed to effectively use soluble fertilizers during the first year after planting. Furthermore, applications of nitrogen to areas that are severely phosphorus deficient may actually suppress tree growth until the phosphorus deficiency is first corrected (Maftoun and Pritchett, 1970; Munevar and Wollum, 1977).

The nitrogen supply of most soils is apparently adequate for early growth of young seedlings. This is particularly true of plantations established on cleared *forest* lands where considerable nitrogen may be mineralized from the easily decomposable fractions of forest floor organic materials. Other significant amounts of nitrogen may be fixed by nonsymbiotic organisms and by symbiotic organisms associated with leguminous or nonleguminous nitrogen fixers, before crown closure. Operational applications of nitrogen to young stands have been made most frequently to short rotation and intensively managed hardwoods

(Capel and Coffman, 1966), and to conifers planted on sands of low organic matter content, such as found in Australia (Waring, 1973) and the flatwoods of the southeastern United States (Pritchett and Gray, 1974). Response of young pine to nitrogen and phosphorus fertilizers as influenced by drainage characteristics is shown in Figure 18.9.

**Potassium Deficiencies in Young Stands.**   On the deep alluvial sands in the northeastern United States and eastern Canada, potassium is often the nutrient that limits maximum growth of *pinus* and *picea* species, and thus it should be applied at time of stand establishment. A single application may last for the entire rotation (Fornes et al., 1970; Stone and Leaf, 1967). Similar long-term responses from potassium applications have been reported for some areas of central Europe (Baule and Ficker, 1970).

Potassium deficiencies occur most commonly in acid sands to loamy sands low in organic matter, in total cation exchange capacity, and in base saturation, such as those of alluvial and aeolian origin. Potassium deficiencies have also been reported on heath lands and on organic soils. Although potassium deficiencies are found in some forests in Yugoslavia, Denmark, Chile, Australia, South Africa, Great Britain, France, Russia, Germany, Canada, and the United States

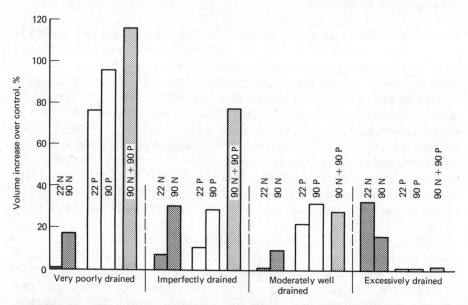

**Figure 18.9**   Percent volume increase over untreated slash pine during the first five years after fertilizing at planting with 22 or 90 kg nitrogen or phosphorus, or a combination of 90 kg nitrogen and phosphorus per hectare on four soil drainage classes in the coastal plains (average of 4 to 6 experiments in each drainage class).

(Leaf, 1968), there are no large areas where potassium fertilizers are applied operationally to forests. Forest trees are apparently able to thrive on relatively low levels of potassium through an efficent system of internal and external cycling of the element.

**Secondary and Micronutrient Deficiencies in Young Stands.**  Calcium, magnesium, and sulfur are absorbed by trees in fairly large amounts, but fetilizers supplying these secondary plant nutrients are seldom used for correcting deficiencies in young trees. Sulfur and limestone are sometimes used for adjusting soil pH to more suitable reactions for seedlings in nurseries, and limestone has been used in midage German forests to hasten decomposition of thick litter layers, increase nitrogen availability, and ameliorate the soil. However, neither limestone nor sulfur is used operationally in young forests.

Where micronutrients are suspected or known to be deficient in soils, they should be applied near time of planting, because of their benefit to the establishment and growth of young seedlings and their relatively long-lasting effect on tree growth. However, in most instances a deficiency will not be suspected until some symptom of abnormal growth or foliage color is noticed. Micronutrient deficiencies have been associated almost entirely with plantation forests and are seldom found in natural forests. Boron is probably the micronutrient most commonly deficient in forests. Deficiencies have been found in plantations of exotic species in Australia, New Zealand, Chile, and some countries of Africa (Stone, 1968). Copper has been reported as deficient in conifers planted on sandy heath lands of northern Germany, in nurseries in England, and on peats and humic sands elsewhere (Stone, 1968). Instances of iron, manganese, and zinc deficiencies in young forests are known, but are not common.

## Fertilizing Established Stands

**Nitrogen Deficiencies in Older Stands.**  Nitrogen fertilizer materials are the most frequently used in fertilization of older stands. Nitrogen deficiency may not become critical until well after crown closure when much of the nitrogen becomes immobilized in the ecosystem organic matter. Deficiencies often result from the accumulation of organic nitrogen reserves in the forest floor, low nitrogen mineralization rates, and strong biological competition for nitrogen after crown closure (Weetman, 1962, Hagner and Leaf, 1973). Therefore, applying nitrogen to older stands on mineral soils is often economically attractive because growth increases can be readily harvested in intermediate or final cuttings. The value of such increases is greater in older trees than in young trees and the carrying costs associated with the investment in fertilizers applied near the end of the rotation is much less than when the fertilizer is applied near time of planting.

The areas of largest use are in the Scandinavian countries and the Pacific

Northwest of North America. For example, Möller (1974) reported that about 100,000 ha of pine and spruce were fertilized in Sweden each year. This annual rate represented approximately 0.5 percent of their total forest area. The area of forests fertilized annually in Finland was approximately the same as that fertilized in Sweden. While most of the fertilizer was applied to established forests on mineral soils in Sweden, about two-thirds of the forests fertilized in Finland were on peatlands. Other central and northern European countries fertilized approximately 50,000 ha of older stands of pine and spruce. In the northwest of North America, some 150,000 to 200,000 ha of predominately Douglas-fir were fertilized annually by 1978. Japan is apparently the only other country fertilizing significant areas of established stands. Hagner (1971) reported that 20,000 to 30,000 ha were fertilized annually in Japan and about 2,000 ha in Oceania.

**Other Nutrient Deficiencies in Older Stands.**   While the great percentage of fertilizers applied to older stands consists of nitrogen materials, there are localized areas where deficiencies of other nutrients are of equal or greater importance than nitrogen. Phosphorus is generally applied in combination with nitrogen to stagnant pine stands in many coastal wetlands. These soils are extremely deficient in phosphorus and significant growth responses can generally be obtained from applications of the combined elements to midage stands where there is adequate stocking (Pritchett and Smith, 1974).

Fertilizers containing nitrogen and phosphorus, and sometimes potassium, are applied to older plantings on some abandoned agricultural soils and on peat and muck lands. Complete fertilizers are perhaps most commonly used in central Europe and Japan. Micronutrient applications are sometimes made to older plantings in Japan, Oceania, and elsewhere, particularly where trees are grown on volcanic soils (Hagner, 1971).

# 19

# FERTILIZER MATERIALS AND THEIR REACTIONS IN FOREST SOILS

The identification of nutrient deficiencies in forest lands has resulted in rapid advances in the science and technology of forest fertilization during the past decade. However, knowledge of the reactions and fate of fertilizer materials applied to forest soils is still fragmentary, and much information developed from agronomic research has been borrowed to fill the void. Unfortunately, much of the standard agronomic fertilizer technology cannot be applied directly to forest situations due to some unique properties of forest soils and nutrient requirements of tree species. Effective and economical use of fertilizers in forestry is largely dependent on an understanding of the properties of the various fertilizer materials and their reactions in forest soils.

The discussion on fertilizer materials for use on forest lands is grouped on the basis of stand age: (1) fertilizers for young plantings and (2) those for midage stands. Some of the following general considerations may help in selecting fertilizer materials for these two conditions.

1.  Phosphorus and phosphorus-nitrogen materials are most often applied to seedlings of plantation forests. Except for special situations, such as in some hardwood plantings and the reforestation of previously cultivated fields, soils of young plantations are quite acid and contain large amounts of partially decomposed organic materials mixed with the mineral soil. During the first year or so after planting, the rather sparse root systems of young trees do not effectively occupy the total soil mass between tree rows. A portion of any soluble fertilizers broadcast during this period may be lost from the site

during heavy rainfall or the fertilizers may stimulate excessive competition between other native species and the young trees. Care must also be taken that the concentrations of soluble fertilizer salts do not become sufficiently high as to damage seedling roots on dry sites. This is a particular concern when the fertilizer material is concentrated in bands.

2.  Nitrogen is the element most often deficient in older stands. Only infrequently have responses to phosphorus, potassium, or other elements been reported under these conditions. Nitrogen deficiency is apparently caused by the gradual immobilization of the element in the forest floor biomass in cool climates or, in temperate or tropical areas, it may be due to the depletion of nitrogen in short-rotation plantations. Fertilization of midage stands usually requires the use of aircraft. Pelletization is important, but solubility of the fertilizer material is of less importance when fertilizing midage stands than when fertilizing young stands.

## NITROGEN EERTILIZERS

Nitrogen is one of the most abundant nutrient elements in the forest ecosystem and is at the same time the element most frequently limiting to tree growth. Deficiencies of this element have been rather widely reported in coniferous stands in cool climates where deep layers of litter and humus materials (with wide C:N ratios) develop as a result of slow rates of decomposition (Weetman and Webber, 1972). At the other climatic extreme, large areas of intensively managed forests in temperate and tropical zones apparently respond to nitrogen fertilizers due to the impoverished conditions of these soils as a result of rapid oxidization of organic matter and loss of mineral elements through leaching.

### Properties of Nitrogen Sources

A wide variety of nitrogen materials is available to the silviculturist. They vary greatly in nutrient concentrations (Table 19.1) and in their reactions with soils. Urea and ammonium nitrate are the most commonly used nitrogen sources in forestry, primarily because of their high nitrogen concentrations. They also possess good physical properties and, as a consequence, they are less expensive to handle and apply than most other sources. However, the forms in which nitrogen is present and the kind of associated ion in the fertilizer compound can produce effects which may be more important than considerations of nitrogen concentration. Smith et al. (1971), McFee and Stone (1968), Pharis et al. (1964), and Benzian (1965) have indicated that ammoniacal nitrogen was physiologically a preferred form to nitrate nitrogen for young conifers. Furthermore, Benzian (1965) found ammonium sulfate superior to nitrochalk and calcium nitrate, especially on near-neutral nursery soils. Ammonium sulfate has sometimes been more effective than ammonium nitrate in reforesting denuded areas, such as old "borrow" pits. In some instances, the superiority of the former material was

**Table 19.1** Composition of Some Fertilizer Materials Used in Forests

| Material | N | P | K | Ca | Mg | S | Available P[a] |
|---|---|---|---|---|---|---|---|
|  |  | percent |  |  |  |  | % of total |
| Ammonium nitrate | 34 | — | — | — | — | — | — |
| Ammonium nitrate-lime | 20 | — | — | 7 | 4 | — | — |
| Ammonium phosphates |  |  |  |  |  |  |  |
|   (18-46-0)[b] DAP | 18 | 20 | — | — | — | — | 100 |
|   (15-60-0) APP | 15 | 26 | — | — | — | — | 100 |
|   (11-48-0) MAP | 11 | 21 | — | — | — | — | 100 |
| Ammonium phosphate-sulfate | 16 | 8 | — | — | — | 14 | 100 |
| Ammonium sulfate | 21 | — | — | — | — | 23 | — |
| Basic slag (Thomas) | — | 4–8 | — | 32 | 3 | 0.2 | 62–94 |
| Calcium nitrate | 16 | — | — | 19 | 1 | — | — |
| Calcium cyanamide | 22 | — | — | 39 | — | — | 98 |
| Dicalcium phosphate | — | 23 | — | 29 | — | — | 98 |
| Isobutylidene diurea | 32 | — | — | — | — | — | — |
| Magnesium ammonium phosphate | 8 | 17 | — | — | 14 | — | — |
| Potassium chloride | — | — | 50 | — | — | — | — |
| Potassium nitrate | 13 | — | 37 | — | — | — | — |
| Potassium-magnesium sulfate | — | — | 18 | — | 11 | 18 | — |
| Potassium sulfate | — | — | 42 | — | — | 18 | — |
| Raw rock phosphate | — | 11–17 | — | 33–36 | — | — | 14–65 |
| Superphosphate (ordinary) | — | 7–9 | — | 18–21 | — | 10–12 | 97–100 |
| Superphosphate (concentrated) | — | 19–23 | — | 12–14 | — | 0–1 | 96–99 |
| Sodium nitrate | 16 | — | — | — | — | — | — |
| Urea | 46 | — | — | — | — | — | — |
| Ureaformaldehyde | 38 | — | — | — | — | — | — |

NOTE: P × 2.29 = $P_2O_5$
       K × 1.20 = $K_2O$
       Ca × 1.40 = CaO
       Mg × 1.66 = MgO

[a]Citrate soluble
[b]Expressed as $N-P_2O_5-K_2O$

thought to be related to its sulphur content. Benzian attributed the larger growth response primarily to the greater residual acidity of the ammonium sulfate. However, it is possible that any advantage of ammoniacal sources over nitrate materials is largely independent of pH (McFee and Stone, 1968) or the associated sulphur ion (Leaf, 1968).

On the other hand, Ebell (1972) reported an apparent superiority of nitrate over ammoniacal nitrogen for stimulation of cone induction in Douglas-fir. Nitrates, and high levels of inorganic nitrogen in general, are suspected of antagonizing pine root mycotrophy, and several pathogens are known to preferentially attack conifers under high nitrogen regimes (Björkman, 1967). Osmotic effects exerted by soluble salts produce varying degrees of physiological drought in young seedlings. Pritchett and Robertson (1960) reported that 150 ppm nitrogen from soluble sources was about the maximum that pine seedlings could tolerate on droughty sands. The possibility of injury depends not only on the amount of fertilizer added and the soil moisture content, but also on the *salt index* of the material. For example, sodium nitrate produces about 3.7 times as much salt in the soil solution as does urea, per unit of nitrogen added. Free ammonia, liberated upon hydrolysis of urea and from diammonium phosphate in alkaline soil, can damage seedling roots and excess ammonium ions can interfere with uptake of other cations. However, the possible beneficial effects of ions associated with ammonium compounds that may help correct multiple deficiencies should not be overlooked.

## Recovery of Applied Nitrogen

The percent recovery of nitrogen applied to an ecosystem depends to a large extent on its retention against leaching and gaseous losses. Nitrates are readily leached from soils and the application of nitrate sources may result in its movement into surface and ground water (Bengtson and Kilmer, 1975). Paavilainen (1972) reported that calcium ammonium nitrate produced better growth of Scots pine *(Pinus silvestris)* on drained peatlands than did ammonium sulfate, calcium nitrate, or urea. Although Paavilainen obtained an initially strong, but short-lived radial growth response to calcium nitrate, the nitrate was rapidly leached from the soil. Ammonium sulfate was only slightly leached and, thus, produced a long-term height growth response. Volatile loss of ammonia may have been the cause for a relatively poor performance of urea. Although ammonium does not move as readily in soils as nitrate, it can be readily converted to the more mobile nitrate form in some forest soils. The activity of most nitrifying organisms is pH dependent and, consequently, the conversion of ammoniacal nitrogen to nitrate nitrogen is believed to be very slow in most acid forest soils.

Historically, urea has been a favored nitrogen source for forest fertilization, particularly for aerial applications to established stands (Nömmik, 1973). However, under some conditions nitrogen recovery from urea can be very low.

Results of field trials in Sweden (Möller, 1973) revealed that ammonium nitrate increased growth of Scots pine stands on mineral soils from 30 to 50 percent more than that obtained with equivalent nitrogen rates applied as urea, after five years. The superiority of ammonium nitrate over urea averaged about 30 percent when applied to Norway spruce *(Picea abies)*. One reason given for the poor showing of urea under Scandinavian conditions was immobilization of urea nitrogen by highly acid humus material of the forest floor. The immobilization and remineralization of nitrogen were found to be positively correlated with temperature and application rates and, to a great extent, governed by the nitrogen carrier itself. Overrein (1970) suggested that changes in the microenvironment associated with urea hydrolysis, besides directly affecting microbial processes, may also affect the nonbiological/biological agencies responsible for the conversion of nitrogen from an active organic phase into a more stable nonextractable passive pool in the humus. There may also be significant changes in the soil microflora following nitrogen applications. Bacterial numbers increase as a result of the decrease in soil acidity brought about by urea hydrolysis. This may lead to greater nitrification and increased leaching losses.

Volatile losses of ammonia from surface-applied prilled urea has ranged from 5 to 10 percent or less (Overrein, 1968; Volk, 1970) to 20 to 40 percent (Watson et al., 1972; Nömmik, 1973). Ammonia losses from urea applied to acid forest soils are explained by the presence of local alkalinity around the urea pellet as a consequence of the enzymatic hydrolysis of urea to ammonia and carbon dioxide. High losses are associated with high initial soil alkalinity, high urease activity, low ammonia-absorption capacity, high temperature, low nitrification capacity, and low soil moisture (Nömmik, 1973). The application of high rates of urea solubilizes a portion of the organic matter that may cause a crust on the soil surface upon drying.

Although Volk (1970) reported gaseous losses of ammonia from prilled urea applied to coastal plain forest soils averaged no more than about 12 percent, due to the low rate of enzymatic activity in those soils, the need for further reduction of these losses is obvious. Timing urea applications to coincide with a period of rainfall usually reduces gaseous losses because the water carries the dissolved urea into the soil where the ammonia can be held on the exchange complex of the soil colloid upon hydrolysis of the urea. Burning the forest floor prior to urea application may slow enzymatic activity and reduce losses. The leaching of unhydrolyzed urea and of dispersed ammonium-saturated humus colloids and associated cations to deeper horizons is a possibility in some acid sands, but such losses are probably not significant under most conditions (Volk, 1970). Results indicate that if fertilizer nitrogen must be applied during the warm, dry summer months, sources such as ammonium nitrate will usually prove superior to urea.

Nömmik (1973) reported that increasing the size of the urea pellet may be a viable method of reducing gaseous losses of nitrogen from surface application.

Volatilization losses from an application of 200 kg nitrogen per hectare, after 31 days of exposure on a *Pinus silvestris* forest floor, averaged 27 percent from small pellets (2-4 mg urea per pellet), but only 15 percent when the same amount of urea was applied in large pellets (2.1 g each). Nömmik (1973) explained that the reduction in losses from large pellets may be due to the increased rate of vertical diffusion of urea and ammonia by increasing the concentration gradient in the zone around the large pellets. Furthermore, the hydrolysis may be retarded by a slower dissolution of the large-pellet urea. The delayed dissolution enhances the chances of the urea being washed into the soil prior to hydrolysis during periods of meager rainfall.

The dissolution of urea may also be delayed by coating the pellet with water-resistant coatings or semipermeable membranes, which prolong the integrity of the pellet. Sulfur or paraffin coatings have been used to delay the release of urea from granules. The length of the delay is a function of the uniformity and thickness of the coating, the kinds and quantities of waxes and microbiocides incorporated into the coating, and the ambient temperature and intimacy of soil contact. Typically, a small percentage of the pellets release their urea soon after application, with a sustained release of from 3 to 9 months. Sulfur-coated ureas have been mostly used in nurseries, mine spoil revegetation, and special forest conditions (Bengtson, 1973). With semipermeable membranes, the urea is released only after the membrane is ruptured following an increase in osmotic pressure resulting from water moving through the membrane into the pellet. Coatings have not been widely accepted for operational use because of their high costs.

A combination of large pellet size with the addition of urease inhibitors, or nonspecific metabolic inhibitors, reportedly reduces the risk of volatile loss of ammonia from applied urea. Bremner and Douglas (1971) reported that phosphoric acid in urea phosphate retarded enzymatic hydrolysis of urea and markedly reduced the volatile loss of urea-nitrogen as ammonia. Nömmik (1973) reported that incorporation of small amounts of either orthophosphoric or orthoboric acid into large urea pellets reduced volatile ammonia losses to half of that from untreated pellets and that further reduction probably could be obtained by increasing the phosphoric acid proportion above the 1:22 phosphoric acid:urea mole ratio that he used. The use of higher ratios of phosphoric acid to urea appears to be practical because many forest soils are deficient in both phosphorus and nitrogen (Bengtson, 1973).

## Other Nitrogen Compounds for Forestry Use

*Ureaformaldehyde* and *isobutylidene diurea* (IBDU), condensation products of urea, have low water solubility and the attendant advantages of slow release and low salt index. They have been used successfully in nurseries and for fertilizing ornamental and Christmas trees, but their high costs per unit of nitrogen have discouraged their widespread use in forests.

Magnesium and other *metal ammonium phosphates* have slow release properties but have not proven sufficiently superior to readily soluble sources to make them attractive for forestry use.

*Anhydrous ammonia* ($NH_3^+$) contains the highest nitrogen concentration of any nitrogenous fertilizer currently marketed and it is one of the least expensive per unit of the element. It is a gas at room temperature and is stored under pressure as a liquid. Its application requires the use of high-pressure tanks and special injection devices. A "forest roller," developed in Germany (Mayer-Krapoll, 1956) for the amelioration of coniferous raw humus, has successfully applied ammonia gas to forest stands. However, it is difficult to apply in most established forests and its application may cause injury to tree roots. It has some potential use in young plantations, particularly hardwood plantations, and in nursery beds, but the difficulty of its application keeps anhydrous ammonia from being truly competitive with solid nitrogen compounds.

*Herbicides* have been applied to young plantings in combination with nitrogen fertilizers, as a means of reducing competing vegetation that is often stimulated by the fertilizer. The herbicide is usually applied in a circle around individual trees or in a band down the tree rows in a separate operation from the fertilization. However, an even more effective procedure appears to be the use of some granular fertilizer coated with a herbicide. For example, diammonium phosphate coated with simazine (2 percent active ingredient) gave good control of herbaceous vegetation when band-applied at the rate of 200 kg per hectare in the southeastern United States coastal plain. Tree heights were greater where herbicide-coated diammonium phosphate was used than where diammonium phosphate was used alone, after two years. Although the effect of a herbicide treatment on tree growth probably will not persist through the rotation period, its influence on seedling survival may be of real value—especially where fertilizers are used on dry sites (Figure 15.7).

## PHOSPHATE FERTILIZERS

The present fertilizer industry traces its beginning to the discovery by Liebig in 1840 that the availability of phosphorus in bones could be increased by treatment with acid. At about the same time, Lawes developed and patented a process for acidulating phosphate rock with sulfuric acid to make ordinary superphosphate. Since the first commercial production in England in 1843, the process for the manufacture of this material has not greatly changed. However, in the last decade considerable progress has been made in developing fertilizer compounds with much higher concentration of phosphorus and often in combination with other nutrient elements.

### Characteristics of Phosphate Fertilizers

The concentration of phosphorus in fertilizers has historically been expressed as the oxide ($P_2O_5$) equivalent. This archaic term will be avoided in this text wherever possible in favor of the simpler elemental expression. However, because most phosphate fertilizers are still sold on the basis of their $P_2O_5$ content, the guarantee (or grade) is expressed as $P_2O_5$ as well as percent phosphorus (P). The following conversion factors are convenient for changing from one expression to the other:

$$\% \, P = \% \, P_2O_5 \times 0.43$$
$$\% \, P_2O_5 = \% \, P \times 2.29$$

A list of the more popular phosphate fertilizers for forest fertilization are given in Table 19.1, along with their phosphorus concentrations.

The plant availability of phosphorus in fertilizer materials is difficult to assess. As a consequence, the solubility of the element, as determined with various extractants, is widely used as a measure of its availability to plants. The fraction of the fertilizer phosphorus extractable in water is expressed as a percentage by weight of the total material and termed *water-soluble phosphorus*. The ammonium orthophosphates are 100 percent water soluble, while super-phosphates are between 90 and 98 percent water soluble. Phosphorus dissolved in normal ammonium citrate, after removal of the water-soluble fraction, is termed *citrate soluble* (Table 19.1). Some familiar materials that are largely water-insoluble, but citrate soluble, are di- and tricalcium phosphate and basic slag. *Available phosphorus* is the sum of the water-soluble and citrate-soluble fractions, while the available phosphorus plus the citrate-insoluble fraction represents the total amount present. Colloidal and rock phosphate have generally been considered as largely citrate-insoluble materials; however, they differ rather widely in citrate solubility according to origin, with some of then being fairly soluble (Bengtson et al., 1974).

## Kinds of Phosphate Fertilizers

**Ground Rock Phosphate.**   Deposits of rock phosphate, which occur in several areas of the world, are the only important sources of fertilizer phosphorus. Milled to a fine powder, the material is sometimes used without further processing, but it is more commonly used as the raw material for a family of phosphate fertilizer compounds. The general formula for phosphate rock is $Ca_{10}(PO_4,CO_3)_6$ $(F,Cl,OH)_2$, with considerable substitution by carbonates and other anions and cations found among various deposits. The concentration of phosphorus in rock phosphate also varies among geographic sources, but it averages about 13 to 17 percent, with fluorine concentrations of 3 to 4 percent. The presence of fluorine in the ground rock greatly influences the availability of its phosphorus component. Ground rock phosphates are essentially insoluble in water and their citrate solubility varies from practically nil to about 30 percent. For example, citrate solubility of several samples from North Carolina, Florida, and Idaho averaged

26, 16, and 11 percent, respectively. High citrate solubility increases the effectiveness of direct applications to annual crops, but there does not appear to be much difference for trees in the long-term effectiveness among sources of rock of varying solubilities. The possible exception is that sources with 25 to 30 percent citrate solubility may provide seedlings with an initial growth advantage which will help them in establishing dominance over competing vegetation (Bengtson et al., 1974).

Ground rock phosphate has not found much favor among agriculturalists because its phosphorus is not sufficiently available to annual crops, which are normally grown in near-neutral soils. It is used to a limited extent in pasture fertilization, but ground rock probably has its greatest potential as a phosphorus source for fertilizing forest trees on very acid soils. Organic acids in these soils slowly dissolve the phosphorus in ground rock, making it particularly suitable for forests on soils with very low or with very high phosphorus-fixing capacities. Acid quartzite sands of many coastal areas contain insufficient iron or aluminum (often less than 40 ppm extractable aluminum) in their surface horizon to retain water-soluble phosphates in the rooting zone of seedlings. Examples are the Aquods (Ground-Water Podzols) of many coastal areas, as exemplified by Leon fine sand in Figure 19.1. While much of the phosphorus leached from the surface of these soils is apparently intercepted by the B2h organic pan, it is doubtful that the phosphorus is significantly exploited from this layer by tree roots. Some clay soils, on the other hand, contain very large amounts of active iron and aluminum in their surface horizons, and water-soluble phosphates applied to these soils are gradually converted into relatively unavailable precipitation products. Consequently, relatively insoluble sources should be used on soil of very high phosphorus fixing capacity. The water-soluble phosphates should be reserved for soils with moderate capacities for phosphorus retention in their surface horizons— soils with 40 to 160 ppm $NH_4OAc$-extractable aluminum, for example. Figure 19.2 illustrates the relative effectiveness of ground rock with a water-soluble phosphate on a soil with a higher phosphorus fixing capacity.

The effectiveness of phosphate fertilizers of varying solubilities is closely related to the size of the fertilizer particles. The percent short-term recoveries of highly soluble phosphates generally increase with increases in granule size or reduction in soil contact. On the other hand, response to sparingly soluble phosphates, such as ground rock phosphate, increases as the particle size is reduced. However, finely ground rock is difficult to spread with ground equipment and is almost impossible to spread from the air. This undesirable physical feature has undoubtedly hindered the acceptance of this otherwise desirable material for forestry use. The need for a finely ground material for improved availability, and a coarse material for ease of spreading presents a dilemma to forest managers. A more soluble raw phosphate, such as north African (Gafsa) rock, may be used effectively after it has been crushed and screened to a

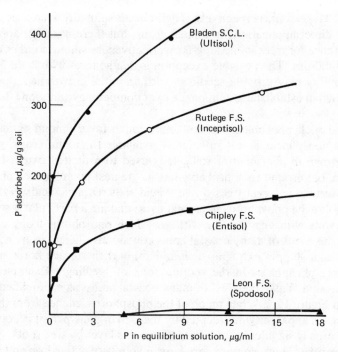

**Figure 19.1** Phosphorus adsorption isotherms of four coastal plain soil types representing four soil orders (Ballard and Fiskell, 1974). Reproduced from *Soil Science Society of America Proceedings,* Vol. 38, No. 2, p. 252, 1974, by permission of the Soil Science Society of America.

relatively uniform sandlike material. It must be emphasized that coarsely ground rock phosphates of low citrate solubility are unlikely to provide adequate phosphorus for tree seedlings even on highly acid soils (Bengtson et al., 1974).

Processes have been developed by TVA National Fertilizer Development Center and others, for the granulation of ground rock phosphate and for cogranulation of rock phosphates with certain fertilizer salts, including KCl and urea, or elemental sulfur. The granulation of partially acidulated rock phosphate provides a mixture of soluble and slowly soluble phosphates in a convenient form for spreading. The ground rock is acidulated with less than the stoichiometric amounts of $H_2SO_4$ or $H_3PO_4$, usually resulting in 25 to 40 percent of the total phosphorus in an available form. The moist matrix can be readily granulated, but it has been reported that pellets thus produced may not slake readily (Bengtson, 1973). It is essential that the granules slake and revert to their original fineness upon contact with moisture, thus increasing their surface area contacts with the soil.

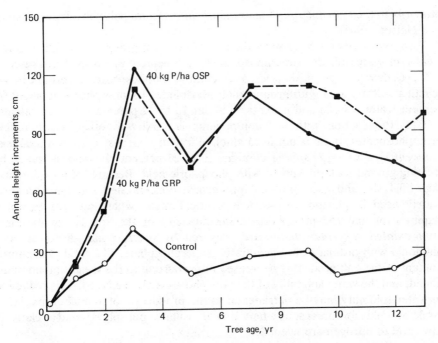

**Figure 19.2** Annual height increments of slash pine as influenced by application of 40 kg phosphorus per hectare as ordinary superphosphate (OSP) or ground rock phosphate (GRP) at time of planting on a Typic Ochraquult.

**Acid-treated Phosphates.**  Most ground rock phosphate, particularly that used in agriculture, is processed to increase its phosphorus availability by an acid or else a heat treatment. The acid-treated compounds, such as phosphoric acid, superphosphates, ammonium phosphates, and nitric phosphates, are the most widely used group of commercial phosphate fertilizers.

*Phosphoric acid* ($H_3PO_4$) is commonly made by reacting rock phosphate with sulfuric acid by the so-called wet process. A much purer form of acid is manufactured by heating phosphate rock in an electric furnace to volatize elemental phosphorus, which is then oxidized to phosphorus pentoxide and combined with water to form phosphoric acid. This latter, more expensive, product is more often used for industrial products than for fertilizers. Phosphoric acid can be applied directly to the soil, where the reaction products are much the same as other orthophosphates. However, agricultural grade phosphoric acid (24 percent P) is most commonly used to acidulate rock phosphate to make concentrated superphosphate, or else it is neutralized with ammonia to produce ammonium phosphates and liquid fertilizers. *Superphosphoric acid,* which contains 34 percent P, is made essentially by dehydrating ordinary phosphoric acid. It is used primarily in the manufacture of ammonium and calcium polyphosphates (Tisdale and Nelson, 1966).

*Superphosphates* have been the most important group of phosphate fertilizers ever since ordinary superphosphate (7 to 9 percent P) was first marketed in 1843. *Ordinary superphosphate,* a calcium orthophosphate, is produced by reacting sulfuric acid with ground rock phosphate. Its phosphorus is about 90 percent water soluble and it contains about 10 to 12 percent sulfur as calcium sulfate. Its reaction with soil components is similar to other water-soluble orthophosphates, but it is not used much in forestry because of its low analysis. *Concentrated superphosphate* contains 19 to 21 percent phosphorus, made by treating ground rock phosphate with phosphoric acid. Because of its solubility, high analysis, and good physical properties, concentrated superphosphate is widely used in phosphorus-deficient young forests, where nitrogen is not a requirement and the phosphorus-fixing capacity of the soil is not extreme. *Ammoniated superphosphates* are prepared by reacting anhydrous or aqua ammonia with ordinary or concentrated superphosphate. The total phosphorus content of the end product is decreased in proportion to the weight of ammonia added, and the water solubility of the phosphorus is also reduced. This reduction in water solubility may be detrimental to the production of annual crops, but it could be a plus for forest soils where slowly available phosphorus and a relatively low level of nitrogen are needed.

*Ammonium phosphates* are commonly produced by reacting ammonia with phosphoric acid. Two popular water-soluble phosphates produced by this process are *monoammonium phosphate,* containing 21 to 26 percent phosphorus with a fertilizer grade of about 11-48-0 ($N-P_2O_5-K_2O$), and *diammonium phosphate* (20 percent P), which is often marketed as an 18-46-0 grade. These ammonium phosphates have the advantage of high plant food content, good handling qualities, and ratios of nitrogen to phosphorus suited for tree seedlings on soils where both of these elements are deficient.

Recently developed phosphate fertilizers that may have value under some forest situations are *ammonium polyphosphate,* a reaction product of superphosphoric acid and ammonia and containing 25 to 26 percent phosphorus with a grade of about 15-60-0; *ammonium phosphate-nitrate; magnesium ammonium phosphate;* and *potassium phosphate.* None of these compounds have yet been adequately tested in forests.

**Heat-treated Phosphates.** Thermal phosphates are made by subjecting phosphate rock to varying degrees of temperature. They do not account for a very large segment of the phosphate market because they are generally more expensive to make and they contain a rather low percentage of available phosphorus. *Basic slag,* or Thomas slag, is a byproduct of the basic open-hearth method of making steel. It is a popular product in Europe where it has been used in nurseries and for fertilizing trees on peatlands and other acid, phosphorus-deficient areas. Thomas phosphate generally contains from 6 to 8 percent phosphorus, but the American product contains only about 60 percent as much. Slags neutralize soil acidity and contain small quantities of a variety of micronutrients that may be of value under special conditions.

*Calcium metaphosphate* is prepared by burning elemental phosphorus in the presence of finely divided rock. Its successful manufacture requires a relatively cheap and abundant source of electric power and, as a consequence, it has not been widely used. It can be used satisfactorily on most acid soils.

### Phosphorus Reactions in Soils

In the acidic environment of forest soils insoluble phosphates are slowly converted to plant available forms. Of equal importance are the reactions of orthophosphates with soil components to form specific compounds, called soil-fertilizer reaction products. For example, when superphosphates are applied to a soil, water vapor moves into the granules forming a saturated solution of monocalcium phosphate and dicalcium phosphate dihydrate. The saturated solution then moves out of the granule leaving a residue of dicalcium phosphate at the original site. This saturated solution is very concentrated with respect to phosphorus and calcium; it is extremely acid (pH 1.5); and it has a reduced vapor pressure. The reduced vapor pressure creates a driving force for further movement of water to the fertilizer and dissolution and movement of fertilizer salts into the soil. The soil moisture content has a marked influence on the rate at which water-soluble phosphorus moves out of the fertilizer granule (Terman, 1971).

The concentrated acid solution moves out in a front reacting with soil constituents. In acid soils, the reaction is largely with compounds of iron, aluminum, and manganese, which are dissolved in the solution and eventually precipitate with the phosphorus (Figure 19.1).

Best results are obtained when slightly soluble phosphate materials are applied in powdered form and thoroughly mixed with a large volume of soil, while water-soluble phosphates give superior results on acid soil when they are granulated and applied in a band.

## POTASSIUM AND OTHER MINERAL FERTILIZERS

### Potassium Sources

Potassium is absorbed by plants in relatively large quantities, but fortunately it is found in most soils in fairly large amounts. Tisdale and Nelson (1966) reported that the earth's crust contained about 2.4 percent potassium, mostly in potash feldspars, muscovite, and biotite minerals. However, the concentration of these primary minerals, as well as that of the secondary or clay minerals of illite, vermiculite, chlorites, and interstratified minerals, is less in the surface soil than in subsoil and their distribution worldwide is far from uniform. Furthermore, much of the total potassium in these minerals is unavailable or only slowly available to plants, and many agricultural soils require the addition of potassium fertilizers for optimum yield. Because trees cycle potassium very efficiently, deficiencies of this element are seldom encountered in forest soils. The exceptions are some peatlands and glacial outwash and coastal sands that are almost devoid of primary potassium-bearing minerals, or they have been depleted of the element through extensive agriculture (Leaf, 1968).

When potassium fertilizers are required in forests, they are generally applied as *potassium chloride* (KCl), which contains about 50 percent K, or *potassium sulfate* ($K_2SO_4$), which contains about 42 percent K and 18 percent S. Potassium chloride is the most widely used of all potassium sources, but potassium sulfate has an advantage in some soils due to its sulfur content. Both of these sources are water soluble, as are all important commercial sources of potassium. When added to soil, they dissolve in the soil solution and dissociate into their respective ions.

A potassium source of minor importance in forestry, but sometimes used in nurseries, seed orchards, and for special situations, is *potassium magnesium sulfate* ($K_2SO_4 \cdot MgSO_4$). This double salt of potassium and magnesium sulfate contains 18 percent K, 11 percent Mg, and 18 percent S. *Potassium nitrate* ($KNO_3$) is an excellent but rather expensive source of potassium (37 percent K), but the associated nitrate ion may not be a desirable source of nitrogen for forest soils. *Potassium metaphosphate* ($KPO_3$) is a concentrated, water-insoluble source containing 33 percent K and 27 percent P that becomes slowly available upon hydrolysis. It has not been widely tested under forestry conditions but should be an excellent source of both phosphorus and potassium for acid soils.

## Potassium Reactions in Soils

Potassium exists in the soil in one of three forms: (1) relatively unavailable, (2) slowly available (fixed), or (3) readily available to plants. These forms are considered to be in equilibrium and any major change in the system can cause a shift in the equilibria. In fact, there is seldom an actual equilibrium existing in the soil due to plant removal and leaching losses and to the continuous but slow transfer of potassium in the primary minerals to the exchangeable and slowly available forms. The unavailable forms occurring in primary minerals and in certain secondary minerals that entrap potassium in their lattice structure account for more than 90 percent of the total soil potassium. When these materials decompose through the weathering process, their potassium is converted to one of the slowly available forms or else it is released as a plant available (exchangeable) form. Figure 19.3 illustrates the equilibrium of the potassium forms in a soil.

The slowly available potassium comes from clay-adsorbed (fixed) potassium

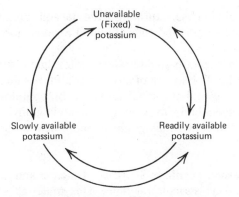

**Figure 19.3** Of the three forms of soil potassium, 90 to 98 percent is in the relatively unavailable (primary and secondary minerals) form, 1 to 10 percent is in the slowly available (nonexchangeable) form, and 1 to 2 percent is in the readily available (exchangeable and soil solution) form.

and is not considered to be in an exchangeable form. Fixation results from entrapment of the $K^+$ ion between the layers of 2:1 clays. This fixed potassium becomes available to plants over long periods of time as a result of wetting and drying of the colloids, temperature fluctuations, and soil acidity changes.

Liming can affect potassium retention in soils through its influence on base saturation. The $K^+$ ion is capable of displacing calcium and other bases from the exchange complex of the soil colloids more easily than it can displace the hydrogen and aluminum ions, which dominate in most acid systems. Losses by leaching of potassium applied to forests are not likely to be large, except in sandy soils and soils subject to flooding. The reduction in leaching losses by forest cover is affected through a reduction in the amount of gravitational water passing through the soil and by absorption of potassium into the plant tissue (Nolan and Pritchett, 1960).

## Calcium, Magnesium, and Sulfur in Soils

These elements are used by trees in relatively large amounts, but they are seldom deficient in forest soils. Calcium and magnesium in soils originate in the decomposition of rocks and minerals from which the soils are formed. Upon liberation into the soil solution, these elements may be absorbed by plants or other organisms, adsorbed by clay particles, lost in percolating water, or precipitated as a secondary mineral. Deficiencies of these bases are most likely to occur in acid, sandy to loamy sand soils, low in organic matter, and low in cation exchange capacity and in base saturation. Alluvial soils are more often deficient than soils of glacial till origin; and soils of humid regions more often than soils of arid areas (Leaf, 1968).

Sulfur is present in soils as sulfides and sulfates, and in organic materials as a component of proteins. The source of most soil sulfur is sulfides of plutonic rock (Tisdale and Nelson, 1966), but a small amount is brought down in the rain. This latter source generally amounts to no more than a kilogram or so per hectare per year but may be as much as 50 kilograms or more near areas of great industrial activity.

**Calcium and Magnesium Fertilizer Sources.** Calcium and magnesium are most economically applied as agricultural lime. This material is used primarily to correct soil acidity, but calcic ($CaCO_3$) limestone is also a primary source of calcium, and dolomitic ($CaCO_3 \cdot MgCO_3$) limestone supplies both calcium and magnesium. Because most forest trees grow well in acid soils and because calcium and magnesium are seldom so deficient as to affect tree growth, lime-stones are seldom used in forests. The exceptions are a few instances where forest managers have used lime to hasten the microbial decomposition of raw humus in order to release nitrogen or phosphorus (Mayer-Krapoll, 1956).

Calcium, and, to a lesser extent magnesium, is contained in some mixed fertilizers or as an associated ion in such fertilizer materials as superphosphate, rock phosphate, gypsum, and ammonium nitrate-lime mixtures. Magnesium is a component of magnesium sulfate and potassium magnesium sulfate. When any of these materials are applied to acid soils they slowly dissolve and the calcium or magnesium is released to the soil solution.

**Sulfur Fertilizer Sources.** Sulfur is seldom applied to soils in the elemental form, except for the purposes of increasing soil acidity. It is sometimes used in forest nurseries to keep the soil reaction acceptably low, but sulfur is more often applied to soils in a number of sulfur-containing fertilizer materials, such as sulfates of ammonium, calcium, magnesium, or potassium, and basic (Thomas) slag and ordinary superphosphate.

Regardless of whether sulfur is applied in the elemental form, as a component of mineral fertilizer compounds, or is derived from the mineralization of sulfur in organic matter, it eventually ends up as sulfate in well-aerated soils. This oxidized form is available to plants, but because of the anionic nature and high solubility of sulfate salts, they are subject to considerable leaching in sandy soils. However, soils with high amounts of clay and hydrous oxides of iron and aluminum have a capacity to retain sulfates in an adsorbed form so that leaching losses are kept to a minimum.

Under waterlogged conditions, in which bacterial reduction takes place, sulfides and even elemental sulfur are formed. Hydrogen sulfide is the product most often formed when sulfates are subjected to reduced conditions and, unless it is precipitated by iron or other metals, it escapes to the atmosphere.

## Micronutrients—Sources and Reactions in Soils

It has been stated (Stone, 1968) that there are essentially no micronutrient deficiencies in natural forests, or, for that matter, in the large areas of moderately disturbed forests still composed of native species growing on intact forest soils. This is due, in large part, to the small quantities of micronutrients required by forests stands and to the efficient and conservative operation of the forest ecosystem. (Thorough and often deep exploitation of the soil, coupled with external and internal cycling of absorbed nutrients, minimize losses and sometimes enrich the surface soil by means of accumulated reserves in the biomass and soil organic matter.) Furthermore, a kind of natural selection has taken place throughout many generations in native stands. Species or individuals sensitive to moderate deficiencies or excesses are handicapped in competition with their more tolerant neighbors; therefore, the vegetation native to an extensive soil area must consist of species and genotypes fitted to the prevailing chemical environment.

Stone (1968) pointed out that natural forests exist in a more stable situation than young forest plantations and, to some extent, volunteer stands of open lands. The planted trees are often exotics, but even when native they may not be truly indigenous genotypes. Their selection or occurrence is determined by factors apart from specific adaptability, and, in many instances, the chemical properties of open lands are quite unlike those of the parent stands. The majority of reports of micronutrient deficiencies in forests concern plantations or disturbed stands, but even among these stands micronutrient deficiencies are not common.

*Boron* occurs in most soils in amounts of only a few parts per million in the form of tourmaline, a resistant borosilicate, or in organic combinations. Upon decomposition of organic materials, boron is released and that part not absorbed by plants is subject to leaching. The amount lost by leaching depends largely on the texture of the soil, movement of water through the soil, and soil acidity. Well-drained sandy or highly weathered soils are most frequently low in boron. Boron deficiency in trees can be induced where soils are marginally supplied with the element by increasing the soil reaction (overliming) or by extremely dry soil conditions (Tisdale and Nelson, 1966).

Boron is the most commonly deficient of the micronutrients in forest plantations. There have been reports of deficiencies in planted wattle, eucalypts, and pine in Kenya (Vail et al., 1961) and in planted pine in Rhodesia, New Zealand, Australia, and Brazil (Stone, 1968). The deficiency, often resulting in dieback of terminal shoots and stunted growth, is commonly corrected by the application of 10 to 20 kg per hectare of commercial borax, or equivalent from colemanite or borosilicate glass. Borax ($Na_2B_4O_7 \cdot 1OH_2O$) contains 10.6 percent boron and is soluble in water and easily leached from the soil or taken up in toxic quantities by plants. For these reasons, colemanite, a naturally occurring calcium borate ($Ca_2B_6O$) of low solubility, is a preferred source, but it is not widely available. Borosilicate glass (frit) is made by fusing a boron salt with glass and then grinding the resultant mass. This product has the extended availability of colemanite, which is so important in sandy soils and under conditions of high rainfall. Boron concentrations among frits normally vary from 3 to 6 percent and the availability of the boron is largely controlled by the fineness of grind. Frits containing a variety of micronutrients are also available.

*Copper* deficiencies and responses in forest trees have been reported for only a few areas. These have generally occurred on sands or highly weathered soils, calcareous sands, or peat and muck. Symptoms are generally noted as disturbance of the terminal and leading lateral shoots, often followed by defoliation and dieback. New shoots of some species are soft and may curve in an S shape (Figure 18.2).

Copper deficiencies can be treated by spraying a solution of soluble salts on the plant leaves or by applying the fertilizer salt to the soil. Copper sulfate ($CuSO_4 \cdot 5H_2O$), a soluble material containing 25.5 percent copper, is commonly used for both purposes. Copper can also be applied as frits or in the chelated form.

*Iron* is one of the most abundant metallic elements in the earth's crust, although its concentration varies widely among soils. It occurs as oxides, hydroxides, and phosphates as well as in lattice structure of primary and clay minerals. Generally, sufficient iron is released from these sources during weathering for normal plant use. Deficiencies of iron are mostly problems of nurseries, shelter belts, shade trees, poorly sited plantations, or pioneer vegetation on calcareous soils (Stone, 1968). Symptoms of deficiency in angiosperms are characterized by green midrib and veins with lighter green, yellow-green, yellow or white interveinal tissue. In conifers, deficiency appears as uniformly pale green or yellowish-green new foliage, contrasting with the darker green of the older foliage. Severely deficient needles brown back from the tips and are lost.

Deficiencies of iron can be corrected in nurseries and in older trees by the application of ferrous sulfate or chelated iron to the soil or directly to the foliage in aqueous sprays. Injections of dry salts directly into trunks of chlorotic trees is effective. Fritted iron can also be used on acid soils. While these iron-containing fertilizers are effective, they are generally too expensive for operational use, except in nurseries or specimen trees. It is more practical to select tree species that are adapted to soils of low iron content or low acidity than to correct the deficiency. Even in nurseries, the problem of iron deficiency can often be best ameliorated through control of soil reaction and fertilization. Increasing the acidity of nursery soils with an application of sulfur is often sufficient to increase the availability of soil iron and overcome a deficiency. The use of acid-forming nitrogen fertilizer is also suggested for such areas.

*Manganese* deficiencies in forest species have rarely been reported, but there have been a few instances of deficiencies in shade trees and shelter belt plantings in regions where the surface soil reaction is high and in forest plantings on eroded limestone materials (Stone, 1968). The amount of the element required by most plants is small and the range between deficiency and toxicity is not large. Tisdale and Nelson (1966) reported that manganese deficiency is severest on high organic matter soils during cool weather months when the soils are subject to waterlogging. As the soils dry out and the season warms up, the symptoms generally disappear. This may result from lowering the soil pH brought about by increased microbiological activity. Manganese availability has been found to increase with increases in soil acidity which may actually result from changes in the oxidation status of the soil rather than acidity per se.

Chlorosis associated with manganese deficiency begins near margins of broadleaf foliage and extends inward toward the major veins. Symptoms can commonly be distinguished from those of iron deficiency by the fine veins remaining green in chlorotic areas, broad bands of normal green color next to major veins, and terminal shoots generally not affected until shoot growth stops.

Manganese sulfate ($MnSO_4$) contains 26 percent manganese and is the most widely used manganese fertilizer material. It can be used as an aqueous solution, included in mixed fertilizers, or spread locally at rates of 2 to 4 kg $MnSO_4$ per tree (Stone, 1968). A manganese ammonium phosphate has been developed, and the element is also available in the chelated and fritted forms. Acidification of nursery soils or irrigation water, as used in correcting iron deficiency, will generally increase manganese availability, except in some acid sands that contain essentially no reserve of this element.

*Molybdenum* deficiencies have not been reported in forest species to date. The foliar concentrations required for good growth are extremely small, usually well below 1 ppm, and symptoms of deficiency are not well established for trees. This element is of interest to forest managers, however, because it is required in relatively large amounts by free-living nitrogen-fixing bacteria, blue-green algae, and organisms involved in symbiotic fixation of nitrogen by both legumes and nonlegumes. It could, therefore, be a determinant in the nitrogen status of forest ecosystems. It has not been determined whether the application of molybdenum fertilizer will stimulate either the symbiotic of nonsymbiotic fixation of nitrogen in forests. However, such a possibility would appear to be a real one because the availability of molybdenum decreases in acid soils, contrary to that of other micronutrients. Applications of sodium molybdate at rates of about 1 kg per hectare have been adequate to overcome molybdenum deficiencies in legumes.

*Zinc* deficiencies in field and orchard crops are fairly widespread in the United States, yet zinc deficiencies in forests of this country are all but unknown.

Symptoms of zinc deficiency in woody plants are well established from reports of the problem with planted pine in western and southern Australia (Stoate, 1950) and with fruit trees in many countries (Stone, 1968). Marked chlorosis or bronzing of young leaves, loss of older leaves from leading shoots, rosetting and dieback of terminal shoots are common in deficient broadleaf plants. In conifers, extreme shortening of branches, needles, and needle spacing in the upper crown, together with general yellowing of foliage, loss of all but the first- and second-year needles, and dieback are symptoms of zinc deficiency (Stoate, 1950).

Zinc is most likely to be deficient in highly leached acid sands or in alkaline soils. The availability of zinc is reduced with increases in soil reaction and, presumably, with increases in organic matter content, although the latter may result from immobilization by microorganisms associated with the organic matter. Zinc is adsorbed by various clay minerals and by the carbonates of calcium and magnesium. As a result of this adsorption, zinc is relatively immobile when applied to most soils. Zinc sulfate ($ZnSO_4$), containing 36 percent zinc, applied to soils at rates of 50 to 100 kg per hectare, has long been a standard recommendation for deficient soils. However, zinc sulfate sprays, soil applications of frits or chelates, and metal implants may be effectively used, especially on specimen trees.

## ORGANIC WASTES AS NUTRIENT SOURCES

The enormous amounts of waste produced by society have the potential for pollution if improperly managed, and yet can be a source of essential plant nutrients and organic matter when applied to agricultural and forest lands. The use of organic wastes as a supplementary source of nutrients for forests appears to be a viable alternative to disposal systems that involve liquid discharge into surface waters, deep wells, or ocean outfalls or to solid waste disposal by landfills or burning. These latter systems of disposal are often expensive and may result in air pollution or damage to water systems as a result of nutrient enrichment and additions of toxic substances. On the other hand, disposal of wastes in forests, at rates tolerated by trees, may save on disposal costs, as well as on energy required to produce an equivalent amount of nutrients as fertilizer. Forests are particularly suited for such a disposal system because there are generally adequate forested areas near centers of waste production so that the need for overloading the system is minimized; yet disposal sites can generally be well removed or buffered from populated areas. Because forest vegetation is not in the human food chain (with the exception of some wild animals hunted for food), the presence of toxic substances or pathogens is of less concern than when these materials are spread on agricultural lands.

Three general types of waste materials have been applied on limited scales to forested areas, perhaps more often as methods of disposal than as sources of nutrients. These materials are conveniently grouped into (1) sewage and other effluents, (2) municipal and mill sludges and animal manures, and (3) solid wastes (garbage) and composts.

## Wastewaters

Secondarily treated sewage effluents and industrial wastewaters have been applied to forest lands as a means of disposal and as a source of nutrients (Sopper and Kardos, 1972). Nutrient concentrations are generally low, but because of the high rates of application, considerable nutrients are applied on an area basis. For example, a 5-cm per week application rate provided 243 kg nitrogen, 48 kg phosphorus, and 143 kg potassium per hectare per year. Reports on disposal of wastewaters on forest lands as early as 1953 (Mather, 1953) indicated no significant groundwater pollution from annual disposal rates as high as 15 m. However, native trees were generally replaced within a few years by shrubs and weeds, mainly as a result of damage to trees from high pressure spray jets and changes in the soil moisture regime (Smith and Evans, 1977).

Researchers in Pennsylvania (Sopper and Kardos, 1972) concluded after eight years that forest ecosystems could be used to alleviate many disposal and pollution problems, and that 90 percent of the wastewater applied at 5 cm per week was recharged to the groundwater reservoir with an average recharge of 15 million liters per hectare during the April-November irrigation period. The forested silt loam-silty clay loam biosystem effectively decreased the phosphorus concentrations of the wastewater by more than 90 percent at the 0.6 m soil depth and from 97 to 99 percent at the 1.2 m depth (Figure 19.4). In the eighth year, the average concentration of nitrate-nitrogen in soil water at 0.6 m depth was as high as 11.7 ppm, but only 1.0 to 5.3 ppm at 1.2 m depth. While these levels are generally below the allowable maximum drinking water limit of 10 ppm, the authors (Sopper and Kardos, 1972) indicated that increasing concentrations of nitrate-nitrogen could become a deterrent to the use of land for liquid waste disposal. This might be overcome by altering the irrigation sequence to create short periods of soil saturation or by using sites with imperfectly drained soils that would provide conditions favorable for denitrification. In their system, red pine responded to an application rate of 2.5 cm per week of secondarily treated effluent, but not to higher rates. On the other hand, white spruce and mixed hardwoods responded to rates of effluent up to 5 and 10 cm per week, respectively.

Smith and Evans (1977) pointed out that the necessary steps in developing satisfactory wastewater disposal systems in forests are to select tree species tolerant of moist, fertile, near-neutral pH conditions, devise methods of application that do not damage the plants, and use rates that are compatible with the species and physical properties of the soil. The land area required is not a major prohibitive factor, according to Sopper and Kardos (1972). At the recommended level of irrigation of 5 cm per week, 52 ha of land would handle 3.78 million liters of wastewater per day.

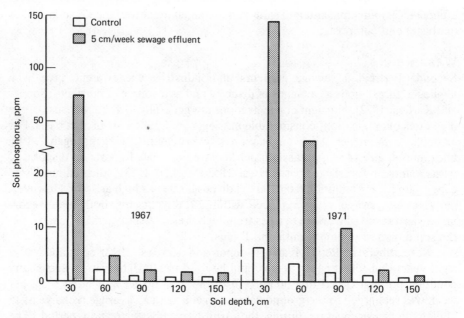

**Figure 19.4** Extractable (Bray) phosphorus at various depths in a deep sandy loam soil in treated (spray irrigated with 5 cm per week of treated municipal sewage effluent) and control plots (Richenderfer et al., 1975).

## Sludges and Manures

**Sludges.**    The separation of solids from wastewater starts with primary settling of easily separable solids in both domestic and industrial sludges. Domestic (municipal) and most industrial wastes are then subjected to a secondary treatment of aerobic decomposition which results in the formation and settling of a biological floc. This thickened slurry (2 to 3 percent solids), together with the primary settled solids, may then be stabilized by a heated anaerobic digestion process. Some 45 million metric tons of solid waste are produced by municipal wastewater treatment plants in the United States each year, and the sludge material contains from 2.6 to 5.6 percent N, 2.8 to 3.4 percent P, an assortment of other essential elements and 35 to 47 percent organic matter on a dry-weight basis (Peterson et al., 1971). The composition of sludge from a settling pond for wastewaters from a chemical cellulose plant, as well as that of digested municipal sludge, is given in Table 19.2.

If properly digested, municipal sludge has little odor and is relatively free of pathogens. The value of digested sludges as a source of nutrients for agricultural crops and for the amelioration of spoil banks has been proven (Peterson et al.,

**Table 19.2** Composition of Digested Municipal Sludge (Peterson et al., 1971) and Industrial Sludge, on a Dry-weight Basis

| Sludge | pH | Conductivity | P | N | K | Ca | Mg | Mn | Na | Zn | Cu | Fe | Al |
|--------|-----|--------------|-----|------|-----|-----|----|-----|------|-----|------|-----|-----|
| | | μmho/cm | | | | | | kg/ton | | | | | |
| Municipal | 7.4 | 6,062 | 34 | 56 | 6.8 | 50 | 16 | 0.7 | 3.0 | 0.7 | 0.4 | 18 | — |
| Industrial | 9.2 | — | 12 | 0.68 | 2.5 | 158 | 15 | 0.9 | 12.7 | 0.1 | 0.03 | 1.4 | 2.6 |

Reproduced from *Fertilizer Technology and Use*, Chapter 18, page 559, 1971 by permission of the Soil Science Society of America.

1971), but they have not been widely tested under forest conditions. The disposal of these wastes in forests at rates of 10 to 25 tons per hectare should be beneficial to tree growth, and may have an added physiological advantage over disposal on agricultural lands in that the forest products are not directly consumed by humans. Furthermore, amounts of nitrates, heavy metal, or fecal coliform filtering into the groundwater should be negligible at the above rates of applications.

**Manures.** Animal wastes vary widely in composition, depending upon the type of animal, kind of feed consumed, and the management and handling of manure during storage and spreading. Fresh manure contains 30 to 85 percent water and all of the inorganic nutrients needed by plants. The compositions of some animal manures are given in Table 19.3.

Manures have long been used as a source of plant nutrient on agricultural lands and probably will continue to be largely disposed of in this fashion. They could be applied to young forest plantations and established stands at rates up to 10 and 20 tons per hectare, respectively. Stephens and Hill (1972) applied 38 tons per hectare of liquid poultry manure in one application, which supplied about 450

**Table 19.3** Composition of Animal Manures (Loehr, 1968)

| Animal | Moisture | N | P | K | S | Ca | Fe | Mg | Volatile Solids | Fat |
|--------|----------|------|-----|------|-----|-----|------|-----|-----------------|-----|
| | % | | | | | kg/ton | | | | |
| Dairy cattle | 79 | 5.6 | 1.0 | 5.0 | 0.5 | 2.8 | 0.04 | 1.1 | 161 | 3.5 |
| Fattening cattle | 80 | 7.0 | 2.0 | 4.5 | 0.8 | 1.2 | 0.04 | 1.0 | 198 | 3.5 |
| Hogs | 75 | 5.0 | 1.4 | 3.8 | 1.4 | 5.7 | 0.28 | 0.8 | 200 | 4.5 |
| Horses | 60 | 6.9 | 1.0 | 6.0 | 0.7 | 7.8 | 0.14 | 1.4 | 193 | 3.0 |
| Sheep | 65 | 14.0 | 2.1 | 10.0 | 0.9 | 5.8 | 0.16 | 1.8 | 284 | 7.0 |
| Broiler chickens | 25 | 17.0 | 8.1 | 12.5 | — | — | — | — | — | — |

**Figure 19.5** Five-year-old slash pine on an Aquod soil; (L) 10 tons per hectare of composted municipal garbage placed in mound prior to planting; and (R) control.

kg nitrogen, to an established white pine plantation. At this rate, there was little odor or enrichment of the groundwater. However, at a rate of 224 tons per hectare (applied in 5 applications at two week intervals) groundwater within 0.9 m of the surface contained 40 to 80 ppm of nitrate nitrogen. The application of very large amounts of manure to soil, unless incorporated immediately by discing, may also create nuisances of flies and odors. Nevertheless, reasonable application rates appear to be environmentally safe and a benefit to tree growth. If discing is a requirement and accessibility to spreading a problem, young plantations may be favored over older stands as disposal sites.

## Municipal Garbage

Solid wastes, particularly the more than a million metric tons daily production of municipal garbage, present a monumental disposal problem. Burning is not permitted in many areas and landfills are often unsatisfactory and difficult to obtain. Composted garbage materials have not proven economically attractive except for lawns and specialty crops (Hortenstine and Rothwell, 1968). Incorporating ground raw garbage into forest soil during reforestation may be a suitable and economical means of disposal. In sandy soils, the organic wastes may improve soil physical properties and moisture retention. Applications of up to 44 tons per hectare on an excessively drained sandy soil in Florida decreased soil acidity, and modestly increased soil organic matter, cation exchange capacity, and exchangeable calcium, magnesium, and potassium, but did not improve growth of planted slash pine after two years (Bengtson and Cornette, 1973). In contrast, ground garbage surface applied and disked into coastal plain flatwood soils before planting slash pine dramatically increased tree growth during the first five years (Smith and Evans, 1977). This latter material contained 0.65 percent nitrogen and 0.18 percent phosphorus (Figure 19.5).

Early results indicate that fairly large amounts of organic wastes can be disposed of in forests without detrimental effects to trees or environment and, in fact, the materials may improve physical and chemical properties of some soils, particularly sandy soils. Because of long rotations common to forest species and the large areas generally available away from centers of populations, forest lands appear to have definite ecological advantages over other means of disposal.

# 20
# OPERATIONAL TECHNIQUES IN FOREST FERTILIZATION

The development of information on techniques of application has probably been slower than any other phase of forest fertilization. There has not been the store of data from agronomic research that could be conveniently adapted and used as a guide to the most appropriate season of the year, stage of tree development, rates, or methods of application that would result in the desired response. Forest trees present different problems from cultivated crops, and even techniques used for fertilization of ornamental and shade trees cannot be effectively used in forests.

## TIME OF APPLICATION

In dealing with perennial crops, timing of the fertilizer application can be in terms of season of the year or it can relate to stage of plant development.

### Season of the Year

**Established Stands.** Fertilizers have generally been applied to forest trees in the spring and early summer to coincide with the period of most active growth. As Viro (1970) pointed out, there is "a periodicity in tree growth at different parts of the growing season." It appears to be particularly important in cool climates to time the fertilizer application in relation to root growth so as to achieve maximum uptake. The period of most active root growth varies from one year to the next, depending on weather conditions, but it is usually greatest in midsummer. Spruce roots start growth in the spring when soil temperature reaches 4 to 5°C and end in autumn when it falls to 6 to 7°C. The roots continue to grow for a considerably longer time than other parts of the tree, storing nutrients for the coming year (Viro, 1970). In warm climates, root elongation may not completely cease during winter months, and growth may be affected more by soil moisture than by temperature (Kaufman, 1968). Thus, season of application may have little influence on fertilizer effectiveness in moderate climates.

Some Swedish results (Friberg, 1974) indicate that fertilization on snow gives a better effect than fertilization on bare ground when using urea, whereas, it is quite the contrary when using ammonium nitrate fertilizers. It is doubtful that either fertilizer material should be applied on snow due to environmental hazards from nitrogen lost to runoff prior to ground thaw. Friberg (1974) and Viro (1970) generally agreed that fertilization of established stands of pine and spruce in Scandinavia should probably be done in the summer or early autumn. Moreover, they also agreed that season of application appeared to have less influence on fertilizer effectiveness than might be expected.

Urea should be applied during the season when there is the greatest likelihood that it will be washed into the soil, thereby reducing the chance of volatile losses. After hydrolysis of urea in the soil there is little danger of gaseous loss, but nitrogen fertilizers are subject to leaching, even in established forest stands (Nömmik, 1973). Applications should be made in fairly small doses and timed to enable the trees to take up the maximum amount of fertilizer possible. Potassium behaves much like the ammonium ion in soils and the same precaution should be exercised in its use, particularly on sandy soils. On the other hand, phosphorus sources can apparently be applied to established stands at any season with equal effectiveness. The exception is applications made to a snow cover. Both soluble and slowly soluble sources of phosphorus appear to be absorbed by mycorrhizal roots as it becomes available in the forest floor with little opportunity for movement out of the surface soil.

**Young Stands.** The seasonal application of fertilizers to young stands may be more critical than for older stands. The root systems of most young trees develop rather slowly and they may not occupy a significant part of the soil mass for several years after planting. Consequently, soluble nitrogen fertilizers applied near time ot planting are subject to leaching from the surface soil, but they may cause salt injury and stimulate excessive vegetative competition before they are leached. McKee and Sommers (1971) concluded from a pot study that season of application had little affect on nitrogen recovery and root carbohydrate reserves, but that their results may not hold under field conditions.

Pritchett and Smith (1972) reported that survival of pine seedlings were significantly reduced by soluble nitrogen sources applied broadcast at planting time at rates in excess of 225 kg nitrogen per hectare, especially on sandy or droughty soils. If nitrogen is to be applied the year of planting, it should be applied in such a way that it minimizes salt damage to the seedling and competition from ground vegetation. This may mean delaying the application for several months, until early summer in temperate zones, in order to allow time for the initiation and development of seedling roots and to avoid excessive early growth of herbaceous plants.

The use of slowly soluble nitrogen sources is an alternative to delayed application, but these sources have not been popular for forestry use due to their relatively high costs per unit of nitrogen. Furthermore, a case can be made for delaying the fertilizer application until after the establishment of a protective ground cover to serve as a sink for mobilized nutrients that might otherwise be lost by leaching from the ecosystem. Ground cover vegetation should be only encouraged on wet sites; it is not a recommended practice on droughty soils.

## Stage of Stand Development

Timing fertilizer application in relation to stand development is influenced by both biological and economic considerations. Trees should be fertilized near time of planting on soils with severe nutrient deficiencies. This is particularly true of phosphorus deficient sites. Growth that is lost by delaying a fertilizer application on these sites may not be recovered by later treatment. In Australia, response by *Pinus radiata* to fertilizer applied shortly after planting was immediate and large on deficient sites, according to Waring (1973). He reported that an early response was partly due to an increase in root:shoot ratio, which permitted roots of the fertilized trees to exploit the site more rapidly. Furthermore, differences in relative growth rate brought about by early fertilization continued to widen until intense competition restricted growth. It appears possible to apply phosphorus fertilizers at planting in such a manner that one application will sustain growth at a maximum for several years, without the necessity for dividing the fertilizer into several applications. Whether the early growth advantage from adding nitrogen fertilizers to young deficient plantations can be sustained for extended periods is not clear. Unlike phosphorus, nitrogen is rather easily lost from the soil, unless there are sufficient plant roots to absorb it. Where soil moisture is not a limiting factor, herbaceous vegetation can be useful as a nutrient sink in retaining nitrogen in the ecosystem until the young trees can use it. The effects of nitrogen on young pine stands on wet sites have been noted to last for 15 years or more. The response to nitrogen fertilizers on dry sites may be of shorter duration because the control of competing vegetation on these sites also reduces the capacity of the site to retain nutrients. Woods (1976) found that the control of competing vegetation by herbicides and cultivation was essential to plantation establishment when fertilizers were used on dry sites in South Australia.

In cool climates, nitrogen deficiencies are generally not encountered in early stages of stand development, but they may develop later as a result of immobilization in the forest floor biomass (Hagner and Leaf, 1973). In Finland, Viro (1966) stated that only when the trees had attained the stage of natural rapid development did they respond to nitrogen fertilizer. This occurs after pine has attained the age of 10 to 15 years and after spruce is 15 to 20 years old, according to Viro. Because of immobilization (or loss) of nitrogen applied to established stands, the effect on growth may completely disappear in 8 to 10 years (Möller, 1974). For this reason, nitrogen applications are almost always delayed to near time of harvest of pine and spruce stands in Scandinavia and the Pacific Northwest. Delaying the nitrogen application until 7 to 9 years prior to harvest reduces the investment period and may increase the relative growth response. In temperate regions, nitrogen may also be most effectively applied to well-stocked pine stands at 12 to 16 years (except in moist and severely deficient sites where the application should not be delayed).

Several researchers have noted that the capacity of older plantation trees to respond to fertilizer is related to free growth and that fertilizing at thinning, particularly with nitrogen, permits the surviving trees to reoccupy the site and regain maximum volume increment faster than would otherwise be possible (Miller and Pienaar, 1973; Waring, 1973). Competition for moisture is apparently a major factor influencing fertilizer effectiveness in dry areas and it may account, in large part, for the fertilizer × thinning interaction.

While phosphorus is normally applied near time of planting, it has been used successfully on stagnated stands of slash pine on the phosphorus-deficient wet savanna soils of the coastal plain (Pritchett and Smith, 1974). Nitrogen is usually applied in conjunction with phosphorus on these sites. There may be a delay of up to three years before a growth response is noted in severely deficient stands because of the necessity of developing adequate tree crowns for effective photosynthesis (Figure 18.3).

## APPLICATION RATES

The rate of fertilizer application must be a balance between high concentrations that could damage young plants or lead to leaching losses, in the case of soluble materials, and the need to apply sufficient materials to maintain a long-term response. Application rates for nitrogen are perhaps the most critical of all fertilizer materials because most nitrogen sources are highly soluble, are subject to losses, and promote excessive growth of competing vegetation.

In tree nurseries, only 25 to 75 kg nitrogen per hectare are generally applied before seedlings have passed the cotyledon stage, so as not to adversely affect germination or result in excessive leaching losses. Additional nitrogen is then applied as needed, as solid material or through the irrigation system, in increments of approximately 25 to 50 kg nitrogen per hectare.

High rates of nitrogen from soluble sources are not effectively used by young plantings, and nitrogen is best applied at moderate rates, even though multiple applications may be needed during the rotation. For young deficient plantations in the U.S. southeast, 60 to 90 kg nitrogen per hectare are recommended (Pritchett and Gooding, 1975). Applications to mid-aged stands of pines in that region are generally 110 to 140 kg nitrogen per hectare. This is only slightly less than the 150 kg per hectare used in Sweden. Möller (1974) explained that while there was a growth response up to 400 kg nitrogen per hectare the lowest cost of production per cubic meter of wood was attained at about 150 kg.

Hardwoods are apparently more nutrient demanding than conifers and when they are grown in short rotation they generally receive a complete fertilizer at rates based on the results of a soil test. Applications of about 80-40-40 (N-P-K kg/ha) may be applied to *Populus* and *Platanus* species soon after planting, with annual applications of 80 kg of nitrogen or of a mixed fertilizer supplying the equivalent amount of nitrogen.

## APPLICATION METHODS

The equipment and logistical support systems for fertilizer application in forestry have undergone extensive evolution during the last decade. This has been possible as a result of some significant advances in solid and liquid fertilizer technology and the development of more efficient handling and delivery systems (Bengtson, 1973).

### Physical Properties that Influence Application

**Granulation.**   Most solid fertilizer materials and mixtures are now granulated to provide dust-free, noncaking fertilizers that are easy to handle and apply. Essentially all equipment for handling and application of solid fertilizer is designed for granular products. Granulation also facilitates bulk-blending, an efficient custom formulation of fertilizers to meet the specific requirements of particular crop-soil situations, without the need for maintenance of large inventories of different fertilizer grades (Bengtson, 1973). Bulk-blending of granular materials permits significant economy and flexibility in application; however, this technique has not been widely used in forestry as yet. It is likely to come into wider use with the development of more reliable technique for diagnosing multinutrient deficiencies and prescribing appropriate nutrient ratios.

"Aerial grade" pellets of urea and ammonium nitrate have been recently developed for forestry use (Figure 20.1). The use of large granules in aerial equipment improves the uniformity of application, increases the swath width, and reduces crown lodging. However, there are indications that crown penetration by different granule sizes varies with tree species. For example, more of the conventional-sized granules of urea and ammonium nitrate lodged in the crowns of spruce than did the large "Skog-AN" granules. On the other hand, the coarse stiff needles of Scots pine retained a greater percentage of the nitrogen applied as large granules than when applied as small granules (Ekberg and Friberg, 1972).

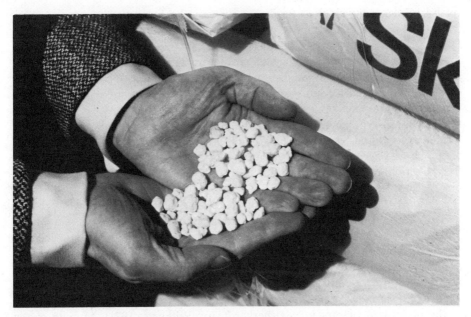

**Figure 20.1** Large pellets of ammonium nitrate for aerial application in forests.

The authors also pointed out that wet foliage considerably increased retention of the smaller granules in both tree species but had little or no effect on retention of the larger forestry-grade granules.

Any reduction in crown retention by using large granules would reduce the chances for gaseous losses of ammonia from foliage-lodged urea. However, Volk (1970) pointed out that nitrogen losses as ammonia from standard-size granules retained in slash pine foliage amounted to only about 3 percent of that applied. Losses could be much higher in other species or if small granules of urea are dropped on a wet canopy. Volatile loss of nitrogen from urea applied to the forest floor may be significantly reduced by the use of larger granules, according to Nömmik (1973). The reduction in acidity and the high concentration of ammonium in microsites associated with the larger granules may reduce urease activity and increase the chances that the urea will be washed into the soil before hydrolysis. Coating fertilizer granules with sulfur or other water-resistant materials should slow the dissolution rate and reduce lodging in wet crowns. It also enables easier handling and application of hygroscopic materials under humid conditions.

**Liquid and Suspension Fertilizers.** The ease with which mixed suspension and solution fertilizers can be transported and handled has resulted in a considerable increase in their use on agricultural crops in recent years. A wide variety of relatively high analysis fluid fertilizers can now be formulated. Some micronutrients and pesticides can also be mixed with the fertilizer for ease of application. However, soil application of liquid fertilizers has not been widely practiced in forestry, mostly because of logistic problems. The weight of liquid materials needed to fertilize a given area is greater than that of most solid materials. This extra weight makes for considerable difficulty in applications to wet or rough terrain. The development of high flotation equipment should help overcome some of these problems, permitting broadcast or localized application of liquid fertilizers to young plantations. For most effective results, urea and anhydrous ammonia may have to be injected into forest soil. This would require that their use be confined to gentle terrain and soils free of surface and subsurface obstructions (Bengtson, 1973).

*Foliar application* of fluid fertilizers should have several advantages for forestry. It is a rapid method of relieving deficiencies and is a means of avoiding reactions in certain soils that could reduce the availability of applied nutrients to the trees. Foliar applications may also reduce the possibility of leaching losses and water pollution by some elements and, thus, reduce the rate of application needed to correct a deficiency. Urea is readily absorbed by most tree leaves, but the use of a spreader-sticker may be required to assure retention and penetration of the liquid into pine needles (Eberhardt and Pritchett, 1971). In spite of the advantages of foliar applications, this method presents the same logistic problems as those encountered with soil application of liquid fertilizers. Except for micronutrient deficiencies that can be corrected with low rates of application, it is unlikely that foliar applications will become economically attractive for operational forest fertilization in the near future. Most inorganic sources of micronutrients must be applied in weak solutions with acidity adjustments to prevent leaf damage. Organic metal complexes are more effective for foliar application of heavy metals, but they are relatively expensive.

*Sprinkler irrigation* systems have been successfully used for the application of liquid fertilizers to tree nurseries and seed orchards. This is a convenient method of soil application of fertilizers where suitable equipment exists. Because of equipment limitations, it is not likely to be used extensively in conventional silviculture. The possible use of irrigation systems for the disposal of sewage effluents and industrial wastes in forests has been mentioned in Chapter 19.

## Application Systems

There are two general systems for fertilizer application in forests. These are based on the use of (1) tractor-powered, ground-traversing equipment, or (2) aircraft. Choosing a particular system depends on such factors as accessibility of the area to ground equipment and the need for selective placement of the fertilizer, as well as the availability and relative costs of spreading a unit of fertilizer by the two systems. For older trees, broadcasting the fertilizer is often the only practical method of application. However, for young stands some selective placement may be desirable. Placement so that young trees can derive maximum benefit from the fertilizer with minimum stimulation of competing vegetation requires the use of ground equipment. Selective placement may also reduce the quantity of fertilizer required for a young stand. There is also the possibility of combining ground application with other operations, such as bedding or planting, and reducing supervision and application costs.

**Ground Equipment.**  The use of ground equipment is largely confined to the application of fertilizers at or near time of planting because of the general inaccessability of machinery in established stands. There are also many steep, rough, and wet cleared sites on which ground equipment cannot be effectively used. However, where ground equipment can be used it is generally less expensive per unit of land area than aerial equipment. A wide variety of ground equipment and systems of application have been devised to replace the time-honored, but generally inefficient, hand application. The latter method is still used in some areas with abundant cheap labor and where specialized equipment is unavailable or excessively expensive, or where a combination of terrain and placement considerations rule out mechanical methods (Bengtson, 1973), as shown in Figure 20.2. However, it appears that machines will become even more commonplace in operational forest fertilization in the future.

Because much of the fertilizer applied by ground equipment near time of planting consists of phosphorus or phosphorus-nitrogen materials, it can be effectively applied either before or after site preparation and planting. Bengtson (1973) reviewed several fertilizer application systems that used tractor-mounted spreaders. These systems involved both broadcast and band applications and they can be grouped into (1) those in which the fertilizer is applied before planting and incorporated into the soil by discing or bedding, and (2) those in which the fertilizer is surface applied at time of planting or soon thereafter.

Broadcast applications of fertilizers are generally made immediately prior to the discing or bedding phases of site preparation. A blower or spinner type spreader, mounted on a conventional farm tractor or rubber-tired articulated skidder vehicle, can spread granulated materials in swaths of 15 to 30 m. The "cyclone" spinning spreader has been most widely used for broadcast applications of granular materials, but the blower spreader may be preferred for soft granules and for finely ground materials, such as rock phosphate or lime. Gravity-flow drop spreaders are sometimes used for spreading ground rock phosphate in an effort to reduce fertilizer drift.

Another system of preplant application involves the simultaneous band spreading of phosphate fertilizer and incorporation into planting beds. A power-driven gravity flow or belt-type dispenser is usually front mounted on a crawler tractor, although it may be mounted immediately in front of the bedding plow.

**Figure 20.2**   Hand application of nitrogen fertilizer to a mature spruce stand in central Norway. Fertilizer is gravity fed from a backpack.

Postplant applications of granular fertilizer can be accomplished with a rear-mounted cyclone spreader. They can be used to broadcast a swath up to six rows wide, but more commonly the spreader is modified with spouts that direct the fertilizer into bands about 1 m wide over each of two rows of young trees (Figure 20.3).

Simultaneous planting and fertilizing is accomplished with a pair of fertilizer hoppers mounted to the front of a tree planter or else at the rear of a wheel tractor. The fertilizer is directed through flexible tubes into or beside the planting slits. The two hoppers permit blending or separate placement of two different fertilizer materials with one-row planters, or they may be used with two-row planters. The fertilizer flow is accurately metered by a drive mechanism operating off an idler wheel that runs along the ground, coordinating flow with ground speed.

**Figure 20.3** A two-row, tractor-drawn distributor for granular fertilizer. Fitted with spouts, it applies a band of fertilizer about 1 m wide over two rows of trees.

Ground equipment has the potential advantage over other systems of application in being flexible, accurate, and relatively rapid. Furthermore, the fertilizing sometimes can be combined with other ground operations, such as bedding or planting. On the other hand, tractor-mounted spreaders are restricted to relatively level and dry terrain. They cannot effectively operate in older stands, except in well-spaced plantations or corridor-thinned stands, and their daily capacity is much less than aerial methods (Bengtson, 1973).

**Aerial Application.**   Aircraft have been widely used for inventory work and, to a lesser extent, for the application of pesticides in forests. In recent years, aircraft have been used extensively for the application of fertilizer to older forest stands in Scandinavia (Hagner, 1971), the Douglas-fir region of North America (Strand et al., 1973), and the southeastern United States (Pritchett and Smith, 1975). Fixed-wing types have been the aircraft most widely used for aerial applications, especially in Scandinavia. However, the increased availability of helicopters with greater lifting power has largely changed this situation in North America (Strand et al., 1973). An example of a helicopter in use in the Pacific Northwest is shown in Figure 20.4.

There are certain problems and limitations in using either system for aerial application of fertilizers to forests. The weather conditions must be good for

**Figure 20.4** Helicopter fertilization of a stand of Douglas-fir with the Bell 212 (courtesy Weyerhaeuser).

relatively long periods for effective operations. Landing strips or heliports must be constructed near the area to be fertilized. Nonuniform distribution patterns are sometimes a problem. Flagging or some method of marking the area is essential for uniform application, especially with fixed-wing aircraft. Frequent landings or hovering over refilling crews and flying at low altitudes over rugged terrain require much attention to safety (Page and Gustafson, 1969). Helicopters can effectively work smaller and more irregular areas than can fixed-wing aircraft, but they generally have a smaller effective working radius from the loading site (Bengtson, 1973). Fixed-wing aircraft require a longer runway, but they may be able to use existing roads as runways in some areas. The length of required runway can be effectively reduced by taking off downslope and landing upslope. In this instance the loading area should be located at the top of the slope.

**Support Considerations.**  In order to take advantage of the large spreading capacities and to minimize the high fixed costs of aerial application systems, substantial and dependable fertilizer transport and loading capacities are required. Some helicopters can carry up to 1.5 tons of fertilizers with the capacity to spread 20 tons of materials per flight hour. At an application rate of 450 kg per hectare, more than 150 ha per day can be treated with adequate logistical support.

In operations near the fertilizer source, the material may be transported directly from the plant to the loading area in bulk transports where it is transferred to front-end loaders and thence into a spreading hopper. More commonly the fertilizer is moved into large tilt bins, with storage capacities in excess of 45 tons, near the area of operation. From these bins, the fertilizer is gravity-fed to trucks that transport it to landing sites (Bengtson, 1973). Where a reliable transport-storage system for bulk materials cannot be devised, bagged fertilizers may be hauled by trucks from a railway siding to the loading zone at any convenient time. The bags can be stored in the open for several months without damage from the weather, although vandalism can be a problem. The 36-kg bags are hand lifted to a front-end loader, opened, and the bulked fertilizer transferred by loader to the aircraft hopper.

Most helicopter systems use a pod or conical tank carried under the helicopter on a sling. The pod has a mechanism to meter fertilizer to a spinner, which broadcasts the fertilizer as the helicopter moves over the forest. The spreader may be driven by an electric or hydraulic motor powered from the aircraft or by a remote-controlled gasoline engine mounted on the side of the pod (Page and Gustafson, 1969). The pod may be refilled by dumping from a front-end loader while the helicopter hovers overhead or else two pods may be used. In the latter instance one pod is refilled while the helicopter is spreading the other load. In fixed-wing aircraft, spreading is generally accomplished by gravity from the hoppers through a venturi distributor beneath the aircraft (Hagner, 1971).

**Distribution Pattern.** Distribution patterns of fertilizer spread by aerial systems can be very erratic, depending on equipment, pilot skill, climatic condition, and terrain. Irregular growth patterns and difficulties in estimating response will result from poor distribution. Aerial photographs and maps are used to locate the drop zone, but flagging, balloons, flares, and previously killed trees are sometimes used to delineate areas and swaths to be fertilized. Flight lines are generally oriented parallel to stream courses to minimize contamination of surface waters. Flight direction, elevation above the ground, and speed all effect the distribution pattern. Two applications at right angles to each other will improve the distribution pattern, but this is not always possible and much depends on skill and care of the individual pilot.

Armson (1972) concluded that "application rates of 124 kg urea per hectare when applied with a Stearman aircarft and venturi type dispenser result in such great variation of pattern of distribution to render this technique unsuitable."

When the rate was doubled, much less variation in the pattern occurred. He used one square foot catchers placed at random in the fertilized area and found that 70 catchers were needed to monitor distribution with an accuracy of ± 11 kg per hectare if a prescribed rate of 112 kg of nitrogen as urea were specified.

With fixed-wing aircraft in New Zealand, the rate of application for single runs at 157 kg per hectare varied from 7 to 500 kg per hectare. A method for testing the ground distribution, termed the "half value," was examined by Ballard and Will (1971). The half value was defined as "the percentage of sampling points which receive less than half the designated rate of application." A value of 10 percent or less was proposed as acceptable. The number of random sampling points required for arriving at a reliable half value was reported to be one for each 0.8 ha fertilized.

Helicopters appear to have the potential for more uniform spread of fertilizer than fixed-wing aircraft, and they offer greater flexibility in landing sites. It is likely that their use in silviculture will increase.

## ECONOMIC CONSIDERATIONS

The uncertainties surrounding the economic aspects of future prices and costs have delayed fertilization in many areas where it is probably needed. Furthermore, much of the operational fertilization presently practiced in the world is done on faith and it largely results from pressure on forest owners to supply more wood to industry without sacrificing requirements for sustained yields. Although forests are exploited for many different purposes in satisfying diverse human needs, it is usually in the context of the effect on wood production that economics of fertilization are viewed. A realistic economic analysis of forest fertilization, like that of most silvicultural practices, is relatively easy to model but is rather difficult to interpret because of the assumptions required by the relatively long investment period and the cost-price uncertainties resulting therefrom.

### Fertilizing Young Plantings

When fertilizer is applied to seedlings near time of planting, a return on the investment may not be realized until the end of the rotation. For pulpwood rotations, this may range from 20 to 40 years in conifer plantations, with the possibility for one or more merchantable thinning as the result of closer initial spacings. In hardwood plantations, fertilization may shorten pulpwood rotations to 5 to 10 years or less (White and Hook, 1975).

When fertilizing at time of plantation establishment there are two general types of site conditions to consider: (1) sites that are so deficient that acceptable tree survival and growth cannot be obtained without the use of fertilizers, and (2) sites where reasonable tree growth can be obtained without fertilizers but where the addition of fertilizers will increase growth rate by a significant amount.

Examples of the first condition are the severely phosphorus- and potassium-deficient soils mentioned previously. Applying fertilizers to these problem soils is essential if the areas are to be used for commercial forestry, and the economics of fertilization are seldom questioned. One application of phosphorus, or potassium, to these sites near time of planting generally improves tree growth throughout the rotation period and serves to convert an unproductive forest site into a productive one at a nominal cost.

Fertilizing young plantations on soils with incipient deficiencies is more difficult to justify from an economic standpoint than plantation on severely deficient soils. First, there must be some means of delineating areas that are most likely to respond well to fertilizers. Soil or site maps and diagnostic tools, such as soil or tissue analyses, are invaluable in forests where deficiencies are not apparent. It is generally assumed that increases in height growth resulting from fertilization are commensurate with increases in site quality. Whether fertilization will alter survival, height to diameter relationships, or stand diameter distribution is not known for most areas. In an economic analysis, Pritchett and Gooding (1975) used, as an example, the application of 50 kg nitrogen and 100 kg phosphorus per hectare at time of planting of slash pine on flatwood soils of the coastal plain. The treatment resulted in a 1.5-m height advantage over the unfertilized trees by age 10 years. The site quality was raised from 65 (19.8 m) to 71 (21.6 m) by fertilization, and the yield from 1730 planted stems per hectare at age 20 years increased from 210 to 261 cubic meters per hectare. Fertilization was credited with a yield increase of 51 cubic meters per hectare at harvest time by these calculations. Based on fertilization costs in 1974, and the projected stumpage price at time of harvest 20 years later, the rate of return on the fertilizer investment amounted to 12 to 15 percent annually.

## Fertilizing Established Stands

Fertilizing older forest stands is generally easier to justify economically because the investment period is much shorter than when fertilizing seedlings. Nitrogen deficiency is the growth-limiting factor most frequently encountered in older stands on mineral soils and the effect of nitrogen fertilizer on growth may persist for less than 9 years in pine and only slightly longer in spruce stands, according to Möller (1974). Therefore, one to three applications are made on deficient sites, from 5 to 20 years prior to harvest. A different procedure may be needed in selecting responsive sites in older stands than in young plantations. Möller (1974) stated that "when the increase in growth is given in cubic meters per hectare, the growth improvement for a given fertilizer dose appears to be related to the pretreatment growth at time of fertilization." This is in contrast to fertilizing at time of plantation establishment where the poorest sites often result in the greatest responses, unless some factor other than the nutrient applied is limiting response.

In calculating the economics of fertilizing older stands the "allowable cut effect" is sometimes used. Under the system, the extra increment attributable to fertilization can be harvested from existing mature stands elsewhere on the forest property, rather than waiting for the full effect to be realized on the fertilized stand. In this way, the benefits from fertilization may then be captured in a short period. For forests managed on a sustained yield basis, the allowable cut for any year can be approximated by using the Hanzlik formula:

$$\text{Allowable cut} = \frac{\text{volume of mature timber}}{\text{length of rotation}} + \text{mean annual increment}$$

When established stands are fertilized, the future yield from these stands, as well as the mean annual increment for the entire management unit, is assumed to be increased and the allowable cut is raised. Hagner and Leaf (1973) pointed out that the allowable cut, or "cost-profit," calculations are probably only applicable to "large forest owners with a long-term yield exploitation policy and with forests supplying the company's own processing industries." Large forest owners who produce wood purely for sale face a rather different calculation picture, although the basic elements are similar. For small forest owners, the economic incentives for fertilization as for all forms of silviculture, are usually small. The same authors (Hagner and Leaf, 1973) also pointed out that some large forestland owners may possess extensive holdings so remote and expensive to harvest that it would be more economical to increase the cut on more conveniently situated lands and, at the same time, stimulate growth there by fertilization.

Möller (1974) concluded that fertilization of older stands may increase average diameter as well as volume. Because there is usually an economic relationship between average diameter and the timber value and logging costs per unit, the gross price of timber rises in step with increasing dimensions while at the same time logging costs are reduced.

While these considerations may not be relevant to economic calculations under all management systems, they are given as examples of possible benefits from fertilization. Fertilization of older stands is a relatively short-term investment, and if the uncertainties surrounding the forecasting of growth effects can be reduced, the economic prospect becomes most attractive.

# 21

# SECONDARY AND OFF-SITE EFFECTS OF CHEMICAL USE IN FORESTS

As a result of the increasing world population, it appears inevitable that pressure will develop to produce maximum amounts of wood fiber, to satisfy an ever-rising demand for land for recreational use, and to yield maximum amounts of clean, pure water from a reduced area of forest lands. These demands can only be met by intensifying the management of certain forested areas beyond levels presently conceived. A variety of techniques, including the use of chemicals, will be needed to achieve these goals. Chemicals can be extremely useful in attacking a multitude of forest problems, such as: controlling diseases, insects, and unwanted vegetation; correcting nutrient deficiencies; fighting wildfires; reducing soil erosion; improving road surfaces; retarding transpiration; speeding up slash disposal; and eliminating undesirable soil pathogens.

Chemical applications, like other management operations, may have both beneficial and adverse effects on the forest ecosystem and its surroundings. They should be used with an understanding of the interactions that occur within forest ecosystems, affecting not only vegetation but soil, air, water, wildlife, and microbiological conditions as well.

Chemicals of various types have been used in forest management for more than a century, but large-scale use of these materials are comparatively recent developments, with the greatest impetus coming after World War II (Tarrant, Gratkowski, and Waters, 1973). Although the peak of pesticide use came during the 1960s, the actual use of these chemicals in forestry was never more than a small fraction of the total used for other purposes. Less than 0.03 percent of the

255 million hectares of forest lands in the United States was treated with insecticides in an average year during the past decade and no more than 5 percent of our forest lands have ever been treated with insecticides at any time (Tarrant et al., 1973).

Fertilizers are the most widely used chemical in forestry and the potential for fertilizer use appears to be increasing. However, only about 0.2 percent of the total forest lands in the United States have been treated with fertilizers to date and it is doubtful that more than half this amount (0.1 percent) will be treated in any given future year.

## FERTILIZER EFFECTS ON FOREST ECOSYSTEMS

Fertilization can be viewed as an energy-additive cultural treatment that triggers a host of interrelated, complex, and dynamic events within the ecosystem. While the most striking result of fertilizer additions is an increase in tree growth rate, this increased wood production presupposes improvement in photosynthetic efficiency or size of photosynthetic apparatus, or both. Thus, improvement in photosynthetic efficiency as a result of the correction of a deficiency of an element involved in chlorophyll synthesis can be considered a primary function of fertilization. Closely related to the primary effects of fertilization are a host of secondary events that result therefrom. Since tree species differ in their nutrient requirement and in their capacity to extract nutrients from soils, fertilizers may affect the balance between species in a mixed stand and between trees and ground vegetation. Survival and growth of ground vegetation is also influenced by the reduction in light accompanying a denser canopy resulting from successful fertilization. Changes in nutrient concentration and physiological conditions in plants can affect animals feeding on the plants, as well as parasites attacking these plants. While fertilized trees may be less (or more) susceptible to insect or disease attack, they may, at the same time, be more palatable for browsing by vertebrates (Baule and Fricker, 1970).

An increase in tree growth as a result of fertilization is like a temporary improvement of site quality. This improvement may be short lived as is the case with nitrogen fertilization of most sites, or it may be profound and long-lasting as with phosphorus fertilization of many deficient soils. Indirect effects of fertilizer applications on soils are less obvious but they can also be of considerable importance. For example, most nitrogen fertilizers are acid forming and an increase in soil acidity may reduce the availability of phosphorus and the bases. It may influence the spectrum of soil microbes, with an increase in certain fungi but a decrease in bacterial numbers. Fertilizer applications may increase or decrease the effectiveness of mycorrhizae, depending on soil conditions; but at moderate rates of application, fertilizers are not likely to greatly affect the numbers of either microflora or mesofauna.

A significant fertilizer response by a stand of trees can effectively lower the water table on wet sites. This results in improved aeration and increased nutrient supply. On well-drained sites, improved tree nutrition may result in an increase in dry matter production per unit of water used, but it may also result in a greater water stress during drought periods. Hence, secondary effects of fertilization on soil properties may be positive or negative, depending on site conditions.

## Fertilizer Effects on Tree Properties

Fertilizer responses are normally expressed in terms of increased tree height, diameter, or volume growth. By definition, growth responses to additions of fertilizers are obtained only in areas deficient in one or more nutrients. However, subtle responses to treatment may be expressed as changes in chemical concentrations, volume or weight ratios of component parts, resistance to disease or insect pests, or in certain wood properties—with or without a change in growth pattern.

**Chemical Composition and Extractives.** The concentration of an element in healthy plants varies among plant parts of a given species and it varies among species for the same plant part. For example, nitrogen concentrations tend to be greater in foliage than in bark, and greater in bark than in wood of trees; and the concentration of nitrogen and bases are higher in foliage of most hardwoods than in foliage of conifers. Nevertheless, concentrations tend to be relatively uniform for a particular type of tissue within a species. It is on this premise that plant analysis and tissue tests for diagnostic purposes are based. The concentration of elements in a tissue, therefore, reflects the nutrient status of the stand and can be expressed schematically as in Figure 21.1.

One should expect fertilization of trees to influence the level of carbohydrates in tissue as a direct result of improved photosynthesis. Such is not always the case, however, due to increased protein synthesis and translocation of sugar or starch to storage organs. Free amino acids, especially arginine, appear to increase in concentration in certain tissues following nitrogen fertilization. On the other hand, optimum potassium supply is often associated with low levels of free nitrogenous compounds. Tests for these compounds may prove to be better indicators of plant nutritional condition than tests for mineral elements, if results can be properly interpreted (Durzan and Steward, 1967).

Fertilizer applications have been noted to influence syrup production in sugar maple *(Acer saccharum)* in New England, but not always in a positive fashion (Yawney and Walters, 1973). Nitrogen has generally increased sap-sugar concentrations for three to five years after application. However, complete (NPK) fertilizers or phosphorus or potassium applied alone have sometimes resulted in little or no change in sap yield and even decreases in sap-sugar concentrations. It is possible that the phosphorus treatment influenced the photosynthetic production of sugar and differential accumulation of starch in the latter part of the summer season.

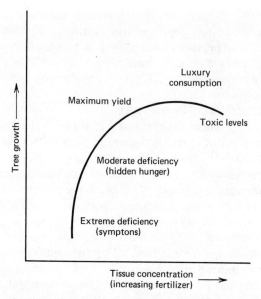

**Figure 21.1** Relationship between nutrient concentration in tissue and growth.

A variety of complex organic compounds, termed extractives, occur in cell walls as surface deposits, or in the cell lumina of both conifers and hardwood. Extractives such as polyphenols, resins, tannins, and latex may be valuable products of the forest, or they may contribute significantly to wood quality by influencing color, odor, taste, permeability, and resistance to insect and fungal attack (White, 1973).

Heavy nitrogen applications have increased resin flow from slash pine by as much as 23 percent (McGregor, 1957). There are also indications that resin soaking of wood can be induced in slash and longleaf pines by removing a small strip of trunk bark and spraying with a dilute solution of paraquat. Increased yields of rosin and turpentine are extracted when the harvested trees are processed at the pulpmill. It is not known whether the fertilizer and paraquat treatments are synergistic.

The yield of latex from *Hevea brasiliensis* can be significantly increased by fertilizer applications (Watson, 1973). Much of the commercial production of rubber is on infertile Oxisols and Ultisols of the humid tropics where the removal of the original tree cover eliminates a major reserve of nutrients. Uptake by the first commercial plantings reduces soil nutrients to growth-limiting levels and liberal use of rock phosphate is needed to encourage vigorous early growth so that the young trees can achieve dominance over competing weeds. Maintenance applications of complete (NPK) fertilizers are routinely used to replace nutrients removed through cropping. Soil and tissue (leaf) tests have proven particularly useful as guides to fertilization in plantations of genetically improved trees capable of yielding more than 2500 kg latex per hectare per year. Latex flow can be increased by the use of stimulants (2,chloro-ethane phosphoric acid), but an increase of 1000 kg per hectare, or more, almost doubles the depletion rate of nitrogen, phosphorus, and potassium from the system and, consequently, higher dosages of fertilizer are required in order to maintain good yields.

**Wood Properties.** Wood quality is determined by such physical properties as density, strength, durability, hardness, grain, color, pulping and machining characteristics. The effect of fertilization on quality largely depends on the condition of the tree before treatment and the particular use to be made of the wood. Changes in a certain property as a result of fertilization many improve the wood for some uses but degrade it for other uses.

Wood *specific gravity* (density) is an important index of wood strength and also an indicator of its general suitability as a raw material for papermaking. For example, more digester capacity is required to produce a given daily tonnage of chemical pulp from wood of low specific gravity than from denser wood, and operating efficiencies are generally higher with high specific gravity wood.

In conifers, high specific gravity is often associated with slow growth and a high proportion of latewood. It denotes higher bulk yield of pulp and greater strength and durability of the paper. However, bursting strength and tensile strength are often inversely related to specific gravity (White, 1973). Fertilizer treatments that significantly increase conifer growth rates may result in a reduction in wood specific gravity due to an increase in the proportion of thin-walled earlywood fibers (in a given annual increment) and a concurrent decrease in the proportion of latewood fibers. Thus, the actual increase in dry matter production from fertilization can be somewhat less than the reported increase in volume increment. However, volume increases usually overshadow any yield loss due to a reduction in specific gravity and, in some instances, no reduction in wood specific gravity or fiber length were detected following the fertilization of three species of young pine trees (Schmidtling, 1973).

It appears likely that fertilizers can be used to increase the growth rate of stagnant trees on impoverished sites to near normal rates without adversely affecting wood quality. Fertilization may also result in more uniform wood in conifers as a result of thickening of earlywood fiber walls and thinning of latewood fiber walls (Gooding and Smith, 1972).

The relationship of specific gravity to hardwood quality parameters is quite complex because of the great variability in anatomy among hardwood species. White (1973) pointed out the distinction between ring and diffuse porous woods. The very large vessels are formed at the beginning of the growing season.

Therefore, if the period of annual growth is prolonged and the total increment increased by fertilization, the proportional amount of large pores in fast-growing ring porous wood tends to be lower and the strength is greater. Thus, fast-growing hickory or ash is preferred over slow-growing trees for tool handles because the wood is stronger.

White (1973) concluded that the deliberate use of fertilizers for the improvement of quality parameters in wood or pulp does not appear to be an important management consideration at the present state of knowledge. Many of the subtle qualities of wood products inherent in the value of manufactured products, such as color, grain, figure, and strength are subject to modification during the manufacturing process. For example, the natural color of woods is frequently altered or masked by staining to either simulate a preferred appearance or to achieve the uniformity required by mass production methods.

**Flowering.**   Fertilizers are widely used to increase flowering in seed production areas and seed orchards. Nitrogen appears to be the critical element for reproductive growth and significant increases in seed yield have been obtained from applications of nitrogen fertilizers on some sites. However, the timing of the application is apparently a critical factor (Schmidtling, 1975), and there are differences in the inherent ability of trees of the same species to flower and produce seed in response to nitrogen fertilization. McLemore (1975) reported that fertilizing longleaf pines increased cone yields and seed size but did not affect number of seeds per cone, percentage of empty or wormy seed, cone size, initial viability, or keeping quality of seed. Cones per tree, sound seeds per cone, percentage of empties, seed weight, and cone size were characteristics of individual trees.

White (1973) reported on a study in which nut production in a 25-year-old plantation of black walnut was increased by 25 percent by NPK fertilization and weed control. This should not be surprising in view of the numerous reports that have shown significant increases in yields of fruits and nuts of orchard species as a result of fertilization.

**Resistance to Disease and Insects.**   The influence of fertilizers on the resistance of trees to attacks of disease and insect pests is generally related to tree vigor and/or the presence of certain organic compounds resulting from the fertilizer treatment. Research in the area has been rather limited and results often appear contradictory. In addition to genetic factors, environmental conditions may affect the host plant, or the parasite concerned, in such a manner to alter the resistance to the pest. The most important factor appears to be nutrition of the host plant. Certain pathogens preferentially attack slow-growing or weak host plants while others are more damaging to fast-growing, succulent tissue (Bjorkman, 1967).

Fertilizer treatments may exert a significant influence on the synthesis of plant constituents that afford the host plant a degree of protection from disease and insect pest. These constituents include lactic acids, alkaloids, nonprotein nitrogenous compounds, such as asparagine, and soluble amides or amino acids, which may be easily utilized by pathogens (Baule and Fricker, 1970). Excessive nitrogen applications may adversely affect phenolic compounds and thereby reduce the resistance of trees to rust fungi. Cuticular waxes and terpenes also play a part in reducing fungal diseases, and Baule and Fricker suggested that the level of these constituents may be manipulated through fertilization.

A balanced nitrogen-phosphorus ratio seems to be important in regulating susceptibility to attacks by parasitic fungi and viruses. However, the ratio of nitrogen to potassium in plant tissue is more likely to hold the key to resistance to many adverse agencies. In some instances, high nitrogen concentrations have reduced resistance while high potassium levels apparently increased resistance to fungi. Potassium has a certain catalytic affect and may act as a regulator on starch formation and moisture relationships in plants (Bjorkman, 1967).

While plant nutrient analysis can provide valuable insight into the relationship between plant nutrition and resistance to pests, one must recognize that changes in the nutrient levels of assimilatory tissue may take place after the attack by insects and diseases. Such changes should be taken into consideration in diagnosis in order to prevent confusion of cause and effect.

*Disease Resistance.* Some examples of effects of fertilization on the susceptibility of trees to fungal attack include increases in infection of the "snow blight" fungus, *Phacidium infectans* (Bjorkman, 1967). Where vigorous pine growth resulted from heavy nitrogen applications, the attack from this fungus has been severe, especially if potassium was kept at a moderately low level. Spruce seedlings in nurseries fertilized with high rates of nitrogen appear to be more susceptible to attack from *Herpotrichia juniperina* than those not fertilized. On the other hand, nitrogen applications have been used to overcome, at least temporarily, the effects of "little leaf disease" of southern pines (Zak, 1964). Nitrogen fertilization may increase pine resistance to the *Phytophthora cinnamomi* fungus and bring about regeneration of the roots.

Phosphorus fertilization and other treatments that stimulate the formation and development of mycorrhizae increase the resistance of the host plant to certain parasitic soil fungi, such as *Fomes annosus*. *Fomes* attack on spruce and pines has been especially severe on soils rich in calcium in northern Germany, southern Sweden, and England. This may result from the fact that high calcium concentration in soils inhibits the absorption of potassium and high soil pH reduces the availability of iron and manganese.

Baule and Fricker (1970) reported that poplar trees in Baden-Wurttemberg treated with nitrogen, phosphorus, potassium, and magnesium grew more vigorously and suffered much less from fungal attack than those fertilized with nitrogen and phosphorus materials only. They also believed that potassium and magnesium treatment played a decisive role in reducing the damage from needle cast fungus *(Lophodermium pinastri)* in central Europe, although a balanced nutrient supply is probably more important than simple additions of fertilizer materials.

Fusiform rust *(Cronartium fusiforme)* of pine generally has been reported to increase after soil fertility improvement. However, Hollis et al. (1975) stressed the complexity of the relationship between tree nutrition and rust incidence, and concluded that increases in rust were generally associated with nitrogen or phosphorus applications to deficient sites. On the other hand, fertilizer application on sites with no detectable nutrient deficiency often resulted in a decline in rust incidence. There are indications that neither early growth initiation nor tree height are primarily responsible for increased infection by fusiform rust. Instead, increases result from physiological changes brought about by fertilization.

*Insect Pests.* Nitrogen applications often result in reductions in numbers of defoliators, bark beetles, and weevils, but they sometimes stimulate populations of sucking insects. Baule and Fricker (1970) reviewed literature which indicated that nitrogen applications reduced populations of the pine looper moth *(Bupalus piniarius),* pine sawfly *(Diprion pini),* black arches moth *(Lymantria monacha),* fir bark beetle *(Ips curvidens),* engraver beetle *(Ips typographics),* and pine shoot moth *(Evetria buoliana).*

Xydias and Leaf (1964) reported that nitrogen, phosphorus, and potassium fertilizer treatments in a white pine plantation were related to tree growth and incidence of white pine weevil damage. In the deep, sandy potassium-deficient soil, potassium fertilizers significantly increased tree growth and weevil damage while nitrogen fertilizer resulted in decreases in both growth and weevil attack.

In young pine plantations, nitrogen applications resulted in only slight reductions in tip moth *(Rhyacionia* spp.) attack, but 22 kg phosphorus per hectare reduced the rate of infestation from 26 percent in the control plots to 17 percent in the phosphorus-treated plots. Where 90 kg potassium per hectare were added with the phosphorus, tip moth infestation was reduced to an average of 12 percent of the trees (Pritchett and Smith, 1972).

Nitrogen fertilization appears to stimulate sucking insects, but potassium additions generally decrease the population of these insects. Mitchell and Paul (1974) evaluated population levels of aphids *(Adelges cooleyi)* on Douglas-fir fertilized with 56 and 224 kg nitrogen per hectare. Aphid fecundity was 11 to 42 percent higher on treated trees than on untreated trees in the first year of the study. Two years after fertilization, the population levels were essentially the same on treated and untreated trees.

There is evidence that moisture stress results in increased attack of certain weevils and leaf-eating insects. Soil water shortages not only influence the water economy of the tree but its nitrogen economy as well. Both disturbances result in a reduction in protein-carbohydrate metabolism in favor of sugar accumulation. Most eating insects apparently depend on sugars more than starches for their carbohydrate requirements and they are, therefore, favored under drought conditions (Baule and Fricker, 1970). The effects of fertilization on water relations and, thus, on insect attack may be both direct and indirect. Fertilizer salts may influence attack through changes wrought in the osmotic pressure of the sap. Furthermore, good nutrition may improve the water use efficiency of trees (i.e., amount of carbohydrates produced per unit of water transpired) by increasing their drought resistance in terms of improved stomatal control. Nevertheless, faster-growing trees, as a result of successful fertilization, use more soil moisture than stagnant trees, and during periods of extended drought they may experience more moisture stress than unfertilized trees and, thus, be susceptible to attack by certain insects.

## Response of Understory Vegetation to Fertilizers

Fertilization of forested areas can have profound effects on understory vegetation and, indirectly, on the fauna that feed on it. In young plantings, the response by the understory vegetation to treatment is often more obvious than the response by the target species, except in extremely deficient areas. If wood production is the primary objective, the increased growth of understory vegetation can be detrimental to the stand because of increased competition for water and nutrients. Furthermore, deer and other herbivores seem to prefer fertilized browse over unfertilized browse and, consequently, they may do considerable damage to fertilized trees.

In older forest stands, there may be little immediate effect of fertilization on understory vegetation, especially where shading has already eliminated most ground cover. The long-term effect of increased canopy shading on a dense ground cover can be a significant reduction in intolerant species. White et al. (1975) reported that cultural treatments that increased slash pine heights from 4.5 m to 7.7 m at age 9 years resulted in a reduction in light intensity 2 m above the soil surface from 30 percent extinction of incoming solar radiation to 85 percent extinction. Light extinction at ground level was not significantly different in the treated and untreated plots, because the herbaceous vegetation just above ground level intercepted light transmitted in plots lacking a significant midstory shrub vegetation. The reduced light intensity at the herbaceous level caused changes both in the amount of forage and in its nutritional quality. A number of studies have shown that site disturbance and fertilization alter the floristic composition of an area. The initial response is often an increase in the number of species, particularly the frequency and density of herbs. However, with a response in tree canopy density, fewer species may be observed on fertilized plots than on untreated plots. Annual species are often shaded out by the fertilized tree crop and other perennial vegetation. While there may be fewer species on fertilized plots, the total biomass of ground cover vegetation may still be greater than in control plots.

In most forested areas, the carrying capacity for deer and other herbivores is increased by fertilization as a result of the stimulation of browse plant growth and improvement in its nutritional quality. Nitrogen and phosphorus, the two elements most commonly applied in fertilizers to increase tree growth, are particularly important in improving the protein levels in food plants.

In the Gulf coastal plain, Wolters and Schmidtling (1975) found that cultivation alone or cultivation with three rates of fertilizers more than doubled pine basal area over the untreated controls by age 12. The low rate of fertilization was 112 kg nitrogen plus 56 kg each of phosphorus and potassium per hectare. Medium and high rates were double and quadruple these amounts. All treatments reduced total browse density and crown cover; however, plots fertilized at medium or high rates contained more browse desirable for deer than the control and had as much crown cover from desirable species as the control plots. On untreated plots, the most abundant browse was *Ilex glabra,* an undesirable species for deer, while on treated plots the most abundant was *I. vomitoria,* a desirable species.

Fertilizers have also been used to improve grazing for domestic animals in young plantations and on forest rangelands. However, in order to minimize damage to young trees, the fertilized areas should not be intensively grazed during the first year or two after planting. The improved carrying capacity may help pay for fertilizer cost, particularly in areas where tree response to fertilizer is marginal. Such conditions apparently exist on a major portion of the rangelands of the U.S. national forests of the west and in some young forest plantations in other areas.

## Fertilization and Water Quality

Fertilizer materials undergo a number of transformations when applied to soils. Transformations in forest soils are not necessarily at the same rate nor do they necessarily yield the same end products as transformation of fertilizer added to cultivated soils, because of general differences in pH, spectrum of microorganisms, and amount of relatively undecomposed organic materials present.

The fate of fertilizers added to forest ecosystem depends on rate and time of fertilizer addition and on a number of site factors, such as types of forest and ground cover, soil physical and chemical properties, and temperature and rainfall regimes. Only relatively small percentages of fertilizers are recovered in the forest crop. Recovery may vary from 5 percent or less for phosphorus and some heavy metals to 15 to 20 percent for nitrogen. This does not mean that the remainder of the material is lost from the forest ecosystem. Some of the fertilizer is taken up by ground cover vegetation and by an expanded population of soil microorganisms, while a substantial portion is retained, at least temporarily, in the soil. Some elements bound in the soil, such as phosphorus fixed in soils high in iron and aluminum compounds and nitrogen and phosphorus incorporated into organic matter, may eventually become available to the tree crop. Nevertheless, a significant part of the added fertilizer appears to be irreversibly lost.

Nitrogen and phosphorus are the two elements whose loss from the site causes the most concern. This is because they are the two nutrients most often applied as fertilizers to the forest and because they are most often associated with eutrophication of surface waters. While the elements added to forests as fertilizers occur naturally in living and dead vegetation and in the soil, losses from these sources are generally quite small. Nitrogen and phosphorus concentrations are particularly low in unpolluted stream and lake waters and these elements often control, or limit, the growth of aquatic plants, just as they often limit the growth of trees on land. Other nutrients are generally considered to be present in concentrations sufficient to meet the biological requirements for optimum growth. With rapid enrichment of lakes by nutrients, algal growth is stimulated and substantial quantities of organic matter may be produced. The decay of this material reduces the water's oxygen supply and the lake may become *eutrophic*.

**Phosphorus.**   Phosphorus is regarded as the element that most often limits the growth of aquatic life because the critical limiting concentrations of phosphorus are so low. For example, the critical concentration for the growth of blue-green algae may be below 0.01 ppm, with profuse growth occurring at concentrations of 0.05 ppm (Stanford, England, and Taylor, 1970). In spite of the small increase in phosphorus concentrations needed to stimulate algal growth, there is little evidence that phosphorus fertilizers applied to forest lands cause any detectable increase in eutrophication of surface waters.

Most mineral soils immobilize phosphorus in iron and aluminum compounds in acid soils and as calcium phosphate in alkaline soils. Essentially no leaching of phosphorus takes place under these conditions. The exceptions are soluble phosphates applied to acid organic soils low in all minerals, and acid quartzitic sands low in iron and aluminum in surface horizons (Humphreys and Pritchett, 1971). Although phosphorus leached from the surface horizon of the latter mineral soils may be retained in a spodic or argillic horizon, phosphates of low solubility, such as ground rock phosphates, are recommended for both situations. Transport by erosion of the soil material on which it is adsorbed is the main way that phosphorus moves from a fertilized area. Fortunately there is very little erosion from normal forested sites. Wildfires, which result in increases in both stormflow and mineralized phosphorus, would appear to be a greater threat to water quality than fertilization. However, there is little evidence that fires increase phosphorus concentrations in streams or lakes in forested areas. McColl and Grigal (1975) reported that phosphorus concentrations in runoff water were elevated for two years after a wildfire in a virgin forest in northern Minnesota, but that there were no detectable increases in concentrations in an adjacent lake or its input stream.

**Nitrogen.**   When added to soil, nitrogen undergoes a complex series of transformations and may exist in the soil in various states of oxidation. Probably 95 percent or more of the nitrogen in forest soils exists in organic (humus) materials. A high proportion of the nitrogen in humus materials is resistant to microbial action. A portion of the added fertilizer nitrogen becomes immobilized in forms similar to those already existing. This takes place during plant residue decomposition by microbial action, during which soluble (mineral) nitrogen is converted to organic forms (Stanford et al., 1970). Urea complexed in this fashion appears to be fairly stable. Urea not immobilized in organic materials is hydrolyzed to ammonium carbonate within a few days after application and some of it may be lost as gaseous ammonia. If applied immediately prior to heavy rainfall, urea can be leached before it is converted to ammonium—particularly in sandy soils. Urea is very mobile and if it does not hydrolyze soon after application, it can serve as pollutant to streams. Once converted, however, the positively charged ammonium ions migrate slowly in soils because the attractive forces of the negatively charged clay and organic colloids restrict their mobility. Unfortunately, the effective cation exchange capacity of most acid forest soils is very low and ammonium is not retained as strongly as in less acid soils. On the other hand, the oxidation of ammonium to nitrate-nitrogen is very weak in these acid soils, and the level of the more mobile nitrate ion in soil water is extremely low, unless nitrate is added as a fertilizer.

If nitrate is added as a fertilizer, such as ammonium nitrate, a part of it may be lost in gaseous form. Biological denitrification occurs in all soils at some time, but it is likely to be important only in soils that contain excessive water and have a high oxygen demand. Oxygen demand is a reflection of an active microbial population with an ample carbon supply to sustain their growth. Saturated soils and ample carbonaceous materials are often found in coastal flats and wetlands and, at least temporarily, in many other areas following timber harvest. Denitrification is a reversion to the atmosphere of nitrogen that might otherwise be available for plant growth, but, from a pollution standpoint, denitrification lowers concentrations of nitrates in ground and surface waters.

Because many factors influence the utilization and fate of nitrogen applied to a forest ecosystem, it is difficult to predict the effects of this fertilizer material on water quality. Nutrients entering streams as a result of fertilization will be mostly absorbed by aquatic plants. They can move downstream, principally as a pulse of algal biomass, if not harvested by fish. The brief nature of the increased productivity suggests that there is not likely to be a substantial change in ecosystem structure associated with the fertility pulse. Moreover, most fertilized watershed studies indicate only minimal increases in nitrogen concentrations in surface waters. Cole and Gessel (1963) found only small amounts of urea nitrogen lost by leaching. Other studies in the Pacific Northwest (Moore, 1975) confirmed that urea applications to Douglas-fir stands pose little threat to water quality unless there is a direct application to stream channels. Moore reported that concentrations of urea-nitrogen never reached 1.0 ppm at downstream monitoring stations following applications of 224 kg nitrogen per hectare. Ammonia-nitrogen increased only slightly above background, and the highest level of nitrate-nitrogen found was 0.12 ppm. Fertilizer nitrogen entered streams only in the form of nitrate after the first three weeks, and 95 percent of the total loss over seven months occurred within the first nine weeks after application (Figure 21.2). Fertilizer nitrogen lost in stream water during the seven-month monitoring period was about 0.25 percent of the total applied.

The use of urea or ammonium sulfate in erosion control fertilization of watersheds burned by wildfire increased urea and nitrate concentrations in streamflow only slightly above that of the control stream. The highest concentration of nitrate-nitrogen measured was 0.2 ppm shortly after application of 55 kg

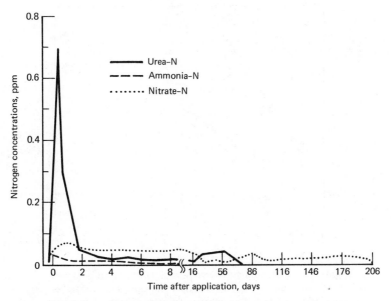

**Figure 21.2** Average concentration of urea-, ammonia-, and nitrate-N in stream water following fertilization of two watersheds in Washington with 224 kg urea per hectare (Moore, 1975).

nitrogen/ha (Klock, 1971). Nitrate-nitrogen loss in grams per hour is shown graphically in Figure 21.3 for a 514-ha watershed that was fertilized with 60 tons of urea. These and other studies in the Pacific Northwest and Alaska indicate that forest fertilization with urea offers minimum potential for pollution of the aquatic environment when properly accomplished.

In contrast with results from the use of urea in the Pacific Northwest, Tamm (1973) reported that heavy applications of ammonium nitrate increased nitrate concentrations of spring water in both Sweden and Germany. In some instances, concentrations exceeded 10 ppm nitrogen and remained high for as long as two years after fertilization. In spite of the high concentrations in spring water, losses of nitrogen from the fertilized area amounted to only about 12 kg per hectare, or 8 percent of that added, during the first two years. Urea fertilization did not result in appreciable leaching of nitrate under Swedish conditions. However, a heavy application of urea (225 kg nitrogen per hectare) to a deciduous forest in West Virginia increased stream nitrate concentrations in connection with stormflow three weeks after application, with 18 percent of the applied nitrogen lost during the first year, mostly during the dormant period. This loss of nitrogen in streamflow after urea fertilization is higher than that in other areas and the reasons for the higher values are not clear. In most fertilized watersheds, urea

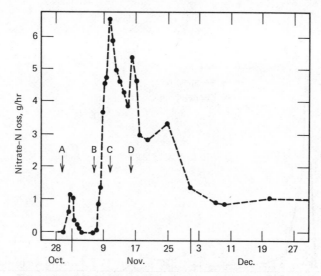

**Figure 21.3** Streamflow nitrate-nitrogen losses from a 514-ha watershed after fertilization with 54 kg nitrogen per hectare as urea. Nitrogen was applied at rates of 4, 13, and 37 kg per hectare at times A, B, and C, respectively. The secondary peak loss (point D) was associated with a period of rainfall (Klock, 1971).

applications have resulted in only small increases in nitrogen concentrations in streamflow, unless the material is applied directly to the stream. Ammonium nitrate offers a greater potential for pollution than urea because of the mobility of the nitrate ion. However, as currently practiced only a fraction of a forested watershed is fertilized in a given year, and evidence does not implicate forest fertilization in significant eutrophication.

## EFFECTS OF PESTICIDES AND OTHER CHEMICALS ON FOREST ECOSYSTEMS

Pests respresent a component of the ecosystem that must be understood and controlled if optimum outputs and benefits from the forest resource are to be obtained. In both ecological and economic terms, insects, diseases, weeds, and other pests limit the productivity, usefulness, and value of forests.

Chemicals have long been the primary defense, particularly against insects; but since World War II striking changes have occurred in the kind and complexity of chemicals used and in the equipment and techniques of application. Until rather recently, the most popular compounds were those toxic to and potentially useful against a wide spectrum of target pests. However, widespread concern for environmental safety of pesticidal chemicals has caused a reexamination of the types of chemicals permitted and the manner in which they are used. This concern involves both their hazards to humans and other nontarget organisms in the environment and the dangers inherent in processing and handling the materials. (Tarrant et al., 1973). Severe restrictions have been imposed on some materials, especially DDT and other chlorinated hydrocarbon insecticides. Extensive efforts are being made by agencies in the United States and several other countries to develop improved standards for evaluating new chemicals. It is probable that present trends toward use of more selective, less persistent chemicals will continue so as to minimize chances of unexpected and adverse changes in the balance and structure of forest ecosystems; although there is little evidence of permanent or temporary damage to the soil from the proper use of present pesticides.

### Herbicides

Many of the commercially important conifer species, such as Douglas-fir, radiata pine, and the southern pines, are unable to compete effectively with most brush species for light, water, and nutrients during the early stages of stand development. Herbicides are effective tools for reducing competition and they offer some advantages over machinery for stand conversion. They control sprouting, a problem with many hardwoods that have been dozed or chopped. The debris left on the surface reduces water and wind erosion and usually does not require burning or windrowing—thus reducing the operational costs as compared to mechanical methods.

**Types of Herbicides.**   Herbicides used in forestry can be grouped into three broad classes based on their mode of activity: (1) hormones, which are rapidly absorbed into the plant and readily translocated; (2) chemicals that kill plant tissues on contact; and (3) soil sterilants.

Hormone-type herbicides, which include such materials as 2,4-D (2,4-dichlorophenoxyacetic acid); 2,4,5-T (2,4,5-trichlorophenoxyacetic acid); and picloram (4-amino-3,5,6-trichloropicolinic acid), can be applied by foliar sprays or injection directly into the tree. Hormones are partially selective in that they kill some species more readily than others, but complete selectivity can only be obtained by injecting the herbicide directly into the unwanted tree or spraying directly on trunks or stumps. Pines are considerably more resistant than hardwoods to phenoxyacid sprays, such as 2,4,5-T, and these chemicals have been widely used in aerial sprays to control brush in pine plantations.

Contact herbicides include paraquat, the arsenates, ammonium sulfamate, and others. Selectivity is achieved by avoiding desirable plants with the application. Many of these materials are toxic to animals, and, consequently, their use in forestry has been rather limited.

Sterilants, such as bromacil and picloram, kill all vegetation in areas to which the material has been applied as pellets on the soil surface. These materials are nonselective and, because soil organic matter and forest litter absorb large amounts of the herbicides, the relatively heavy applications required for control make them expensive to use.

**Hazards from Herbicide Use.**   There is increasing public concern about environmental quality, particularly as it may be affected by widespread use of chemicals. Because of their obvious effects on forests and woodlands, herbicides receive a large share of the concern. The hazards associated with the use of any chemical depend largely on the toxicity of the material and the rate and care exercised in its application.

The direct effect of a chemical on an organism is the result of the intrinsic toxicity of the chemical and the potential for exposure of an organism to the chemical. Toxicity may be acute—rapid response to fairly large doses administered over a short period of time; or chronic—slow response to exposure of relative small doses administered over a relatively long period. The potential for exposure depends upon the behavior of the chemical in the environment, such as movement, persistence, and fate.

There is only limited absorption and very little translocation of many herbicides by vegetation. A large percentage of the intercepted herbicide that is not absorbed will be washed from the surface of leaves to the forest floor by rainfall. Herbicides remaining on leaf surfaces and any pesticide not translocated to other plant parts will also enter the environment of the forest floor due to leaf fall. The greatest concentration of herbicides in foliage is generally found shortly after application. Growth dilution, weathering, and metabolism of the herbicide by the plant are important ways by which the concentration is rapidly decreased. Norris (1975) reported residues of 2,4,5-T applied by helicopter at the rate of 2.24 kg/ha as isooctyl ester of 0.5, 0.6, 3.3, and 11.0 ppm in vine maple, blackberries, grass, and Douglas-fir, after one month. After six months the same vegetation contained residues of 0.2, 0.02, 0.10, and 0.5 ppm. He concluded that long persistence of high level herbicide residues will not occur in forest vegetation. This conclusion is supported by the resprouting of lush vegetation on sprayed areas within a year after application.

The forest floor is a major receptor of aerially applied spray materials. Herbicides in the forest floor may be volatilized, adsorbed on soil mineral or organic matter, leached through the soil profile by water, or degraded by chemical or biological means. Degradation is the principal means by which most environmental pollutants are reduced. However, herbicides vary greatly in their rate of degradation. For example, on red alder forest floor material 80 percent of applied amitrole and 94 percent of the 2,4-D were degraded in 35 days. About 120 days were required to degrade 87 percent of the 2,4,5-T, and only 35 percent of the picrolam degraded in 180 days, as shown in Figure 21-4 (Norris, 1970).

**Damage to Forest Streams.** Contamination of streams is one of the principal concerns for herbicide use in forests. The chemicals may enter forest streams by direct application to surface waters, leaching through the soil profile, or in mass

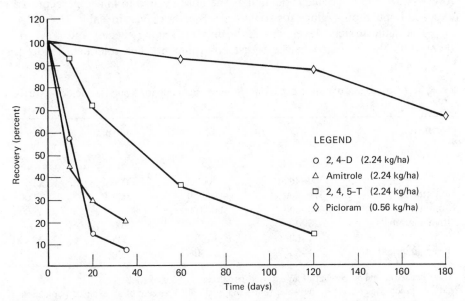

**Figure 21.4** Percent recovery of four herbicides from red alder forest floor material (Norris, 1970). Used with permission.

overland flow. All streams in or near target areas generally contain some herbicide residue, with peak concentrations usually occurring shortly after application. However, peak concentrations seldom exceed 0.1 ppm in streams in sprayed areas or 0.01 ppm in streams that run adjacent to, but do not actually enter, target areas. These levels are not acutely toxic to fish or animals (Norris, 1975). Since water usually reaches streams through subsurface flow, contamination of streams can be minimized by leaving an untreated buffer strip along each side. Rapid dilution with downstream movement reduces concentrations below detectable limits. Thus, use of herbicides does not appear to constitute a water pollution hazard when reasonable care is taken to protect streamside vegetation and the stream from direct contamination.

**Toxicity.** The potential exists for a major part of the herbicides applied to vegetation to reach the forest floor or mineral soil. The toxicity of herbicides to soil microorganisms and their subsequent influence on soil fertility has received only limited study. Norris (1975) reviewed current literature and concluded that 500 to 1000 ppm of most herbicides are required to produce an adverse effect on soil organisms. He pointed out that an application of one kilogram of herbicide per hectare of land is equivalent to only about 3 ppm in the surface 2.5 cm of soil and well below the level shown to influence microbial populations.

Norris compiled tables of the acute and chronic toxicity characteristics, including the median tolerance limit for four herbicides commonly used in forestry. Median tolerance is defined as the dose needed to kill 50 percent of an exposed population. Values for some organisms are given in Table 21.1.

The acute toxicity levels are fairly uniform among species and are higher than would be expected in the forest environment. The tolerance levels for

**Table 21.1**  Median Tolerance Limits[a] of Some Organisms for Four Herbicides (Norris, 1975)

| Organism | 2,4-D | 2,4,5-T | Amitrole | Picloram |
|---|---|---|---|---|
| | | mg/kg | | |
| Birds | 360–2000 | 300 | 2000 | 2000 |
| Rodents | 375–800 | 400–950 | 5000+ | 2000+ |
| Ruminants | 400–800 | 500–1000 | — | 2000– |
| Other mammals | 100 | 100 | 1200 | — |
| Fish[b] | 1–60 | 1–30 | 325 | 13–90 |
| Other aquatics[b] | 1–5 | 0.5–50 | 20 | 1+ |

[a]Dose needed to kill 50 percent of an exposed population.
[b]Concentration of herbicide in the water (ppm) that will kill 50 percent of an exposed population of aquatic organisms in 48 hours.

chronic exposure to herbicides are considerably lower than the levels for acute toxicity. They average from 50 to 250 mg per kilogram of body weight per day for most mammals. The short persistence of most common herbicides prevents chronic exposure to doses of these magnitudes under field conditions.

The acute toxicity of 2,4,5-T is only slightly greater than the toxicity of 2,4-D, and early reports that it was teratogenic (fetus-deforming) were apparently prompted by high levels of a highly biologically active contaminant, dioxin, in 2,4,5-T experimental materials. Dioxin is formed as a byproduct of one of the steps in the manufacture of 2,4,5-T, and it is now legally restricted to less than 0.1 ppm of the herbicide. Commercially available formulations of 2,4,5-T do not show teratogenic effects in mice or rats in dosages up to 50 mg per kilogram of body weight per day when administered on days 6 through 15 of the gestation period, according to Norris (1975). He calculated these animals should be able to tolerate levels of 2,4,5-T near 500 ppm in their diet with no teratogenic effect if the daily intake is 10 percent of the body weight. Application rates of 2 to 3 kg per hectare and the rapid reduction of the herbicide residues in vegetation preclude acute or chronic exposure to these levels in the environment.

The acute toxicity of amitrole and picloram is very low and the short persistence and high tolerance for these chemicals make chronic toxicities an unlikely problem under forest conditions. Furthermore, the four herbicides used most commonly in forests have a very short residence time in the body of animals. They are nearly completely eliminated in the feces and urine within a few days after feeding and chronic toxicity can develop only by continuous fresh exposure to large quantities of the chemical. This is unlikely to occur because of the rapid decline of the materials in the environment.

## Insecticides

The careful use of insecticides poses no real threat to plant life, but because of the persistence and toxicities to fauna of many insecticides, their use generally causes greater concern for the environment than use of herbicides. Like herbicides, the initial distribution of insecticides will be among the four components of the environment: air, vegetation, forest floor, and surface waters. The amount of chemical entering each part will depend on the chemical and equipment used and a number of environmental factors (Figure 21.5).

Significant quantities of aerially applied pesticides do not reach the target area. Only 30 to 40 percent of malathion and DDT have generally been recovered on vegetation and the forest floor after spraying in forested areas. Airborne chemicals can be degraded, taken up by plants, adsorbed on various surfaces, or transported in the air as droplets or vapors to other locations. Chemicals intercepted by vegetation may be volatilized, washed to the soil by rain, or adsorbed on the leaf surface and later transported to the forest floor by falling leaves. The small amount that enters the plant and is translocated may be

**Figure 21.5** Insecticide use in forests is generally confined to localized applications to prevent spread of serious pests. Local applications do minimum damage to the soil and its environs. Large-scale applications of insecticides, such as sometimes used in the control of spruce budworm, are controversial.

degraded by plant metabolism, excreted by roots into the soil, or stored in plant parts. Volatilization from plant and forest floor surfaces is apparently not an important mechanism except for a few chemicals (Norris and Moore, 1971).

Some insecticides, such as DDT, chlordane, heptachlor, toxaphene, and other chlorinated hydrocarbons, are quite persistent in soils, while others, such as the carbamates and pyrethrum, have relatively short half-lives. The half-lives for carbamates is about eight days while that of pyrethrum is only a few minutes. The adsorption of chemicals is an important part of their behavior in the forest floor and soil. Adsorbed molecules of chemicals are not available for volatilization, biological degradation, or leaching. The persistence of Bidrin, an organophosphorus systemic insecticide, applied at the rate of 85 g per square meter, was measured in organic and mineral soils in North Carolina over a 90-day period (Werner, 1970). Significant residues from this high rate of application of the insecticide were present in the upper 15 cm of soil for only about 15 days following application. The rate of downward movement was faster in the mineral soil than in organic soil because of greater adsorption by organic colloids, but little movement was noted beyond 15 cm in either soil (Table 21.2).

DDT is extremely resistant to movement in soil due to its very low water solubility (1.2 ppb). Any appreciable movement of DDT through soils must, therefore, be the result of movement of colloidal particles of the free or adsorbed pesticide. Resistance to degradation and leaching exposes DDT to overland transport for extended periods of time. Such chemicals as the carbamates and pyrethrum are susceptible to overland flow for only short periods because of their rapid degradation. Furthermore, overland flow is relatively uncommon in most forested watersheds.

Contamination of surface waters by insecticide is due almost entirely to direct application or drift of the material during application. Norris and Moore (1971) reported surface water contamination due to direct application of DDT reached a maximum concentration of 0.28 ppb in samples taken a few hours after spraying. Most samples contained less than 0.01 ppb DDT.

**Table 21.2** Bidrin[a] Residues Found in the First 15 cm of Two Forest Soils During a 90-day Period After Treatment (Werner, 1970).

| Soil Layer, cm | Days After Treatment | | | |
|---|---|---|---|---|
| | 15 | 30 | 60 | 90 |
| | ppm | | | |
| | Organic Soil | | | |
| 0–2.5 | 0.068 | 0.048 | 0.024 | 0.012 |
| 2.5–5 | 0.066 | 0.019 | 0.000 | 0.000 |
| 7.5–10 | 0.012 | 0.000 | 0.005 | 0.004 |
| 12.5–15 | 0.007 | 0.006 | 0.011 | 0.000 |
| | Mineral Soil | | | |
| 0–2.5 | 0.066 | 0.040 | 0.033 | 0.008 |
| 2.5–5 | 0.067 | 0.032 | 0.017 | 0.001 |
| 7.5–10 | 0.045 | 0.007 | 0.002 | 0.001 |
| 12.5–15 | 0.045 | 0.000 | 0.000 | 0.003 |

[a]Bidrin (3-hydroxy-N, N-dimethyl-*cis*-crotonamide dimethyl phosphate), an insecticide of Shell Chemical Company, was applied at rate of 26 g/m$^2$.

Concentrations of insecticides reaching surface waters are rapidly reduced due to volatilization, adsorption on stream sediments, adsorption by aquatic organism, degradation by chemical, biological, or photochemical means, or dilution with downstream movement (Norris and Moore, 1971). Concentrations of carbamates exceeding 0.1 ppm will rarely be encountered in forest streams. Malathion, an organophosphate, is rapidly degraded in soil and water and enters water only by stream channel interception and limited surface runoff. Ultra-low-volume aerial applications will rarely produce more than 0.5 ppm malathion in streams.

In spite of the evidence that judicious use of most insecticides in forests poses only minor hazards to stream pollution and the environment, the use of more selective and less persistant chemicals is to be encouraged. Furthermore, alternatives to toxic chemicals, such as insect attractants, repellents, and feeding deterrents, hold considerable promise as components of integrated pest control programs. Such pest management programs are attractive means of eliminating the indiscriminate use of resistant and broad spectrum insecticides that have caused much of the concern for pesticide use in forests.

## Other Chemicals

A number of other chemicals are used in forests for various purposes. The total amounts of these chemicals used are generally quite small and their impact on the forest ecosystem is equally small. However, damage to the environment can be very real in localized situations.

Forest fire retardants rank next to fertilizers in volume of use. Several million gallons of retardant chemicals are annually applied to wildfires in the United States. Diammonium phosphate, monoammonium phosphate, and ammonium sulfate are used as retardants and even more widely used as fertilizer. They are not in themselves toxic to plants when used at rates recommended for fertilization. Localized high rates applied during fire suppression activities could damage vegetation, if the vegetation escaped from the wildfire. However, most vegetation will have been destroyed, and without plant uptake and with increased water movement in burned areas, much of the chemicals may be leached into lower horizons or lost to overland flow.

The use of borates in fire suppression represents a greater hazard to the environment than that of ammonium compounds, because high levels of boron are toxic to plants and could retard rate of revegetation. However, such damage is localized and is generally minor in comparison to the environmental damage from unsuppressed wildfires.

Sodium and calcium chloride are used for reducing icing on roads in cold climates. The salt-charged snow melt causes considerable salt burn to roadside plantings. In imperfectly drained areas the salt may concentrate and kill significant numbers of trees. Wider use of sand and frequent snow removal may partly substitute for salt in some sensitive areas.

Tarrant et al. (1973) listed other chemicals used in small amounts in forests. These include alkylpolyoxyethylene ethanol as a soil wetting agent to reduce erosion; hexadecanol, a saturated fatty alcohol for use as a transpiration retardant on planting stock and forest stands and to reduce soil water loss; and asphalt and wax emulsions as agents for speeding slash disposal by fire. None of these chemicals pose a serious threat to the environment at their present limited use. However, no chemical should be used in forests without an understanding of its immediate and long-term effects on nontarget organisms in the environment. Most pesticides are rapidly degraded in the soil and, when properly used, they cause no long-term damage to the soil biological communities or its physical and chemical properties.

# 22
# EFFECTS OF FIRE ON SOILS AND SITE

Forest fires have a history almost as ancient as the forests themselves. Unconstrained wildfires, ignited by natural forces, such as lightning and volcanos, exerted a profound influence on the world's vegetation types long before humans made their appearance. It is likely that all forests, with the exception of those that are perpetually wet, have been burned at one time or another (Spurr and Barnes, 1973). Fires and secondary succession following forest fires have played a dominant role in maintaining species diversity and in shaping the composition and structure of the earth's forests for thousands of years.

## TYPES OF FIRES
Forest fires can generally be grouped into wildfires, planned fires used in the destruction of forest for the sake of agricultural or grazing operations, and prescribed fires used in the management of natural ecosystems.

### Wildfires
Wildfires are a natural force in the evolution of plant communities in many parts of the world. In some regions they are infrequent, erratic, and often so severe as to force plant communities back to early successional stages. Lightning fires of this type are not uncommon in the Douglas-fir regions of the Pacific Northwest (Figure 22.1). Frequent, mild fires were probably more common where primitive people had a role in the initiation of the fires to aid in game hunting. These mild fires were prevalent in the mid-west and eastern United States where they helped maintain fire-conditioned "natural" communities (Spurr and Barnes, 1973) and shaped the development of coastal plain vegetation. Plants with a wide variety of adaptations to protect them from fire are found in frequently burned ecosystems.

**Figure 22.1** Lightning is the cause of many wildfires in the Pacific Northwest and, to a lesser extent, in some other forested areas.

These adaptations include fire resistant bark, heat resistant seeds, and protected buds.

Wildfires, whether ignited by people or some natural force, can be extremely destructive of the forest stand, ground-cover vegetation, and the forest floor. Wildfires and fires used for agricultural landclearing burn with sufficient heat to kill most or all of the vegetation above the soil surface and to initiate a new succession of plant life. Such fires often degrade the surface soil which, under some conditions, can be disastrous. Peat bogs and other organic soils may be virtually destroyed by burning after drainage or extreme drought. Burning surface litter and humus layers on rocky or shallow soils can be equally devastating to site productivity. Wildfires that destroy the protective forest canopy and reduce transpiration result in decreased infiltration rates and greater runoff, and may cause serious soil erosion problems, especially on steep terrain.

## Use of Fire in Land Clearing

In more recent times people learned to use "planned" burns to attain the beneficial effects of fire without the devastating effects of conflagrations. Fire was used in converting forests to agricultural lands by early Romans, and "swaling" agriculture, a practice based on the use of fire to clear forested area for crops, was used in many parts of central and northern Europe until the end of the last century (Viro, 1969). In North America, natives and early settlers alike used fire in their farming and livestock activities and these fires undoubtedly accounted for the openness of eastern forests and the existence of prairie islands in the midst of forest lands. Fire is an essential component of shifting cultivation, or "swidden" agriculture, in many tropical countries today. It appears certain that man has long contributed to nature's impact on the maintenance of fire-conditioned plant communities. Examples of these fire-climax communities in the United States are the red, jack, and pitch pines in New England and the Lake Region; lodgepole and ponderosa pines in western states; Douglas-fir in the Pacific Northwest, and southern pines in the southeastern states.

## Prescribed Fires

Of particular interest from the silvicultural standpoint is the use of controlled or "prescribed" fires by the modern forester. Such fires are applied to natural fuels under weather, fuel moisture, and soil moisture conditions that will allow confinement of the fire to a predetermined area and at an intensity of heat and rate of spread required to accomplish certain planned benefits.

The objectives of prescription fires vary according to the different ecosystems in which they are used, but they generally include one or more of the following: (1) to reduce fuel accumulations that contribute to high-intensity wildfires; (2) to increase reproduction by stimulating seed production, triggering cone opening, and improving seedbeds; (3) to reduce disease and insect problems; (4) to control undesirable and competing species; and (5) to remove unpalatable material and encourage desirable plants for grazing animals or wildlife habitat (Cooper, 1971).

Fire is probably the most economical tool available for disposing of logging slash (Figure 22.2) and for retarding competition from weed species. It is used for regenerating harvested species, as well as converting existing cover to another species. Its traditional use in some northern areas of Europe and North America has markedly declined in recent years, particularly in heavily populated, industrialized nations in which the use of fire has been associated with intensive forest management. The development of rugged machinery and powerful tractors, more complete tree harvest, fractionation of land holdings, and objections to smoke have contributed to this decline. Fire is still a major management tool in Australia, New Zealand, and western and southern United States. In Australia, fire is used for waste disposal, for the conversion of forest composition from indigenous to exotic species, and for the regeneration of native eucalypts (Rennie, 1971).

In the United States, prescription burning of forests probably had its origin in the southeastern coastal plain. It has been an accepted forest management tool of the region for generations and was first promoted by naval stores operators

**Figure 22.2** Burning is an economical method of disposing of logging slash in preparation for a new crop of trees (U.S. Forest Service photo).

who burned intentionally as a means of reducing the danger of uncontrolled fires (Cooper, 1971). The approximately one million hectares of forests that are burned by prescription each year in the region present a classical example of the use of fire in maintaining less competition-tolerant species (pines) as a major component of the forest vegetation. Pines exhibit a higher degree of fire tolerance than do most hardwoods (Langdon, 1971). Without fire or some similar disturbance most upland forests of the area would ultimately be composed of heavy-seeded hardwood species. Consequently, prescribed burning was developed as a management tool that is used to interrupt the successional trend toward a climax oak-hickory forest, a trend heightened by the exclusion of wildfires. The reduction of wildfire hazard in forests is perhaps of greater importance than the reduction of competing species. In the absence of fire, up to 25 tons per hectare of fuel may accumulate on the forest floor (Heyward and Barnette, 1936). Natural decomposition processes are not rapid enough to keep fuel loads at a safe level. Wildfires in areas of heavy fuel loads can rapidly develop into catastrophic conditions during dry periods.

In pine forests of the United States, hazard reduction burns are generally carried out during the dormant season when air temperatures are low (less than 10°C), upper litter moisture is relatively low (8 to 12 percent) and lower litter moisture moderately high (more than 20 percent), relative humidity is between 20 and 50 percent, and winds are steady (Cooper, 1975).

The quantity of heat required to raise the temperature of living vegetation to a lethal temperature of about 60°C is directly proportional to the difference between the lethal temperature and the temperature of the vegetation initially. For this reason summer fires are more damaging to vegetation than winter fires of the same intensity. Snow cover generally precludes controlled burning at higher elevations in the western United States during the winter season; hence most fire activity is scheduled for the fall months. Repeat burning is used to keep the fuel load at reasonable levels. On most pine sites, natural fuels build back to a critical level in about five years.

*Backfiring* (forcing the fire to advance against the wind) is the most common technique in hazard reductions, as seen in Figure 22.3, but strip, spot, or flank firing may be appropriate in special situations (Cooper, 1975). Although backfires are more time consuming, they are easier to manage, produce less smoke, and are generally more effective in reducing fuel accumulations to an acceptable level, than are other types of controlled fires. A well-executed prescribed backfire can consume about two-thirds of the understory fuels with minimum damage to the soil, overstory, or other amenities of the forest (Figure 22.4). The actual damage to the soil and to long-term site productivity depends on the type of fire, soil and climatic conditions, and the steepness of the terrain.

Slash disposal burning can be accomplished during any month of the year if fuel is right and favorable weather prevails. Two to four months are generally required for logging debris to reach its optimum combustion condition.

**Figure 22.3** Control burning using a backfire with a light fuel load.

**Figure 22.4** Periodic burning under controlled conditions results in parklike stands of southern pines.

## EFFECTS OF FIRE ON SOIL PROPERTIES

The effects of forest fires on the physical, chemical, and biological properties of soils are directly related to the severity of the burn. The intensity and duration of soil heating by prescribed fires, such as used for hazard reduction, are significantly less than those generated by slash burning or wildfires and, consequently, the effects of controlled burning on soil properties are generally less than the effects of wildfires.

## Soil Physical Properties

Wildfires may consume a large part of the forest floor, eliminating many of the beneficial effects of this organic layer on soil physical properties. On the other hand, prescribed fires seldom remove more than 50 percent of the surface organic layers. Sweeney and Biswell (1961) reported that 76 percent of the litter (01) and 23 percent of the duff (02) horizons were consumed by four test fires in ponderosa pine in California. The soil organic fraction (A1 horizon) is not generally affected by light burns. Some nondestructive distillation of volatile substances and abnormal drying of organic colloids may occur at shallow depths, but this happens only with unusually hot fires.

**Soil Temperature During and After Burning.**   Changes in temperature at different depths in soils, as a result of burning, are functions of the thermal conductivity of the soil and the temperature and duration of the fire. Controlled fires in longleaf pine flatwoods and well-drained upland sites with fuel accumulations of up to 15 years seldom generated temperatures above 52°C for more than 15 minutes at shallow (3 to 6 mm) soil depths, and except for brief intervals of two to three minutes, the maxima were below 121°C (Heyward, 1938). The highest temperature observed 2.5 cm below the surface was 66°C. Viro (1974) reported that the temperature of mineral soil below the burning humus does not increase greatly in boreal forests—seldom exceeding 100°C beneath a 3-cm humus layer. Ralston and Hatchell (1971) reviewed studies on increases in soil temperature during burning and some of the results in their review are illustrated in Figures 22.5 and 22.6. Cromer (1967) reported that where eucalypt logs were pushed into heaps or windrows the soil was heated to a considerable depth. While temperatures of 100°C were not common below 10 cm, 50°C was occasionally recorded at depths of more than 30 cm.

Soil temperatures following a burn are influenced by the alteration in the insulating capacity of the litter layer and heat absorption pattern of the surface layer as a result of a dark ash deposit.

Forest floor material acts as an insulator against soil temperature and soil moisture changes and the removal of this material exposes the mineral soil to the vagrancies of the weather. Partial removal by a light burn does not greatly affect this mulching property in temperate zones (Ralston and Hatchell, 1971). Furthermore, Viro (1974) reported that improved thermal conditions from thinning the

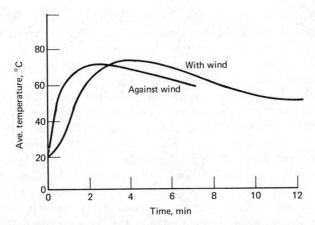

**Figure 22.5** Soil temperatures (3 to 6 cm below surface) during first 12 minutes of two types of controlled burns in five-year rough in longleaf pine in Florida (Heyward, 1938).

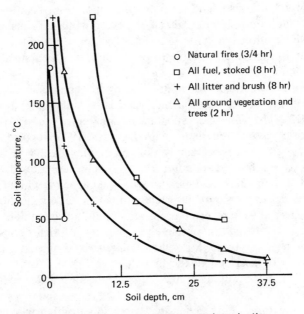

**Figure 22.6** Soil temperatures at varying depth subjected to different type fires in eucalypt forests (Beadle, 1940).

humus layer was the main purpose of prescribed burning in Fenno-Scandian forests. He cited mineral soil temperatures of 12.5°C at 10 cm depths in burned forests and 8.7°C at similar depths in unburned areas during summer months. The thickness of the humus layer averaged 3 cm in the burned sites while the humus and moss layers were 9 cm thick in unburned areas.

The darker surface of a burned site effectively absorbs solar radiation; consequently the surface layers of soil in burned sites are warmer than those of unburned sites, especially during the growing season. Viro (1974) reported average summer temperatures of 31.3°C at the surface of burned humus layers in boreal forests, but only 18.0°C at the surface of unburned humus layers. It appears that unless the canopy shade is also removed, as might occur in a wildfire, the effects on soil temperature of the dark color associated with ash deposits on burned surfaces may be of minor consequence in well-stocked or dense stands. However, in open pine stands, the influence of a dark ash layer on the temperature of the underlying mineral soil can be significant during summer months.

**Soil Moisture.** The effects of fire on soil moisture are indirect and are often ill defined. If the majority of the forest floor is removed by burning, water absorption and retention by the humus layer may be significantly reduced. The elimination of this organic mulch results in increased evaporation. The higher temperatures of a burned site also tend to increase evaporation. Ash and fire-charred materials may filter into the mineral soil on severely burned sites and reduce the rate of water infiltration and increase runoff. The net result of these changes can be a reduction in moisture available to plants and, on steep sites, a loss of soil materials.

*Water-repellent* soils can develop from hydrophobic substances vaporized during burning of surface litter layers on sandy soils (DeBano, Mann, and Hamilton, 1970). The vaporized substances apparently move downward, condense, and form a water-repellent layer. Dyrness (1976) reported that burning increased water repellency of sandy soils at depths of 2.5 to 23 cm, which persisted for up to five years after the fire. This condition reduces water infiltration rate (Figure 22.7) and moisture storage capacity, which are of particular importance on the dry sandy sites where repellent layers most often develop. While many sandy soils in forested areas are difficult to rewet once they become excessively dry, they seldom completely dry under a forest floor. Water repellency has not been reported for regularly prescribed-burned soils of coastal plains. This may be ascribed to the fact that the forest floor is not drastically reduced and the mineral soils seldom completely dry during a successful burn.

**Soil Erosion After Fire.** Fire generally affects soil erodibility if the mineral soil is exposed, either by hot wildfire or by repeated burning over long time periods. Porosity and infiltration rates decrease and bulk density increases following fires

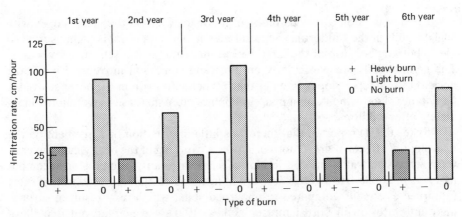

**Figure 22.7** Mean infiltration rates in heavily burned, lightly burned, and unburned areas for the first six years after the fire (Dyrness, 1976).

on many soils. Soil aggregates are dispersed by beating rains and pores may become clogged by fire particles that decrease macropore spaces, infiltration, and aeration (Ralston and Hatchell, 1971). The action of worms and other soil fauna contributing to soil porosity also may be significantly reduced on frequently burned areas. Where grazing is permitted, trampling of cattle on burned areas increases soil bulk density more than equivalent trampling on areas with an accumulated litter layer. However, the amount of erosion following a fire depends on the inherent erodibility of the soil; steepness of slope; time, amount, and intensity of rainfall; severity of fire; and plant cover remaining on the soil.

Where surface organic horizons are not completely consumed in prescribed fires, changes in the pore space and infiltration rates often are too small to be detected. This is particularly true of prescribed burns on sandy soils such as found in much of the southeastern coastal plain of the United States. Ursic (1970) reported annual increases in stormflow from two burned watersheds supporting hardwoods on hilly terrain in northern Mississippi ranged from 2.5 to 7.5 area-cm during the first three years, representing increases of 16 to 50 percent over the expected flow. In these calibrated watersheds, a controlled backfire consumed the *L* layer but only a fraction of the *F* layer. Nevertheless, sediment production exceeded the expected values by 48 to 100 percent the first year.

In some mountainous areas, erosion may accompany increases in peak runoff, especially where the forest floor is substantially consumed by fire. In a 100-hectare wildfire in the central Rocky Mountains, observations (Striffler and Mogren, 1971) indicated that some erosion occurred during the first summer after the burn, but that erosion was not considered a serious problem. The amount of exposed mineral soil surface was more important than the degree of slope in initiating particle movement during low-intensity storms, but slope was the controlling factor during high-intensity storms.

Erosion is often a serious problem following fires on the steep slopes with shallow soils in the California chaparral region. The chaparral is highly susceptible to burning because of the large fuel accumulations and seasonal dry winds. The granitic soils are particularly erodible and some soil movement may take place on the steeper slopes (greater than 70 percent) even in the absence of fire. Gravitational movement, or slippage, of debris often increases severalfold immediately after wildfires.

Sykes (1971) reported that there was little information or agreement on the hydrological effects of fire in northern forests. Burning of the heavy moss layer in these forests may alter the distribution of summer runoff, with erosion accompanying flash floods. However, there are indications of increased infiltration rates in burned-over soils compared to unburned soils, which may result in surprisingly little erosion in boreal forests (Sykes, 1971). He pointed out that these infiltration data are in contrast to those reported by workers in temperate zones where infiltration rates on burned-over areas have been slower than on unburned areas.

## Soil Chemical Properties

Changes in the chemical properties of soils during burning are primarily related to the rapid conversion, or oxidation, of nutrients contained in the organic materials of the living vegetation and litter on the soil surface. Burning differs from natural processes of biological oxidation mainly in the speed with which nutrients are released. In biological decomposition, mineralization of nutrients in the forest floor material is slow but steady and most of the nutrients are absorbed by plant roots as they are released. On the other hand, burning releases nutrients rapidly and some are lost by volatilization, while a portion of the soluble minerals may leach into the soil beyond tree roots.

Cromer (1967) reported that burning of slash piles, or windrows, released readily available nutrients to the soil and partially, or completely, sterilized the surface soil horizon, depending on the severity of the fire. *Pinus radiata* grown in the ash beds were superior to the surrounding trees and contained higher concentrations of phosphorus than fertilized trees on unburned soil. He attributed the greater uptake of phosphorus to a reduction in antagonistic soil organisms (Figure 22.8).

**Organic Matter.**   While large percentages of the standing biomass and organic matter may be destroyed in a severe wildfire, the materials consumed in a controlled burn are confined to the understory vegetation and forest floor debris, and only a small part of the total of these may be burned. The amount of material consumed in a prescribed fire depends primarily on the kind and amount of fuel present and the weather conditions. It may be as little as 2000 to 4000 kg per hectare in an annually burned forest and 4000 and 9000 kg per hectare in infrequently burned areas (Wells, 1971).

**Figure 22.8** "Ashbed" effect (left) on growth of *Pinus radiata* in Victoria, Australia, may result from soil sterilization, as well as increased nutrient availability.

Annual burns consume approximately the same amounts of materials as in annual litter falls. Litter falls in mature stands of southern pines have commonly been estimated at 4000 to 6000 kg per hectare per year. Bray and Gorham (1964) reported an average annual litter fall of 5500 kg per hectare in warm temperate forests. These values are probably not significantly different from litter fall amounts in most well-stocked temperate zone deciduous and mixed forests. A loblolly pine plantation in the southeastern coastal plain of the United States accumulated 26,000 kg per hectare of forest floor humus when left unburned for 20 years (Wells, 1971). This is comparable to the 21,000 kg per hectare of forest floor litter accumulated in a 12-year-old slash pine plantation in Florida (Mead and Pritchett, 1975). From these data one could conclude that the half-life of litter in unburned pine forests of the coastal plain is about 2 to 3 years. Fire speeds up the oxidation rate and after 20 years of annual summer burns or annual winter burns the forest floors of loblolly pine stands averaged 7800 or 14,500 kg per hectare, respectively (Wells, 1971).

In spite of obvious losses of organic materials during prescribed burns, reports generally have indicated no significant long-term decreases in total organic matter to the soil-forest floor system (Viro, 1969; Wells, 1971). Heyward and Barnette (1934) reported that there was no evidence that fires depleted soil nitrogen or soil organic matter in longleaf pine stands of the coastal plain. Viro

(1969) came to similar conclusions for boreal forests in Finland, although much longer periods were required for the sites to return to the preburn state than in temperate-zone forests. The weight of the humus layer proper ($F + H$ layers) diminished by about 24 percent (from 33 to 25 tons per hectare) during a prescribed burn in the boreal forest and this difference between burned and unburned areas had been reduced by half (12 percent) during the first 13 years after burning.

The explanation for the apparent maintenance of organic matter in the ecosystem, in spite of the obvious losses during prescribed burning, lies in increases of organic matter in mineral soil layers equivalent to that lost from the forest floor. For example, organic matter in the 0 to 5 cm zone of a mineral soil in annually burned loblolly pine was about 30 percent more than in nonburned areas after 20 years, as shown in Figure 22.9 (Wells, 1971). The increase in organic

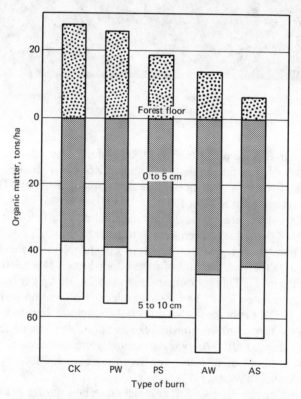

**Figure 22.9** Organic matter in the forest floor, 0 to 5, and 5 to 10 cm of mineral soil for check (CK), periodic winter (PW), periodic summer (PS), annual winter (AW), and annual summer (AS) treatments after 20 years (Wells, 1971).

matter content in the surface mineral soil may result from (1) increased growth of fibrous-rooted plants, such as grasses, following burns; (2) movement of colloidal-sized charred material from the burned floor into the mineral soil by gravity or water; (3) isoelectric precipitation of alkali humates produced during burning; and (4) an accumulation of decomposition-resistant organic residues mixed in surface mineral soil. None of these explanations appears completely satisfactory for all situations. Furthermore, maintenance of soil organic matter should not be expected in all burned areas. Certainly, hot fires from burning windrows may significantly reduce both organic matter and total nitrogen in the underlying mineral soil as well as in the above-ground materials (Cromer, 1967).

**Total Nitrogen.** As pointed out (Wells, 1971; Viro, 1974), total nitrogen is highly correlated with organic matter and, when the forest floor is burned, nitrogen is decreased in relation to the severity of the fire. Wells (1971) reported that while total nitrogen content decreased in the forest floor during controlled burning, it accumulated at about the same rate in the 0 to 10 cm of the mineral soil (Figure 22.10). As a matter of fact, Heyward and Barnette (1934) reported that there was no evidence that repeated controlled burnings of virgin longleaf pine forests depleted soil nitrogen. While 100 to 300 kg nitrogen per hectare are generally volatilized during a controlled burn, it appears that increased biological nitrogen fixation may to a significant extent replace the nitrogen lost from burned areas. There are suggestions (Jorgensen and Wells, 1971; Wells, 1971) that increased soil temperature, moisture, nutrient supply, and pH conditions resulting from a reduction in ground cover and deposition of ash on the burned surface may favor both symbiotic and nonsymbiotic fixation of nitrogen. Leguminous plants are often more prevalent in burned areas (Cooper, 1971) and the activities of some nitrogen-fixing bacteria and blue-green algae are thought to increase as the pH of the surface soil increases.

Viro (1974) reported that the average loss of 320 kg nitrogen per hectare due to burning of boreal forest floors required nearly 50 years for full recovery. However, he considered this loss to be "unimportant because the nitrogen in living vegetation is totally unavailable to other plants. Mineralization of the nitrogen in the logging waste and humus layer is very slow. On the other hand, burning decreases acidity of the humus layer and in this way encourages mineralization of nitrogen. So, despite the loss in the total amount of nitrogen, burning greatly increases the mineral nitrogen."

From the above discussion it appears obvious that the effects of control burning on the long-term total nitrogen content are complex and vary among ecosystems. There can be little doubt that considerable amounts of nitrogen are lost during any burning operations, with only small amounts of the gaseous nitrogen bound as ammonia in the unburned humus. The loss of nitrogen during burning of the forest floor may have little adverse effect on tree growth because

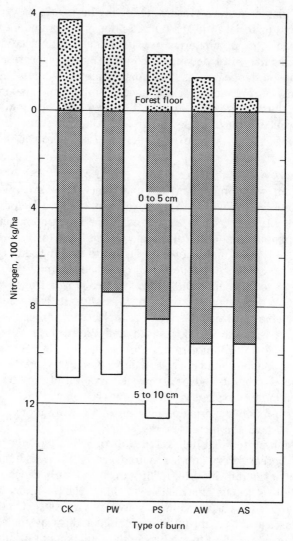

**Figure 22.10**  Nitrogen in the forest floor, 0 to 5, and 5 to 10 cm of mineral soil for check (CK), periodic winter (PW), periodic summer (PS), annual winter (AW), and annual summer (S) treatments after 20 years (Wells, 1971).

of increased nitrogen fixation and improved nutrient availability, but the net loss to the system may be greater than generally reported. Most investigations (Heyward and Barnette, 1934; Viro, 1974; Wells, 1971) have not taken into consideration the possible differences in the amounts of nitrogen immobilized in the living biomass of unburned areas compared to that of burned areas and any conclusions must be based on a relative few long-term studies.

**Soil Acidity.**   In a very acid, poorly drained sandy soil of the lower coastal plain, 20 years of burning decreased the acidity from pH 3.5 to 4.0 in the *F* and *H* layers of the forest floor, and from pH 4.2 to 4.6 in the 0 to 5 cm of mineral soil (Wells, 1971). There was no significant change in acidity below 5 cm. Most of the change occurred during the first 10 years of burning, according to Wells. Since an increase in alkalinity of the surface layer results from the base elements contained in the ash residue, the magnitude of the change depends on the amount and base content of the ash, as well as the texture and organic matter content of the soil. Greir (1975) reported that an intense fire in central Washington produced an average ash weight on the soil surface of 2900 kg per hectare. The ash layer contained 23 kg N, 314 kg Ca, 54 kg Mg, 70 kg K, and 22 kg Na per hectare. However, ash deposition from controlled fires probably amounts to less than half of this value. The destruction of organic matter by burning also reduces the formation of organic acids that are normally formed during biological decomposition and that contribute to soil acidity.

Heyward and Barnette (1934) reported that soil pH values of the surface mineral horizon in frequently burned longleaf areas averaged 5 percent higher than in unburned areas. Since these sandy soils were sampled to a depth of 10 cm, their values do not represent the true changes in acidity at the surfaces of the mineral soil. Viro (1974) reported that the oxides or carbonates released during burning in Finnish forests reduced acidity of the remaining humus layer by 2 to 3 pH units, but that the effect did not persist because of leaching of these alkali compounds and the formation of new humus. The acidity of the humus layer reverted to its original level 50 years after burning. During the first 20 years after burning, acidity of the surface of the mineral soil was on the average 0.4 pH units higher on burned than on unburned sites, and a difference of 0.2 units persisted for at least 50 years.

**Available Nutrients.**   The forest floor contains a substantial portion of the nutrient reserve in most forests. It has been pointed out that in unburned forests, forest floor organic matter slowly decomposes and nutrients are made available for use by higher plants through biological oxidation. Fire drastically speeds up the process of oxidation and some of the mineral nutrients thus released are dissolved and rapidly leached into the mineral soil. Thus, it is not surprising that the ash deposit from a fire increases available phosphorus, potassium, calcium, and magnesium. These increases in available nutrients are greatest directly after a fire, but increases in some elements often persist for 5 years or longer. Wells (1971) reported significant increases in extractable calcium and magnesium in the 0- to 5-cm layer of mineral soils, and significant increases in extractable phosphorus in the 0- to 10-cm layer, after 20 years of annual burning in lower coastal plain sands. Potassium levels in these surface soils were not significantly influenced by burning, presumably because of the rapid leaching of this monovalent cation (Table 22.1).

**Soil Organisms.** Most soil organisms are sensitive to changes in soil temperature, soil moisture, and nutrient supplies; yet there appears to be little agreement among researchers on the effects of fire on these important contributors to the forest ecosystem. Reasons for the differences largely relate to differences in environmental requirements among the various microflora and mesofauna and to differences in experimental techniques.

**Table 22.1** Amounts of Phosphorus, Potassium, Calcium, and Magnesium in the Forest Floor (FF), 0 to 5, and 5 to 10 cm of Mineral Soil in Plots Under Different Burning Regimes for 20 years (Wells, 1971)

| Layer | Annual Summer | Annual Winter | Periodic Winter | Control |
|---|---|---|---|---|
| | | kg/ha | | |
| | | Phosphorus | | |
| FF | 4.0 | 9.0 | 16.6 | 18.1 |
| 0 to 5 | 4.1 | 4.6 | 3.2 | 3.2 |
| 5 to 10 | 1.7 | 2.0 | 1.6 | 1.6 |
| | | Potassium | | |
| FF | 6.0 | 14.7 | 26.5 | 27.7 |
| 0 to 5 | 18.6 | 21.2 | 17.4 | 17.5 |
| 5 to 10 | 10.0 | 12.6 | 9.4 | 10.0 |
| | | Calcium | | |
| FF | 45 | 86 | 150 | 157 |
| 0 to 5 | 159 | 83 | 57 | 32 |
| 5 to 10 | 62 | 30 | 26 | 18 |
| | | Magnesium | | |
| FF | 7.6 | 15.6 | 27.2 | 30.8 |
| 0 to 5 | 24.5 | 18.8 | 12.4 | 10.6 |
| 5 to 10 | 15.0 | 11.2 | 7.4 | 7.4 |

*Soil Microbes.*   Heat can cause immediate reductions in bacterial populations (Ahlgren, 1974), but the extent of the changes in microbial populations depends on the intensity and duration of the fire, soil moisture and texture, and depth at which the organism resides within the soil. Heating the soil for one hour at 100°C produces an initial depression in numbers of bacteria, followed by a sharp increase. The population decline may last only until the first postfire rainfall and after a few months populations in the upper 4 cm of soil on burned land may be higher than on unburned areas. Jorgensen and Hodges (1971) reported few indications that prescribed burning of a loblolly pine plantation adversely altered the qualitative or quantitative composition of fungi or bacteria plus actinomycetes to the extent that soil metabolic processes were impaired. In fact, annual burns had no effect on microbial populations in the mineral soils and only fleeting effects on bacteria plus actinomycete populations in the forest floor (Table 22.2). However, Wright and Tarrant (1957) reported that there were fewer fungi in the upper 4 cm of soil on recently burned Douglas-fir areas than on areas burned six months previously, but only on severely burned areas was this decrease detected at soil depths below 8 cm.

Ahlgren (1974) suggested that fungus species comprising the population on burned land might be expected to differ from those on unburned land because of the heterogeneous nature of the habitat requirements of the organisms. Since surface soil pH varies from alkaline soon after a fire to a more acid condition as the ash minerals are leached out, the species of fungi found in an area probably vary with age of burn.

Some pathogenic fungi occur in or on the soil for a portion of their life histories and fire is of value in purging the forest of certain diseases. A classic

**Table 22.2**   Numbers of Fungi and Bacteria + Actinomycetes in Soil Layers by Burn Treatment (Jorgensen and Hodges, 1971)

| Soil Layer | Burn Treatment | Fungi | Bacteria + Actinomycetes |
|---|---|---|---|
| | | | millions/g |
| F + H | No burn | 1.51 | 51.1 |
| | Periodic burn | 3.28 | 70.8 |
| | Annual burn | 1.18 | 28.2 |
| 0 to 5 cm | No burn | 0.12 | 4.1 |
| | Periodic burn | 0.14 | 3.0 |
| | Annual burn | 0.13 | 6.5 |
| 13-18 cm | No burn | 0.03 | 1.3 |
| | Periodic burn | 0.02 | 1.1 |
| | Annual burn | 0.02 | 1.1 |

example is brown needle spot of longleaf pine caused by *Septoria alpicola*. Winter burning eliminates the disease for about a year, permitting better seedling development during the two to three years before the disease recurs in abundance (Ahlgren, 1974). Prescribed burning before and after thinning may reduce losses to *Heterobasidion annosum* of slash and loblolly pines on sandy sites.

Little is known about the effects of fire on algae, but one would suspect that they are reduced in surface soils in a manner similar to other microorganisms. However, this reduction is apparently quite temporary. They generally thrive under alkaline conditions and are often the first organisms to colonize burned wetlands. Increased populations of algae should be expected under these conditions.

Soil biological processes, like soil organisms, are affected by forest fires in a number of ways. Nitrogen mineralization in forest soils is not believed to be greatly affected by burning, although this may depend on initial soil acidity and texture. The proportion of mineral nitrogen to total nitrogen was similar for annually burned and nonburned areas in the lower coastal plain of the United States (Jorgensen and Hodges, 1971). However, Viro (1969) reported a threefold increase in ammonium nitrogen content of the humus layer during burning in Finland. The nitrate contents of all horizons of the mineral soil were higher on burned sites than on unburned sites as a result of increased nitrification in the former areas and subsequent leaching into the mineral soil. The activity of blue-green algae and nitrogen-fixation by other organisms may be enhanced by increases in available nutrients, decreases in soil acidity, and increases in light intensity at the soil surface, due to burning and reduction in ground vegetation.

*Soil Animals.*　Soil fauna includes those animals living in the forest floor and the mineral soil for all or part of the year. Many of them move back and forth between the two strata; the preferred stratum depending on stage of development or environmental conditions. The effects of fire on soil animals, therefore, depend on their habitat and mobility, as well as the tolerance of the organisms to heat and desiccation. In prescribed burns, the heat of the fire may be less important in reducing insect populations than later environmental changes brought about by the fire (Ahlgren, 1974). In fact, most organisms in the top 2.5 cm of mineral soil apparently survive moderate fires. Buffington (1967) attributed the decreases in soil animals after fires to loss of both incorporated and unincorporated organic matter that reduces food supply for the smaller organisms and, in turn, for their predators.

Fellin and Kennedy (1972) reported that there were generally more arthropods present in older prescribed burns than recent burns in forest clearcuts in northern Idaho. In duff samples, the *Acarina, Chilopoda, Thysanoptera, Protura,* and *Thysanura* were most numerous in more recent burns. Mineral soil samples collected before mid-July contained about 90 percent of the total number of individuals, mostly immatures. In sandy soils of the lower coastal plain, Metz and Farrier (1971) found that annual burning caused a reduction of the soil mesofauna immediately after the burn. Since there were no significant differences in population between control plots and the periodically burned plots, they concluded that the recovery period for this type of burn was less than 43 months. Mineral soil in the annually burned plots consistently had higher populations of mesofauna than did the soil of periodically burned and control plots, although these differences were not significant. Metz and Dindal (1975) collected more Collembola on the controls than on the annually burned plots. However, there were no significant differences between control plots and those burned periodically (last burned about 43 months previously).

Ants are less affected by fire than other groups of insects because of their adaptations to the hot, xeric conditions of early postfire topsoil (Ahlgren, 1974). Furthermore, their cryptic habits enable them to survive fire below the level of intense heat and their social organization adapts them to rapid reestablishment on burned land.

Earthworm populations are apparently markedly reduced by most fires. A 50 percent reduction in earthworms resulted from litter burning in the Duke Forest (Pearse, 1943), and Heyward and Tissot (1936) reported populations were four times greater in unburned than in burned surface (0 to 5 cm) mineral soil in longleaf pine stands. There are some indications that the earthworm population may be more influenced by postburn adverse moisture conditions and food supply than by excessive temperatures during burning.

## EFFECTS OF FIRE ON AIR AND WATER QUALITY

### Smoke Pollution

Forest burning contributes substantial amounts of emission to the atmosphere. A survey of the primary air pollutants indicated that in 1968, forest fires (including wildfires and prescribed burnings) produced about 8 percent of the total atmospheric pollution (Dieterich, 1971). Forest fires produce a variety of combustion products, most of which are not unique to forest fuels. The major components listed in decreasing order of amounts produced are carbon dioxide, water vapor, particulate matter, hydrocarbons, other organics, nitrogen oxides, and minute quantities of many other substances. One of the most objectionable of the combustion products is smoke, the particulate fraction of the emission material. It is important from the standpoint of esthetics and atmospheric degradation (Figure 22.11). When present in large quantities, smoke can cause drastic reductions in visibility and create locally hazardous conditions for surface and air transportation, as well as cause damage to exposed materials and result in human discomfort. Smoke particles, most of which are less than one micron in diameter, contain some chemical compounds that may pose health hazards if they combine with other pollutants to form harmful chemical products (Dieterich, 1971).

The use of prescribed fire for land management and protection carries with it

**Figure 22.11** Smoke causes a reduction in visibility and can create hazardous conditions for air and ground transport systems (U.S. Forest Service photo).

an obligation to conduct burning operations in such a manner as to eliminate or minimize adverse environmental impacts. McLean and Ward (1976) described an air-curtain combustion device that provided an environmentally acceptable technique for disposal of concentrations of forest residues. In situations where complete disposal of large-size residue is required in a smoke-sensitive airshed, this mobile burner provides a practical method of disposal. However, more complete utilization of tree components may be the only satisfactory method of reducing the need for slash burning and its accompanying smoke (Figure 22.12). Unfortunately, no mechanical means has been devised for eliminating smoke associated with prescribed fires needed for fuel reduction in forests. Nevertheless, the principal pollutant, smoke, is subject to some degree of manipulation by

**Figure 22.12** More complete utilization of tree components, as illustrated by this whole-tree chipper may be a practical alternative to burning in smoke-sensitive areas (U.S. Forest Service photo).

timing the burns when favorable weather and fuel conditions exist to minimize smoke production and encourage smoke dispersal.

Effective management of smoke is largely dependent on an ability to forecast weather for specific areas and to utilize those forecasts in developing plans to keep the smoke away from major highways, airports, and metropolitan areas. The National Weather Service issues periodic "air pollution potential" forecasts on a synoptic scale for the entire United States. These forecasts warn of conditions conducive to poor smoke dispersion generally associated with stagnating high-pressure systems. They can help identify localized problems with stagnating air and contribute to more effective management of smoke from prescribed fires.

Most states have regulations concerning open burning, but they also have exemptions that permit controlled use of fire for purposes of forest and game management, although the rules differ widely from state to state. Some states specifically prohibit any type of open burning during periods of air stagnation. Other states have regulations that specify that prescribed burning can be done only when prevailing winds blow away from cities, towns, and smoke-sensitive areas or at varying distances from primary and secondary highways and airports. Enforcement of regulations on controlled burning also varies from state to state and it is not always vigorously pursued. Much controlled forest burning takes place in areas remote from population centers and, unless there are significant reductions in visibility in smoke-sensitive areas, the only damage appears to be subtle additions to those of other air pollution sources. Nevertheless, some improvement in current regulations on prescribed burning, and the enforcement of those rules, are needed if national standards on air quality are to be achieved. It is in the interest of the public and forest managers alike to achieve this goal with regulations that will not unduly restrict prescribed burning, because resultant wildfires may contribute more to air pollution than the prescribed fires.

## Water Quality

Information on the effects of fire on water quality is rather limited and the results of research on the problem are sometimes conflicting due to differences in environmental conditions under which they were obtained. There have been several reports on the influence of wildfires and slash burning on increases in suspended sediments, dissolved chemicals, and water temperature of streams flowing from watersheds (Fredriksen, 1970; Levno and Rothacher, 1969; McColl and Grigal, 1974). Essentially no research has been reported on the effects of prescribed burning of forests on water quality.

**Sedimentation.**   Increases in sedimentation following slash burning have generally been less than that produced by either road construction or clear-cuts. No significant increase was noted in prevailing sedimentation rate of one stream in western Oregon after a moderately intense slash burning of logged areas that covered 25 percent of the watershed (Fredriksen, 1970). Accelerated surface soil erosion after burning can probably be expected only on steep land surfaces bare of vegetation. These conditions can be found in mountainous areas following extremely hot slash fires or wildfires. Turbid water, which results from abnormal sedimentation, blocks light transmission and reduces primary production of aquatic plants. It detracts from the recreational values of streams and lakes and increases the treatment costs of municipal water supplies. However, there is little evidence that sedimentation increases significantly in streams from forest lands that have been control burned for hazard reduction.

**Dissolved Chemicals.**   Materials dissolved in stream water often represent a loss of nutrients from some sites as well as a reduction in water quality. Water from undisturbed forest streams is generally quite low in dissolved chemicals and a large part of those chemicals present consist of silica, sodium, calcium, and magnesium (Fredriksen, 1970). These chemicals are released by mineral weathering in quantities in excess of the requirements of the forest stand for these elements. On the other hand, nitrogen and phosphorus are present in streams in much smaller concentrations than the bases. As nitrogen and phosphorus are mineralized by decomposition of organic compounds in soils, they are rapidly taken up by trees and microorganisms and only very low concentrations are found in the ground water.

In a clear-cut and slash burned Douglas-fir forest in Oregon, the concentrations of nutrients in forest streams were considerably greater following clear-cutting and slash burning than following clear-cutting alone. Nitrate-nitrogen concentration rose to a maximum of 0.43 ppm one year after burning (Figure 22.13). Although increases were large, the streams' concentrations and total amounts lost were still quite low. Ammonia concentrations in streams were higher than nitrate concentrations for a short period following burning, but virtually the entire ammonia loss occurred within two weeks following the slash fire. The total loss of nitrogen amounted to about 2.2 kg per hectare per year for the first two years following burning, after which the loss of nitrogen approximated the 0.05 kg per hectare from the control area.

Phosphorus concentrations in stream water from a clear-cut area were almost the same as those from an undisturbed forest, but increased concentrations were evident the first year following slash burning. Although the inorganic phosphorus concentrations in streams during the year following slash burning were 3.8 times that of the concentration of streams from the control area, the actual increases were very small considering the low average levels of the control streams (about 0.02 ppm). The losses of potassium, calcium, and magnesium the first year following slash burning averaged 6, 81, and 26 kg per hectare as

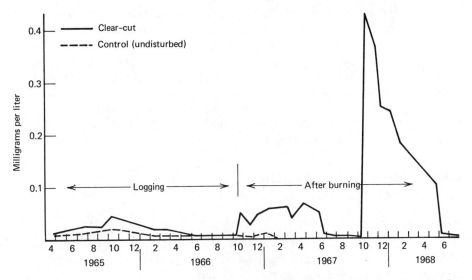

**Figure 22.13** Nitrate-nitrogen concentrations of streams from an undisturbed forest and a clear-cut and slash burned area (Fredriksen, 1970). Used with permission.

compared to 3, 26, and 8 kg per hectare from the undisturbed forest (Fredriksen, 1970). These increased nutrient loads rapidly declined as the area became revegetated.

Following a wildfire in a virgin forest on a lake watershed region of northeastern Minnesota, phosphorus concentration in the runoff was significantly elevated for two years but decreased to normal in the third year (McColl and Grigal, 1975). However, there was no detectable increase in the phosphorus concentrations of an adjoining lake and its input stream. The authors concluded that under similar circumstances, controlled burning would not damage streams or lakes by elevating phosphorus levels. While it is likely that nitrate and phosphorus concentrations in the soil solution increase following controlled burning, the filtering action of the soil should prevent these nutrients from reaching open streams, in absence of overland flow.

**Stream Temperature.** Increases in stream temperature as a result of shade removal and stream surface exposure to sunlight are generally associated with timber harvest. However, Levno and Rothacher (1969) reported that surviving understory shrub vegetation and logging debris adjacent to a stream may provide shade that cushions the solar radiation impact after canopy removal. In their study, clear-cut logging in Oregon increased maximum water temperature only 2°C after all merchantable trees had been removed but before slash burning and stream clearance. However, maximum temperatures increased by as much as 6°C for short periods in logged and burned areas.

The vigor and disease resistance of prized game fish may be affected as much by the duration of the increased water temperature as by the peak temperature of the water. Maximum temperatures decline as clear-cut areas revegetate. Fredriksen (1970) reported a mean monthly maximum temperature increase of 6°C decreased to 1°C within 2 years because of the regrowth of vegetation over the stream.

The impact on stream temperatures of the removal of ground vegetation in a successful prescribed burn should be minimal because the shade provided by the canopy of the overstory vegetation is not altered.

## MANAGEMENT IMPLICATIONS

The use of fire in the management or manipulation of forest ecosystems involves a host of complex and interacting factors. The prescription for wise and intelligent use of fire is not easily derived, because fire may benefit some components of the ecosystem while, at the same time, degrading other components. These advantages and disadvantages must be weighed in each situation because no prescription will fit all environmental conditions.

There is fairly conclusive evidence that periodic burning, when properly carried out, has little influence on the physical and chemical properties of soils or on long-term site productivity—with the exception of fires on fine-textured soils over steep terrain where infiltration may be reduced and erosion increased. Loss of nutrients is minimal except for the volatile loss of nitrogen. This loss of nitrogen may be largely replaced by increased fixation on burned areas. Viro (1974) concluded that the advantages of prescribed burning of Fenno-Scandian forests mostly outweigh the disadvantages. In those forests, the important advantages result from the rapid oxidation of forest floor accumulations. Advantages include improvement in nutrition and thermal conditions of the site, assured regeneration, and the opportunity to select species of trees more suitable for the site.

Fire has been used as a management tool in the Douglas-fir region of western North America for half a century. Kayll (1974) listed the times to burn as (1) when slash and weather conditions are safe, (2) when slash accumulations are extremely heavy or slash areas become so large that fire control is impracticable, (3) when competing undesirable species invade, (4) when necessary in site preparation for seeding or planting, or (5) when disease or insect infestations threaten. In the southeastern United States and parts of Australia, fire is often prescribed to improve wildlife habitat as well as for hazard reduction in forests. Wildfire that results from excessive accumulation of fuel on the forest floor can do much more damage to the soil and the environment than controlled burns.

There appears to be a trend toward increased use of fire in managing natural parks and other resources, a recognition of the role of fire as a natural component of wilderness, and an allowance for fire to follow its natural course under carefully specified conditions (Kayll, 1974). While it must be recognized that prescribed fire will never become a management tool to the total exclusion of other techniques, it will continue to be an effective one and its use will probably diversify and increase in many forest ecosystems.

# 23
## INTENSIVE MANAGEMENT AND LONG-TERM SOIL PRODUCTIVITY

Forests have covered much of the earth's land surfaces for thousands of years without any overt evidence of yield decline. Crucial to this continued productivity is the near equilibrium eventually established in all mature forests in the absence of harvest or major disturbance. This equilibrium involves cycling, a process by which nutrients stored in forest vegetation are released for reuse after the litter falls to the forest floor and decomposes. With tree harvest or other disturbance of this cycle of events, some changes in soil and water properties and loss of nutrients from the ecosystem can be expected. Whether any of these changes or losses have detrimental effects on long-term soil productivity depends largely on the extent of the disturbance and the recuperative capacity of the soil.

The use of plantation forestry as a means of improving the relatively low fiber yields of natural forests and meeting increased timber demands has been widely accepted. The advantages of uniformity in species, size, age, and, hence, standardization of product and orderliness of management are apparent. Perhaps less obvious are the effects of uniformity of forest operations on soil properties. Clearfelling not only removes a large proportion of the total biomass, representing a loss to the ecosystem of a significant quantity of plant nutrients, but also alters dramatically the microclimate at the forest floor and may be detrimental to the physical properties of the soil. There may also be changes in soil properties as a result of converting mixed hardwood vegetation to a coniferous cover. The possibility of coniferous monocultures causing site degradation has been widely reported and the subject of some controversy (Lutz and Chandler, 1946). In fact, most early silvicultural texts advocated the use of mixed stands as a means of avoiding the dangers associated with conifer monocultures (Evans, 1976).

## SECOND ROTATION PRODUCTIVITY

By the first part of the last century large areas of spruce *(Picea abies)* were being planted in Germany and Switzerland, and as early as 1869 a reduction in growth of young spruce as compared to growth in the previous rotation was observed in parts of Germany (Weidemann, 1923). The second-rotation problem generally implicated soil deterioration or fungal attack on the replanted crop with the old stumps acting as a source of infection. However, there was little mensurational data available until Weidemann did a detailed study and concluded that the growth of spruce had declined considerably in the second and third rotations due to adverse effects on the soil from clear-cutting and replanting of pure stands. There were similar reports of declines in Switzerland and Norway.

More recently, Holmsgaard et al. (1961) questioned the evidence of universal reduction in increment by successive crops of spruce. Where declines had occurred, they were believed to result from severe droughts aggravated by the thick mor humus accumulated under the Norway spruce. The changes in soil and humus types beneath the spruce stands may have reflected more the unsuitability of the species on some sites and the poor water regimes rather than soil degradation and loss of fertility.

Miehlich (1971) compared properties of soils supporting natural oak-beech vegetation with soils planted to spruce for 120 years in south Bavaria. There was less total manganese, magnesium, calcium, potassium, and phosphorus in soils supporting spruce than under the native vegetation, but except for manganese the differences were less than 10 percent. Whether reported changes in C:N ratio, soil acidity, aeration, moisture conditions, and microbial activity were indicative of declining growth of the species that induced them was not clear.

Evans (1976) reviewed largely unconfirmed reports of reduced growth in the second rotation of pure teak *(Tectona grandis)* plantations in India and Java and confirmed reports of reduced growth of *Pinus pinaster* in the Landes of France and of *Pinus sylvestris* growing on sandy soils in The Netherlands. A decline in productivity of second and subsequent rotations was also reported for some plantations of *Larix kaempferi* in Japan. Perhaps best known of the more recent reports of second rotation declines are those from Australia and New Zealand associated with the use of exotic species. Thousands of hectares of *Pinus* spp. have been established in relatively short rotations (15 to 45 years) in these countries, as well as in South America and southern and eastern Africa, and nagging questions regarding the maintenance of productivity in successive crops of coniferous monoculture have persisted.

Isolated reports of growth declines in the second rotations of *Pinus radiata* began to surface in South Australia about 1950, according to Evans (1976), and detailed evidence of the losses was presented by Keeves (1966) and Bednall (1968). Keeves presented data comparing basal area growth from permanent sample plots maintained through the second generation that confirmed a substantial drop in productivity, regardless of the initial site quality. Bednall (1968) reviewed the problem and suggested that it was due to one or more of the following factors: (1) depletion of nutrient reserves; (2) reduction in soil water supply by the first rotation that was not adequately replenished before replanting (alteration in pore space and nonwetting characteristics of the soil may also have influenced the moisture regime); or (3) inhibition of root growth and penetration due to a buildup of toxic substances and a deterioration in soil physical properties.

A decline in productivity of the second rotation crops of *Pinus radiata* was also noted in the South Island of New Zealand (Stone and Will, 1965). The authors reported a widespread nitrogen deficiency in the second generation, even though the first rotation crops on the same sites were apparently well supplied with nitrogen. They suggested that the root complex and associated microflora of this exotic pine had a capacity to break down some fraction of soil organic matter that was inaccessible or resistant to the native vegetation originally growing on the site. However, this source of nitrogen was apparently depleted by the first pine crop.

Mensurational investigations into the magnitude, persistence, and extent of possible decline in productivity of radiata pine in its second rotation on Moutere gravel soil in New Zealand was reported by Whyte (1973). He concluded that there was evidence of declines in productivity on some soils in the second rotation. The magnitude of the decline, when it occurred, was sufficient to extend the rotation periods by about two years to attain the same height and by about eight years to attain the same total stem volume as that of the first rotation. He stated that declines in productivity were mostly transitory, representing longer initial periods of stagnation in second crops than in the first. The greatest reductions often occurred on ridges and upper slopes. There was less of a reduction as one proceeded down the slope, and on the lower slopes and valley bottoms growth was often better in the second crop. The author did not speculate on the causes of growth decline on certain sites, but he did suggest that measures should be taken to shorten the period of stagnation at the beginning of the rotation and to reduce sheet erosion of topsoil on ridges and upper slopes.

While the declines in second rotation productivity are apparently isolated instances in New Zealand, India, The Netherlands, and some other countries, the problem is undoubtedly real with spruce in central Europe and with pine in South Australia. Lack of evidence of a problem in other countries with infertile soils may lie with the difficulty of comparing yields of two crops 20 to 50 years apart (during which time changes in cultural practices may negate meaningful comparisons of available data). Woods (1976) pointed out that once second generation trees on dry sandy soils have experienced an initial period of stagnation, they appear incapable of significantly altering their growth rate after canopy closure. Figure 23.1 is a stylized presentation of differences in type of growth obtained from successive rotations under these conditions (Keeves, 1966; Bednall, 1968).

Evans suggested that the rate of litter decomposition and consequent rate of nutrient cycling may be too slow to maintain good growth of second rotation trees, particularly on relatively impoverished soils. He also raised the question of possible changes in soil microbiology, particularly mycorrhizal associations, in successive generations. Woods (1976) reported that early weed control and correct fertilizer placement can result in good "getaway" in the first 12 months after planting, which, along with subsequent fertilization and weed control, may do much to overcome the second rotation stagnation of *Pinus radiata* in Australia. It is clear that the problem of second rotation decline is complex and that the extent of the problem varies according to soil properties. An examination of the changes that take place in the soil (and soil solution) as a result of harvesting and the site disturbances associated with reforestation may help in planning management systems that will assure long-term productivity.

## SOIL COMPACTION AND EROSION

Forest harvesting can have detrimental effects on soil physical and chemical properties. These effects are sometimes expressed in terms of reduced forest productivity.

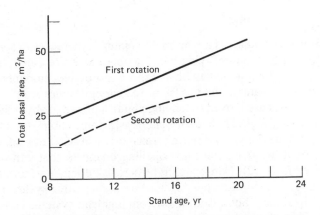

**Figure 23.1** Stylized presentation of composite data on trends in basal area production with age of first and second rotations of radiata pine on dry sands.

## Soil Compaction

Of particular concern is the structure of soils containing a relatively high percentage of fine-textured materials. When these soils are wet they can be damaged by compaction from the use of heavy logging equipment or the dragging of logs. Compaction reduces porosity and the rate of water infiltration. Natural recovery of porosity may require a decade or more (Patric and Reinhart, 1971). Rubber-tired skidders reduced soil bulk density by 20 percent and macropores by 68 percent in northern Mississippi and the rutted soil required 10 to 12 years to recover (Dickerson, 1976), as shown in Figure 23.2. Compaction and deep disturbance can cause damage to residual trees in partially cut stands, even though the watersheds are relatively unaffected. Severe soil compaction may retard subsequent growth of young trees on the affected area.

Stone (1973) pointed out that procedures for minimizing compaction and related damage vary with soil properties. They include avoidance of wet weather logging, shift of activity to nonsusceptible areas when susceptible soils are wet, concentration of main haul traffic on a few major trails, choice of logging methods or use of low bearing-pressure equipment. Loosening and revegetating or mulching the disturbed area may hasten its recovery (Hatchell, 1970). According to Hatchell, one vehicle trip can do almost as much damage as multiple trips over moist, medium-textured soils. Therefore, on these soils logging traffic should be confined to a few primary trails that later may be restored to a productive state by discing or subsoiling. On the other hand, little damage is done by one or two trips over dry, sandy soils and logging traffic can be effectively dispersed over a number of trails on these soils. This may reduce the need for restorative practices and the resulting increase in exposure of mineral soil may improve seedbed conditions.

## Soil Erosion

Soil losses from forested areas are normally minimal. Increases in soil movement and stream turbidity accompanying harvest operations are chiefly due to road construction and other activities that expose excessive amounts of soil, or concentrate surface waters, and not from tree canopy removal in itself. In most timber harvests, logs are moved to loading points along roads by tractor skidding or else by cable (highlead, skyline, or balloon). In some cable systems part or all of the log is held off the ground during transport; whereas, in skidding, the soil surface may be scarred by the logs, opening channels that can develop into gulleys (Rice et al., 1972). Consequently, tractor skidding is generally restricted to gentle or moderate slopes, while highlead and other cable systems are used on extremely steep slopes. Soil disturbance from highlead systems can be held to a minimum if yarding is conducted uphill. The resultant pattern of yarding paths radiates down and out from the landing, thus dispersing runoff and reducing erosion potential. The cable systems were developed for clear-cutting of large timber, but they may also be used for shelter-wood or seed-tree harvests on some

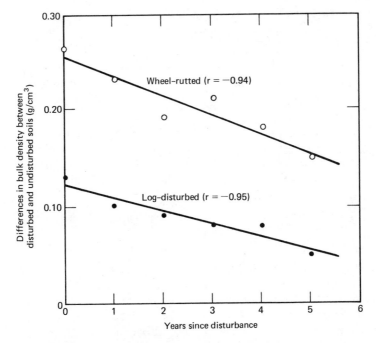

**Figure 23.2** Changes in soil bulk densities with time after tree-length skidding in northern Mississippi (Dickerson, 1976). Used with permission by the Soil Science Society of America.

terrain. From the standpoint of soil disturbance, techniques such as balloon or helicopter harvesting are preferred in unstable areas. However, until such harvest methods become more economical, cable logging will undoubtedly be used in place of the tractor on most steep slopes with helicopters reserved for sites not suited for conventional logging systems. The effects of harvest and logging systems on the percent of mineral soil exposed are shown in Table 23.1 for the Pacific Northwest.

Increased hazard of surface erosion from tree yarding and skidding comes about from litter movement and soil surface exposure, deep disturbance of the soil, and soil compaction. Some surface soil exposure brought about by mixing of litter and dry soil as a result of movement of logs and equipment is found in all intensities of harvest and is of little consequence, except on the most erosive sites. On the other hand, deep soil disturbance from log plowing, cutting of skid roads and landings, and use of heavy equipment during wet weather can reduce water penetration and produce channels that promote gully erosion on slopes. The worst effects of compaction are in skid trails and other deeply disturbed areas. They add to the danger of runoff and erosion unless subsequent treatments are applied to stabilize the surface and deflect drainage concentrations.

**Table 23.1**  Percent Mineral Soil Exposed Using Various
Timber Harvest Methods and Logging Systems
in Oregon and Washington (R. M. Rice, J. S.
Rothacher, and W. F. Megahan, 1972)[a]

| Harvest Method/Logging System | Percent Bare Soil |
| --- | --- |
| Clear-cut/tractor | 26–29 |
| highlead | 16–19 |
| skyline | 6–12 |
| balloon | 6 |
| Selection/tractor | 16 |
| cable | 21 |

[a]Used with permission "Erosional Consequences of Timber Harvesting: An Appraisal;" *Watersheds in Transition* published by the American Water Resource Association.

Rice et al. (1972) stressed that erosion rarely occurs uniformly in a forested
watershed. Most of the surface is undisturbed by logging, but because erosion is
localized, it is often deep and includes a large proportion of subsoil. Consequently, site degradation may be much less than would be the case if erosion was
more or less uniform over the whole surface, although it may appear to be much
worse. Because of the diversity of species within a natural forest ecosystem,
bared areas are quickly invaded by pioneer species, and the initial high rates of
sediment production rapidly decline.

Erodibility of steep sites in the western United States appears to be influenced by the nature of the parent rock. Soils derived from acid igneous rock tend
to be more erosive than soils derived from other parent materials. Soil chemical
properties may influence erodibility as it affects the nature and extent of the
vegetative cover and the soil organic matter content.

Accelerated erosion is a possible undesirable side effect of use of fire
following a timber harvest, according to Rice et al. (1972). The effects of fire are
most harmful on steep slopes (slopes exceeding 70 to 75 percent), where they
may induce dry ravelling, and on coarse-textured soils where a hydrophobic
layer may develop. While severe fires may reduce soil organic matter content
and change surface soil properties, the overall influence on soil moisture properties is minor since severe burns usually cover a very small portion of the total
surface area (Figure 22.2). Road building to facilitate harvesting far exceeds
logging or fire as a cause of accelerated erosion.

## HARVEST REMOVALS AND NUTRIENT BUDGETS

A major avenue of loss of nutrients from forest ecosystems is through removal in
the harvested crop. Estimates of such losses have been monitored for many
years, especially by European foresters, and they are known for many species

and sites. A forest stand annually absorbs almost as much nutrients from a hectare of soil as some agricultural crops. However, less than a third of the absorbed nutrients are immobilized in the merchantable stem wood and bark, while the remainder returns to the soil reserve in foliage, branches, fruits, and roots.

## Nutrient Removal in the Harvested Crop

The removal of accumulated nutrients during conventional logging averages no more than a kilogram, or less, per hectare per year for phosphorus and sulfur, up to about 10 times this amount for nitrogen and potassium, and 5 to 15 kg per hectare per year for calcium, depending on the species, stocking, age, and site (Rennie, 1955).

Generally the harvest of hardwoods removes more nutrients from a site than the harvest of an equal volume of conifers (Table 23.2). Within similar species, nutrient removal is essentially proportional to the volume harvested. Thus the more productive sites normally suffer the greatest nutrient losses during harvest. However, nutrient losses by conventional harvests from even the most productive sites are relatively small on an annual basis. Such losses can probably be replaced by soil weathering and natural imputs, except for some particularly infertile soils. Rennie (1955) studied the effects of nutrient removal in harvested timber on the capacity of a number of British forest soils to replenish harvest losses. He reported that only about 29 percent of the calcium, potassium, and phosphorus in the harvested tree was to be found in the bolewood of 50-year-old pines, and an additional 17 percent was in the bolebark. The remaining 54 percent was in the slash and needles that generally remained on the site at harvest. He concluded that most soils could replace the nutrients in the harvested timber without a long-term decrease in productivity. Of the soils he studied, only the very shallow and highly silicious moor soils were so low in mineral reserves that fertilizers were presumed to be needed to assure a permanently productive timber forest.

While there are many sandy soils in parts of Australia, France, Great

**Table 23.2** Nutrient Removal within Stem Wood and Bark of Thinnings Plus Main Crop by Clear-cutting After 50 Years (Rennie, 1955)[a]

| Stand Type | Ca | K | P |
|------------|-----|-----|-----|
| | | kg/ha | |
| Pines | 85 | 32 | 6 |
| Other conifers | 107 | 59 | 9 |
| Hardwoods | 495 | 116 | 22 |

[a]Used with permission.

Britain, New Zealand, and the southeastern United States that do not contain sufficient soil reserves of critical elements to sustain good tree growth under intensive management (Hagner, 1971), these soils are apparently the exception rather than the rule. On the other hand, the trend toward more complete utilization of tree components and the adoption of shorter rotations could significantly reduce the capacity of many marginally deficient soils to replace nutrients removed in harvests. Logging systems where whole trees are moved to a central landing for delimbing deprive the soil of nutrients contained in the slash just as efectively as do whole-tree chippers (Figure 22.12). Problems relating to more complete utilization result from recent advances in processing methods which make it possible to use slash and the whole of small or defective trees that were formally left on the forest floor.

A number of studies have been conducted to determine the effects of different systems of harvest involving varying degrees of tree utilization on nutrient removal. Some of these studies are summarized in Table 23.3. For example, Weetman and Webber (1972) calculated that 79 kg nitrogen per hectare would be removed from the site by the conventional shortwood method of harvest (to a 7.6-cm top) of an uneven age stand of red spruce-balsam fir in Quebec while a full-tree method removed about 387 kg nitrogen per hectare. Approximately four times as much of each element was removed in the full-tree as in the shortwood method of harvest of this stand. Will (1968) reported that the removal of whole trees in a 26-year-old radiata pine plantation about doubled the soil nutrient depletion rate over that experienced by the harvest of stemwood and bark only. Boyle and Ek (1972) came to essentially the same conclusions regarding the harvest of Wisconsin-mixed hardwoods. Research with southern pines supports the conclusion that the harvest of the total above-ground tree removes about twice the amount of nutrients from the site as are removed in a conventional harvest of stemwood and stembark only. Whether the harvest removal of nutrients will eventually deplete the soil of nutrients depends on soil reserves and recuperative powers and natural inputs. A number of attempts have been made to develop nutrient budgets as a means of analyzing the problem.

## Nutrient Balance Sheets

Studies that equate nutrient removal in the harvested wood with nutrient reserves in the surface soil (Rennie, 1955) often fail to consider important natural inputs and losses of nutrients from the ecosystem that can influence long-term site production. Weetman and Weber (1972) attempted such a nutrient balance sheet and concluded that full-tree logging in Quebec would not result in any significant reduction in growth of the second generation due to nutrient removal, except on some swampy sites without lateral soil water movement and on dry sandy sites with low organic matter contents. They calculated that, with these exceptions, the soils of the Canadian Shield were mineralogically rich enough and had sufficient cation exchange capacity to support the nutrient losses associ-

**Table 23.3** Nutrient Removal (above ground only) by Different Harvesting Systems

| Species, Age, and System of Harvest | Total Removal (kg/ha) | | | | Average Annual Removal (kg/ha/yr) | | | | Reference |
|---|---|---|---|---|---|---|---|---|---|
| | N | P | K | Ca | N | P | K | Ca | |
| **Slash pine (15 yr) Florida** | | | | | | | | | Pritchett and Smith, 1974 |
| (a) Whole tree | 345 | 24 | 137 | 221 | 23.0 | 1.6 | 9.1 | 14.7 | |
| (b) Stemwood with bark | 182 | 11 | 74 | 128 | 12.1 | 0.7 | 4.9 | 8.5 | |
| **Loblolly pine (16 yr) N. Carolina** | | | | | | | | | Wells and Jorgensen, 1975 |
| (a) Whole tree | 257 | 31 | 165 | 187 | 16.0 | 1.9 | 10.3 | 11.7 | |
| (b) Stemwood with bark | 115 | 15 | 89 | 112 | 6.5 | 0.9 | 5.0 | 7.0 | |
| **Radiata pine (26 yr) N. Zealand** | | | | | | | | | Will, 1968 |
| (a) Whole tree | 224 | 28 | 224 | 129 | 8.6 | 1.1 | 8.6 | 5.4 | |
| (b) Stemwood with bark | 128 | 18 | 157 | 105 | 4.9 | 0.7 | 6.0 | 4.0 | |
| **Black spruce (65 yr) Quebec** | | | | | | | | | Weetman and Webber, 1972 |
| (a) Whole tree | 167 | 42 | 84 | 276 | 2.7 | 0.65 | 1.3 | 4.2 | |
| (b) Stemwood with bark | 43 | 12 | 25 | 98 | 0.7 | 0.18 | 0.4 | 1.5 | |
| **Red spruce-fir (Uneven age) Quebec** | | | | | | | | | Weetman and Webber, 1972 |
| (a) Whole tree | 387 | 52 | 159 | 413 | | | | | |
| (b) Stemwood with bark | 79 | 11 | 47 | 150 | | | | | |

ated with any type of tree harvest, when one takes into consideration additions of nutrients in dust and precipitation.

Boyle and Ek (1972) were of the opinion that nutrient element reserves of most forest soils of the Great Lake region, when considered in conjunction with natural nutrient additions, were adequate to maintain established growth rates for several generations—even where boles, tops, and branches are harvested.

Data presented in nutrient balance sheets, as exemplified in Table 23.4, generally indicate that nutrient reserves of soils are adequate for several rotations of trees harvested by either conventional or full-tree systems. However, extrapolation of results from these types of calculations to other forest ecosystems is hazardous due to the general lack of information on the rate and extent of availability to trees of nutrients contained in the soil and parent rocks, atmospheric inputs, and losses from the soil following clear-cut harvesting. Nevertheless, estimations of both long and short-term nutrient availabilities can be particularly important to tree nutrition studies in intensively managed forests on sandy and infertile sites.

Long-term available nutrients can be defined as those soil reserves represented by the total quantity of essential elements in the soil, usually including the humus layers plus the mineral soil to the effective rooting depth; while the short-term available nutrients are those in the surface soil extractable by some standard solution and readily available for plant use. Total nutrients contained in the humus plus exploitable mineral soil were calculated by Weetman and Webber (1972) for two Quebec sites, but only the short-term available mineral nutrients plus total nitrogen were given for the Wisconsin sites (Table 23.4). Both methods

**Table 23.4** Losses of Nutrients by Conventional Harvest of 45- to 50-year-old Stands of Hardwood, Compared with Soil Contents and Annual Inputs (Boyle and Ek, 1972)[d]

| Nutrient Element | Removed in Logging | Soil Nutrient Content[a] | Site Reserves Rotation Equiv.[b] | Annual Inputs[c] |
|---|---|---|---|---|
| | | kg/ha | | |
| N | 120.3 | 1803 | 15.0 | 13.1 |
| P | 12.1 | 179 | 14.8 | 0.3–0.4 |
| K | 60.2 | 185 | 3.1 | 1.0–4.0 |
| Mg | 24.1 | 207 | 8.6 | 0.5–1.0 |
| Ca | 240.6 | 765 | 3.1 | 2.0–7.0 |

[a]Expressed as soil total N; while P and K were extracted with Bray solution, and Mg and Ca with neutral $N$ NH$_4$OAc.
[b]Number of 45-year rotations possible with present soil contents but without inputs.
[c]Annual inputs from dust and precipitation estimated from literature.
[d]Used with permission by the National Research Council of Canada from the Journal of Forest Research, Vol. 2, pp. 407-412.

of expressing reserves have serious limitations due to the difficulties in calculating nutrient availability and projecting these values to future stand conditions. However, to date there are no other satisfactory methods available for estimating nutrient reserves and one or the other (or a combination) of the above methods must be used in developing nutrient balance sheets. A problem with short-term calculations is that the measure of "available" soil nutrients is largely dependent on the strength of the extracting solution. The amounts of nutrients extracted may not be related to the amounts of nutrients actually available to the trees during a given period of time. While quantative methods of determining total soil reserves are generally accurate, they too are difficult to interpret. Since forest trees apparently have only a limited capacity to extract nutrients from unweathered minerals, the rate of replenishing the supply of available nutrients from this reserve is difficult to estimate. The replenishment rate depends largely on the types of mineral present and the various soil weathering processes. Therefore, the values often reported for total soil reserves are probably too large, because only a fraction of total amounts becomes available during a specified rotation period. In contrast, the estimates for available soil nutrients are no doubt quite conservative for long-term supply estimates. An integration of the two values may prove to be the best method of reporting soil reserves until some accurate system of determining rate of nutrient release is developed.

The amounts of nutrient inputs from precipitation and dust and from biological fixation of atmospheric nitrogen, as estimated in Table 23.4, are highly variable among sites and largely unknown for most areas. It appears evident that atmospheric inputs into the forest ecosystem are considerably greater than losses by leaching from forested areas and they may largely replace nutrients removed in the harvested trees during a rotation period. Annual fixation of atmospheric nitrogen by symbiotic and nonsymbiotic organisms is probably low in most forest habitats, but it may amount to several hundred kilograms per hectare during a rotation (Richards and Voigt, 1965). Considering all factors, it would appear that most temperate forest soils have the capacity to recover from natural disturbances and timber harvests by nutrient replacements through mineral weathering and natural inputs. In support of these conclusions, one must consider that some Indians of North America in their agricultural and hunting pursuits burned large areas of forest, perhaps repeatedly, and that most of these areas now support excellent timber. In fact, a sizable portion of nonmountainous forest lands of western Europe and eastern North America was once used for agriculture. The forests that have revegetated abandoned agricultural lands, plus the great areas of even-aged forests growing on old burns, demonstrate the capacity of soils to supply nutrients to new forests after the elimination of the original cover. In fact, one can only be awed by the restorative power of most soils and the essentially conservative nature of forest ecosystems. This primarily results from their large biomass and development of protective humus layers, their intensive and often deep root systems, and their effective retention and cycling of absorbed nutrients (Duvigneaud and Denaeyer-DeSmet, 1970).

## HARVEST EFFECTS ON WATER QUALITY

Perhaps of greater concern than the removal of nutrients in the harvested wood are the changes in microclimatic conditions, activity of soil microbes, and mobilization of soil nutrients following canopy removal. These changes, in conjunction with increased water yields and runoff, can lead to stream pollution and a reduction in soil nutrient reserves on disturbed sites. The effects of forest harvesting on water quality are of concern, therefore, for long-term soil productivity as well as for environmental values.

All natural streams contain varying amounts of dissolved and suspended matter, although streams issuing from undisturbed watersheds are ordinarily of high quality. Waters from forested areas are not only low in foreign substances, but they are relatively high in oxygen content and low in temperature. Nonetheless, some deterioration of stream quality can be noted during and immediately after clear-cutting, even under the best of logging conditions (Aubertin and Patric, 1974). The potential for water quality degradation following timber harvest may involve turbidity, as well as increases in temperatures and nutrient content. Sediment arising from logging roads is the major water-quality problem related to forest activities in many areas.

Roads cannot be constructed without some soil disturbance. Roads on gentle to moderate slopes in stable topography pose few problems. Much of the disturbance is localized and objections arise primarily from the visual aspect before revegetation occurs. Logging roads do not constitute a great hazard in the coastal plain and Piedmont of the southeastern United States. Difficulties and hazards mount when roads are pushed onto steep terrain, cut into erosive soils or unstable slopes, or encroach on stream channels (Stone, 1973). Improper location, design, or construction can result in movement off-site of soil and debris and cause changes in turbidity, bedload, sediment deposits, or spawning success of certain fish.

### Water Turbidity

*Water turbidity* has long been identified with water quality, and has probably been measured more frequently than other properties (Douglass and Swank, 1972). Hoover (1952) reported 7000 ppm of particulate matter in stream water, following large storms in watersheds where logging methods and roads were not subject to control, as compared to 80 ppm in an undisturbed watershed. Furthermore, turbidities resulting from as much as 56,000 ppm of suspended materials have been reported in streams arising from severely disturbed watersheds, while maximum turbidity of water from a nearby uncut forest was only 15 ppm. Douglass and Swank (1972) pointed out that these high turbidities demonstrate the potential damage that can occur without controls to protect the water resource. They emphasized, however, that when proper logging methods and road location and construction procedures are followed, only small and temporary increases in water turbidity occur. Data on increases in sediment production from nonpoint sources due to harvesting and site preparation are not well documented nor easily obtained. Most of the existing data pertain to soil movement, or relocation, and not to sediment production. Undisturbed forests seldom experience significant overland flow and, consequently, little sediment production. Ursic (1975) reported sediment yields from small coastal plain catchments of pole-sized loblolly pine planted for erosion control and from mature southern pine seldom exceeded 224 kg per hectare annually when rainfall ranged from 106 to 168 cm and stormflows were up to 30 area-cm. Annual sediment production from small, pine-covered headwater catchments in north Mississippi averaged less than half this amount, most of which came from channels created by previous farming operations.

While tree felling in itself does not produce an increase in suspended materials, harvesting usually increases forest litter and significantly alters the microclimate of the forest floor, which hastens its deterioration. These latter changes can reduce water infiltration, promote release of nutrients into the ground water, and change stream temperatures. Soil movement was noted on small catchments of upland hardwoods deadened with a herbicide and underplanted with pine. Sediment production was increased by about 450 kg per hectare during the first year (Ursic, 1970). However, sediment yield had been cut by 75 percent by the fourth year. Skidding can create serious disturbances and result in considerable soil movement on steep slopes. It does not necessarily increase sediment load of streamwater in gentle terrain. Ursic (1975) reviewed studies in the Ouachita Mountains in central Arkansas where yields after harvesting were 11 times greater than the seven-year pretreatment average of 15 kg per hectare per year. Sediment returned to prelogging levels in three years.

## Stream Temperature

*Stream temperature* is normally affected only by removal of stream-side shade. Complete removal of the forest canopy from small streams by clear-cutting can increase water temperature from a few degrees to as much as 10 to 15°C and result in greater fluctuation between maximum and minimum temperatures (Figure 23.3). The effect on individual streams varies according to the amount of canopy removed, length of exposure time to full sunlight, stream width and volume of water, and initial water temperature. Temperature increases are minimal on cloudy days and in large streams. In some mountainous and cold-climate areas, slight warming of streams may enhance their productivity. On the contrary, warming of small streams following clear-cutting at lower elevations may exceed the tolerance of trout, salmon, and some other aquatic life.

Swift and Messer (1971) measured effects of various types of management on maximum stream temperatures at Coweeta watershed in North Carolina. When hardwood trees in a cove site were killed, summer maximum temperatures

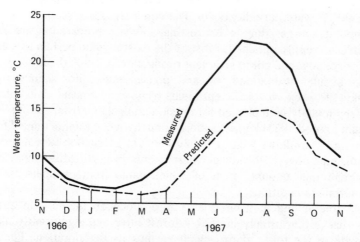

**Figure 23.3**  Measured average maximum stream water temperatures on a clear-cut watershed in Oregon and the predicted temperatures had the watershed not been logged and burned (Levno and Rothacher, 1969).

increased by 2 to 3°C, but winter temperatures were little affected. Subsequent clear-cutting of the dead timber raised summer maximums 3 to 4°C and winter maximums about 2°C above the temperatures of streams flowing from undisturbed forests. An understory cut had little effect on either summer or winter maximum temperatures. Studies have generally shown that forestry practices that open up the stream channel to direct insolation are the only practices that increase stream temperatures. These increases can generally be avoided by leaving a strip of shade from 20 to 40 m wide along each side of the stream. This buffer strip may also reduce many kinds of streambed and streamside disturbances and trap soil and debris washed downslope toward the stream following harvesting. Such a strip provides esthetic and wildlife values and contributes to the timber supply if judiciously logged by the selection method. As a matter of fact, some thinning within the strip may be desired to reduce windthrow hazard to tall trees exposed by cutting of surrounding trees.

## Water Chemistry

The effects of various methods of harvest on *water chemistry* have been measured on small-gauged watersheds in a few coniferous and deciduous forests. The release of plant nutrients from the forest floor is generally related to the degree of canopy removal. The increase in nutrient availability results from the decay of residues of the harvested trees, as well as an increase in the decomposition rate of the soil organic matter. The increased radiation warms the forest floor above its former temperature regime and a reduction in transpiration following canopy removal often increases soil moisture content. These changes in the microclimate of the forest floor have the potential for increased microbial activity and nutrient leaching.

Partial harvests, as in the selection or shelterwood methods, seldom result in a measurable decrease in water quality. Under these methods, the residual trees are stimulated to greater than normal growth rates by the increase in light, water, and nutrient availabilities. This increase in tree growth, combined with the regrowth of ground vegetation, provides an effective sink for the released nutrients, and the expanding crowns and roots of the residual trees eventually fill the growing space created by thinning. On the other hand, clear-cut harvests may have a significant influence on stream water chemistry under some site conditions.

One of the best-documented studies of the effects of vegetation removal on water quality was that of the Hubbard Brook watershed in the White Mountains of New Hampshire (Likens et al., 1970). All vegetation in this northern hardwood forest was cut during the winter and left in place on a well-calibrated watershed. Vegetation regrowth was inhibited by periodic applications of herbicides during the first two years. Annual streamflow was increased 33 cm or 39 percent the first year and 27 cm or 28 percent the second year, above the values expected if the watershed had not been deforested. Over 90 percent of this increase occurred during the four-month growing season.

Large increases in streamwater concentration were observed for all major ions, except $NH_4$, $SO_4$, and $HCO_3$, approximately five months after deforestation of the Hubbard Brook watershed. Concentrations in bases increased several hundred percent during the two years subsequent to deforestation. For example, the excess three-year losses of calcium, potassium, and magnesium were 206, 86, and 46 kg per hectare, respectively. However, the most surprising development was the loss of about 340 kg per hectare of nitrate-nitrogen from the treated watershed during the first three years. Nitrate concentrations in stream water exceeded the health levels of 10 ppm $NO_3$-N recommended for drinking water during much of the three-year period. While the high nitrate concentrations in the clearfelled watershed streams were diluted by streamflow from undisturbed areas so that the concentration from the total forest averaged only about 0.5 ppm, the losses from this experimental clearfelled area were unacceptably high. The high values resulted, in part, from allowing all vegetation to decompose in place and to repeated treatment of the watershed with a broad-spectrum herbicide.

Later, the authors (Likens and Bormann, 1972) reported the concentrations of nitrates and associated bases in streams from nearby areas that had been operationally clear-cut were less than one-third that of the experimental watershed. Nevertheless, these latter values were still considerably higher than those reported for watersheds in other regions. For example, Aubertin and Patric

(1974) reported no significant change in nutrient concentration following the clear-cutting of a hardwood watershed in West Virginia. Gessel and Cole (1965) found that the concentration of nitrogen, potassium, and calcium in water moving through the forest floor of a 35-year-old Douglas-fir plantation was fairly large (Table 23.5) and the amount almost doubled following the disturbance associated with clear-cutting. However, only very small amounts leached beyond the root zone (90 cm) in the sandy soil. Furthermore, during the first year after cutting, only about 0.5, 0.1, and 4.7 kg per hectare more nitrogen, potassium, and calcium were leached below the 90 cm depth in the clear-cut area than in the undisturbed area. Fredriksen (1970) confirmed the relatively small losses after clear-cutting in the Douglas-fir region. Douglass and Swank (1972) reported results from experimental watersheds at Coweeta, North Carolina, that did not show accelerated loss of ions to streams. No significant increases in concentrations of calcium, magnesium, potassium or sodium were found in streamwater following a partially clear-cut aspen watershed in Minnesota, and only a slight increase in nitrates, from 0.12 ppm up to 0.16 ppm, was detected (Verry, 1972). Tamm et al. (1974) reported nutrient losses from clear-cut watershed in Sweden that were not significantly different from those obtained for most areas in the United States. Increases in ground water concentrations of nitrate-nitrogen following clear-cutting averaged about 1 ppm and total nitrogen losses during one to six years after cutting was estimated at 3 to 5 kg per hectare per year.

The causes of the unusually large losses of nitrate nitrogen and associated cations from operational clear-cuts of watersheds near Hubbard Brook in the White Mountains of New Hampshire are not entirely clear. Fortunately, the areas have been rather well monitored (Likens et al., 1970) and the data reveal some properties of these watersheds that are somewhat different from other gauged watersheds in the United States. The 6- to 15-cm thick humus layers that develop under northern hardwoods of cool regions in the northeast United States and Canada are thicker than forest floors of temperate region mixed and coniferous forests. These surface organic layers often contain more than 1000 kg nitrogen per hectare (Likens et al., 1970) and the C:N ratio is generally narrower than that of coniferous forest floors. These properties contribute to a greater potential for nitrification under favorable conditions.

Removal of the forest canopy by clear-cutting decreases transpiration and generally improves soil water conditions during the growing season. Maximum soil temperatures during the warm season are appreciably lower under a forest canopy than in the open. This is particularly true in cooler regions. At Hubbard Brook there was an average summertime increase in the surface soil temperature of about 8°C after canopy removal. These conditions promoted rapid decomposition of the forest floor material and mineralization of organic nitrogen. Nitrates thus produced were leached from a watershed in the absence of a well-established understory vegetation to absorb it. Increased water yields as a result of

**Table 23.5** Movement of Elements Through Soil Under Forested and Clear-cut Conditions During First Year After Cutting of Douglas Fir (Gessel and Cole, 1965)[a]

| Plot Treatment | Soil Depth | Nitrogen | | Potassium | | Calcium | |
|---|---|---|---|---|---|---|---|
| | | Amt. in Leachates | Leached from Soil | Amt. in Leachates | Leached from Soil | Amt. in Leachates | Leached from Soil |
| | cm | kg/ha/yr | % | kg/ha/yr | % | kg/ha/yr | % |
| Forested | 2.5 | 4.80 | 2.7 | 10.0 | 27.0 | 16.6 | 11.8 |
| | 90 | 0.63 | 0.02 | 1.0 | 0.4 | 4.5 | 0.5 |
| Clear-cut | 2.5 | 11.09 | 4.7 | 17.6 | 42.0 | 22.3 | 14.6 |
| | 90 | 1.11 | 0.04 | 1.1 | 0.5 | 9.2 | 1.0 |

[a]Used with permission from *Journal* American Water Works Assoc.

reduced evapotranspiration in the clear-cut area enhanced the opportunity for leaching. It generally takes two or more years following clear-cutting for ground cover to become established in northern forests. Furthermore, in the Hubbard Brook watershed, shallow bedrock prevented deep seepage so that excess water rapidly appeared in downslope streams.

While the results from the Hubbard Brook studies probably apply only to clear-cut watersheds of northern hardwoods, they nonetheless serve to point up the potential for nutrient loss in clear-cut areas with similar forest floors. In such areas, harvest methods should be used that retain some shading of the soil surface. This may be accomplished by redistribution of harvest slash and a minimal disturbance of the forest floor, by rapid regrowth of ground cover vegetation, and by partial cutting methods or clear-cutting in narrow strips.

## FURTHER CONSIDERATIONS FOR INTENSIVE FOREST MANAGEMENT

While there is evidence of soil damage and environmental degradation associated with intensive management operations under some combinations, it is probable that plantation forestry will continue to gain acceptance in many parts of the world. Only through an increase in the intensity of management can the ever-rising demand for wood products be met. Furthermore, where intensive forest management is possible it has great appeal from a biological as well as from an economical point of view. For example, relatively intolerant subclimax species that grow best in full sunlight, such as most pines, are best managed as even-age monocultures. They generally have the seed supply, fast growth rate, and other characteristics that make them well suited for plantation management. It is widely conceded, for example, that early attempts at the use of selection systems of harvesting southern pines proved to be a kind of high grading that caused stands to degenerate rather than improve (Smith, 1972). Present populations of these species are being replaced by progeny of genetically superior trees—a practice that can best be accomplished by clear-cut harvesting and planting in prepared sites.

Some degree of site preparation is necessary for plantations on most soils in order to reduce weed competition and assure satisfactory establishment and early growth of the planted trees. Such preparation is hardly possible except in clear-cut areas. It has been reported that site disturbances associated with partial cuts may damage residual trees and increase the risk of *Fomes annosus* root rot attack on pines, unless the cut is properly timed (Rehfuess and Schmidt, 1971). Controlled burning for hardwood suppression, fire hazard reduction, and improvement in wildlife habitat is a management practice in many pine forests. However, fire cannot be used in young stands and it is often lethal to young trees in uneven-age stands.

   In spite of its attractiveness from the standpoint of efficiency of operations to meet a growing demand for its products, plantation forestry is not suited to all climatic regions, terrains, soils, or species. In forests near populated areas or with high recreational use, esthetics can be quite important. In these areas, the visual impact of harvesting should be controlled to the extent possible. If clear-cutting is used, the size of the harvested area should be kept to a minimum and a cutting pattern adopted that blends with the landscape, takes advantage of natural features, and avoids the harshness of straight lines and color contrasts. More complete tree utilization will help eliminate unsightly logging residues (although small branches and foliage are best left on the site to protect the soil). Avoiding unstable sites and proper planning of access roads can reduce earth slippage on steep slopes and timing of site preparation and reforestation operations can shorten the period of the unsightly condition of the area.

   More important than aesthetic limitations are some biological constraints on intensive management that should be considered. In most areas clear-cut harvest of a mature stand of trees has a dramatic impact on the soil. The sudden removal of the forest canopy opens the floor to the full effects of the sun's radiation. Mean surface soil temperatures in harvested areas rise several degrees above those in nearby forested areas during the summer. While surface soil moisture content may decrease due to the higher temperatures, seasonally perched water tables are often higher in harvested areas, as a result of the near elimination of transpirational water losses. The subsequent increase in water yield may help the watershed manager, but it can also create a potential for nutrient leaching on well-drained sites and for excessively high water tables on poorly drained sites. Swamping can occur on moist, level to gently sloping sites where lateral drainage is restricted and impervious layers prevent downward movement of water (Trousdell and Hoover, 1955). In cool, moist climates as found in Scandinavia, an increase in surface soil moisture following clear-cutting of large areas may accelerate growth of sphagnum mosses that inhibit seedling establishment and growth. These latter sites are best managed by partial harvest techniques in order to keep increases in surface soil moisture to a minimum.

   Increases in biological activity associated with clear-cut harvesting and site cultivation are generally beneficial in cold regions. With a reactivation of soil biological activity, undecayed litter continues to accumulate with accompanying podzolization and soil deterioration through stronger leaching and loss of fertility (Goor, 1974). Even in temperate regions, increased biological activity as a result of regeneration activities may have little or no impact on long-term productivity on level and reasonably fertile soils. However, such activities may decrease soil productivity in mountainous areas, in subtropical and tropical regions where the soils are old and highly weathered, and in other areas where soils contain very low nutrient reserves. Furthermore, such land-clearing operations as blading, rootraking, and windrowing—which removes obstacles to harrowing, bedding,

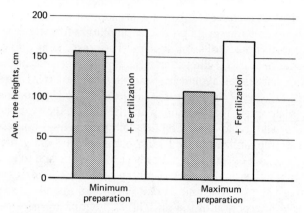

**Figure 23.4** Slash pine heights three years after planting on "minimum" (logged and chopped only) and "maximum" (logged, bladed, and bedded) site preparation treatments on a coastal plain Ultic Haplaquod. Fertilized plots received 56 kg phosphorus at planting and 118 kg nitrogen per hectare one year later (Pritchett and Wells, 1978).

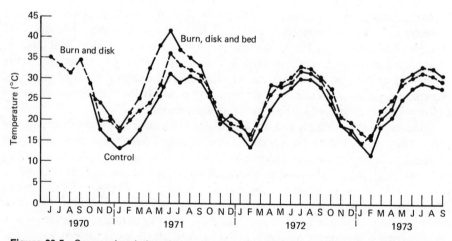

**Figure 23.5** Seasonal variations in maximum soil temperatures 2.5 cm below surface of a coastal plain Aeric Haplaquod. Longleaf pine was harvested in 1968 and the area subsequently left without further treatment (control) or site prepared as indicated and planted to slash pine in 1970 (Schultz, 1976).

or planting—also remove the protective litter and much of the topsoil, the center of biological activity and nutrient reserves in most forest soils. In an attempt to devise methods of reducing mobilization and losses of nutrients, a test was established in the coastal plain flatwoods with varying intensities of site preparation (Pritchett and Wells, 1978). Slash pine planted in clear-cut and chopped-only plots were 30 percent taller after three years than trees planted on beds following the removal of all vegetation, litter, and debris by blading into windrows (Figure 23.4). Forest floor litter and vegetation reduced fluctuations in surface soil temperature, as indicated by results from a similar test by Schultz (1976) in Figure 23.5. Nutrient concentrations in soil solution samples indicated that surface debris in the chopped-only plots decomposed more slowly and released nutrients over a longer period than in bedded plots and, thus, resulted in lower leaching losses and more effective use of mineralized nutrients.

One should conclude from this discussion, therefore, that while plantation forestry may result in a long-term reduction in site productivity, such instances are rather rare and the cause of the reduction varies among sites depending on soil, site, and climatic factors. It is noted that the removal of trees per se does not cause soil erosion, except on the most sensitive sites, and even here methods of harvest can be used that will result in a minimum of soil disturbance. Nutrient removals in harvest are modest, on an annual basis, and can generally be replaced by natural inputs into the ecosystem. Soil nutrient losses following clear-cut harvesting are generally quite small, with the apparent exception of some northern hardwood areas where partial or strip cuts should be used so as to provide continuous shade for the forest floor and to reduce surface runoff. However, site preparation operations may cause significant losses of fertility and they should be carefully designed to fit site conditions. As a general rule, techniques that result in minimum disturbance of the forest floor, consistent with good tree regeneration and growth, are recommended.

With wise planning and responsible management, the forest soil and site can be judiciously matched with the appropriate silvicultural system, harvest method, site preparation operation, species selection, and stand-tending procedures. When this is properly done, damage to site productivity can be minimized or avoided. Today's silviculturist has an obligation to preserve the fertility of our forest soils so that they will continue to produce forests for the future.

# APPENDIX A

## COMMON AND SCIENTIFIC NAMES OF NORTH AMERICAN TREES

| Common Name | Scientific Name |
|---|---|
| Alder, red | *Alnus rubra* |
| Ash, black | *Fraxinus nigra* |
|   green | *Fraxinus pennsylvanica* |
|   white | *Fraxinus americana* |
| Aspen, bigtooth | *Populus grandidentata* |
|   quacking | *Populus tremuloides* |
| Baldcypress | *Taxodium distichum* |
| Basswood, American | *Tilia americana* |
| Beech, American | *Fagus grandifolia* |
| Birch, gray | *Betula populifolia* |
|   paper | *Betula papyrifera* |
|   river | *Betula nigra* |
|   yellow | *Betula alleghaniensis* |
| Blackgum (tupelo) | *Nyssa sylvatica* |
| Boxelder | *Acer negundo* |
| Butternut | *Juglans cinerea* |
| Cherry, black | *Prunus serotina* |
|   pin | *Prunus pennsylvanica* |
| Cottonwood, black | *Populus trichocarpa* |
|   eastern | *Populus deltoides* |
|   swamp | *Populus heterophylla* |
| Cypress, Arizona | *Cupressus arizonica* |
|   Monterey | *Cupressus macrocarpa* |

| Common Name | Scientific Name |
|---|---|
| Dogwood, flowering | *Cornus florida* |
| western | *Cornus occidentalis* |
| Douglas-fir | *Pseudotsuga menziesii* |
| coastal | *Pseudotsuga menziesii (var. menziesii)* |
| Elm, American | *Ulmus americana* |
| slippery | *Ulmus rubra* |
| winged | *Ulmus alata* |
| Eucalyptus | *Eucalyptus spp.* |
| Fig, Florida strangler | *Ficus aurea* |
| Fir, balsam | *Abies balsamea* |
| grand | *Abies grandis* |
| noble | *Abies procera* |
| Pacific silver | *Abies amabilis* |
| subalpine | *Abies lasiocarpa* |
| white | *Abies concolor* |
| Hackberry | *Celtis occidentalis* |
| Hawthorne | *Crataegus spp.* |
| Hemlock, eastern | *Tsuga canadensis* |
| mountain | *Tsuga mertensiana* |
| western | *Tsuga heterophylla* |
| Hickory, bitternut | *Carya cordiformis* |
| pignut | *Carya glabra* |
| shagbark | *Carya ovata* |
| water | *Carya aquatica* |
| Holly, American | *Ilex opaca* |
| Hornbeam, American | *Carpinus caroliniana* |
| Juniper, California | *Juniperus californica* |
| western | *Juniperus occidentalis* |
| Larch, subalpine | *Larix lyallii* |
| tamarack | *Larix laricina* |
| western | *Larix occidentalis* |
| Locust, black | *Robinia pseudoacacia* |
| Madrone, Pacific | *Arbutus menziesii* |
| Magnolia, southern | *Magnolia grandiflora* |
| Mahogany, West Indies | *Swietenia mahogoni* |
| Maple, red | *Acer rubrum* |
| silver | *Acer saccharinum* |
| sugar | *Acer saccharum* |
| Mesquite | *Prosopis juliflora* |
| Oak, black | *Quercus velutina* |
| blackjack | *Quercus marilandica* |
| bur | *Quercus macrocarpa* |
| Chapman | *Quercus chapmanii* |

| Common Name | Scientific Name |
|---|---|
| California live | *Quercus agrifolia* |
| cherrybark | *Quercus falcata (var. pagodaefolia)* |
| chestnut | *Quercus prinus* |
| chinkapin | *Quercus muehlenbergii* |
| laurel | *Quercus laurifolia* |
| live | *Quercus virginiana* |
| northern red | *Quercus rubra* |
| nuttall | *Quercus nuttallii* |
| overcup | *Quercus lyrata* |
| pin | *Quercus palustris* |
| post | *Quercus stellata* |
| scarlet | *Quercus coccinea* |
| southern red | *Quercus falcata (var. falcata)* |
| turkey | *Quercus laevis* |
| water | *Quercus nigra* |
| white | *Quercus alba* |
| Palmetto, cabbage | *Sabal palmetto* |
| Pecan | *Carya illinoensis* |
| Persimmon | *Diospyros virginiana* |
| Pine, bristlecone | *Pinus aristata* |
| eastern white | *Pinus strobus* |
| jack | *Pinus banksiana* |
| limber | *Pinus flexilis* |
| loblolly | *Pinus taeda* |
| lodgepole | *Pinus contorta* |
| longleaf | *Pinus palustris* |
| pinyon | *Pinus edulis* |
| pitch | *Pinus rigida* |
| pond | *Pinus serotina* |
| ponderosa | *Pinus ponderosa* |
| red | *Pinus resinosa* |
| sand | *Pinus clausa* |
| Scots | *Pinus sylvestris* |
| shortleaf | *Pinus echinata* |
| slash | *Pinus elliottii (var. elliottii)* |
| south Florida slash | *Pinus elliotti (var. densa)* |
| spruce | *Pinus glabra* |
| sugar | *Pinus lambertiana* |
| Virginia | *Pinus virginiana* |
| western white | *Pinus monticola* |
| Poplar, balsam | *Populus balsamifera* |
| Redcedar, eastern | *Juniperus virginiana* |
| southern | *Juniperus silicicola* |
| western | *Thuja plicata* |

| Common Name | Scientific Name |
|---|---|
| Redwood | *Sequoia sempervirens* |
| Sequoia, giant | *Sequoiadendron giganteum* |
| Spruce, black | *Picea mariana* |
|   Engelmann | *Picea engelmannii* |
|   Norway | *Picea abies* |
|   red | *Picea rubens* |
|   Sitka | *Picea sitchensis* |
|   Siberian | *Picea sibirica* |
|   white | *Picea glauca* |
| Sweetgum | *Liquidambar styraciflua* |
| Sycamore, American | *Platanus occidentalis* |
| Tanoak | *Lithocarpus densiflorus* |
| Tupelo, swamp | *Nyssa sylvatica (Var. biflora)* |
|   water | *Nyssa aquatica* |
| Walnut, black | *Juglans nigra* |
| White-cedar, northern | *Thuja occidentalis* |
| Willow, black | *Salix nigra* |
| Yellow-poplar | *Liriodendron tulipifera* |

# APPENDIX B
## CONVERSION FACTORS
### (For English and Metric Units)

| A | B | To Convert A to B<br>Multiply by: | To Convert B to A<br>Multiply by: |
|---|---|---|---|
| **Length** | | | |
| Kilometer, km | Mile, mi | 0.621 | 1.609 |
| Meter, m | Chain, ch (66 ft) | 0.05 | 20.117 |
| Meter, m | Feet, ft | 3.281 | 0.305 |
| Meter, m | Yard, yd | 1.094 | 0.914 |
| Centimeter, cm | Inch, in. | 0.394 | 2.54 |
| **Area** | | | |
| Kilometer$^2$, km$^2$ | Mile$^2$, mi$^2$ | 0.386 | 2.590 |
| Kilometer$^2$, km$^2$ | Acre, acre | 247.1 | 0.00405 |
| Hectare, ha | Acre, acre | 2.471 | 0.405 |
| Hectare, ha | Section (640 acre) | .0039 | 259.00 |
| Meter$^2$, m$^2$ | Chain$^2$, ch$^2$ | .0025 | 404.686 |
| Meter$^2$, m$^2$ | Feet$^2$, ft$^2$ | 10.764 | 0.093 |
| **Volume** | | | |
| Meter$^3$, m$^3$ | Acre-inch | 0.00973 | 102.8 |
| Hectoliter, hl | Cubic Foot, ft$^3$ | 3.532 | 0.2832 |
| Liter, l | Quart (liquid), qt | 1.057 | 0.946 |
| Liter, l | Gallon, gal | 0.265 | 3.78 |
| Centimeter$^3$, cc | Fluid Ounce, oz | 0.034 | 29.57 |
| Meter$^3$, m$^3$ | Foot$^3$, ft$^3$ | 35.31 | 0.028 |
| Meter$^3$, m$^3$ | Cords (90 cu ft) | 0.39 | 2.55 |
| **Mass** | | | |
| Ton (metric), t | Ton (English), t | 1.102 | 0.9072 |
| Kilogram, kg | Pound, lb | 2.205 | 0.454 |
| Gram, g | Ounce (avdp), oz | 0.035 | 28.35 |
| **Pressure** | | | |
| Atmosphere, atm | Lb/inch$^2$, psi | 14.70 | 0.06805 |
| Bar | Lb/inch$^2$, psi | 14.50 | 0.06895 |
| Bar | Atmosphere, atm | 0.9869 | 1.013 |
| Kg (weight)/cm$^2$ | Atmosphere, atm | 0.9678 | 1.033 |
| Kg (weight)/cm$^2$ | Lb/inch$^2$, psi | 14.22 | 0.07031 |
| **Yield or Rate** | | | |
| Ton (metric)/ha | Ton (English)/acre | 0.446 | 2.240 |
| Kilogram/ha | Pound/acre | 0.891 | 1.12 |
| Meter$^3$/ha | Feet$^3$/acre | 14.291 | 0.07 |
| Meter$^3$/ha | Cords/acre | 0.159 | 6.298 |

# LITERATURE CITED

Ahlgren, I. F. 1974. The effect of fire on soil organisms. p. 47–72. *In* T. T. Kozlowski and C. E. Ahlgren (ed.) Fire and Ecosystems. Academic Press, Inc., New York.

Alban, D. H. 1972a. An improved growth intercept method for estimating site index of red pine. USDA For. Serv. Res. Pap. NC-80. 7 p.

Alban, D. H. 1972b. The relationship of red pine site index to soil phosphorus extracted by several methods. Soil Sci. Soc. Amer. Proc. 36(4):664–666.

Alexander, M. 1977. Introduction to Soil Microbiology. 2nd ed. John Wiley and Sons, Inc., New York. 467 p.

Allen, E. K., and O. N. Allen. 1965. Nonleguminous plant symbiosis. p. 77–106. *In* C. M. Gilmour and O. N. Allen (ed.) Microbiology and Soil Fertility. Oregon State Univ. Press, Corvallis.

Allison, F. E. 1965. Decomposition of wood and bark sawdusts in soil, nitrogen requirements, and effects on plants. U.S. Dep. Agr. Tech. Bull. 1332. 58 p.

Alway, F. J., J. Kittredge, and W. J. Methley. 1933. Composition of forest floor layers under different forest types on the same soil. Soil Sci. 36:387–398.

Anderson, T. D., and A. R. Tiedemann. 1970. Periodic variation in physical and chemical properties of two central Washington soils. USDA For. Serv. Res. Note PNW-125. 9 p.

Appelroth, S. E. 1974. Work study aspects of planting and direct seeding in forestry. p. 202–249. *In* IUFRO Symp. Stand Establishment. Wageningen, The Netherlands.

Armson, K. A. 1972. Fertilizer distribution and sampling techniques in the aerial fertilization of forests. Univ. Toronto Fac. For. Tech. Rep. 11. 27 p.

Armson, K. A., and V. Sadreika. 1974. Forest tree nursery soil management and related practices. Pub. Min. Natur. Resour., Toronto. 177 p.

Arneman, H. F. 1960. Fertilization of forest trees. Advance. Agron. 12:171–191.

Attiwill, P. M. 1967. The loss of elements from decomposing litter. Ecology 49:142–145.

Aubertin, G. M., and J. H. Patric. 1972. Quality water from clearcut forest land? North. Logger Timber Proc. 20(8):14–23.

Aubertin, G. M., and J. H. Patric. 1974. Water quality after clearcutting a small watershed in West Virginia. J. Environ. Quality 3(3):243–249.

Auten, J. T. 1945. Prediction of site index for yellow-poplar from soil and topography. J. For. 43:662–668.

**Baker, H. P.** 1906. The holding and reclamation of sand dunes and sand wastes. For. Quart. 4:282–288.

**Baker, J. B.** 1973. Intensive cultural practices increase growth of juvenile slash pine in Florida sandhills. For. Sci. 19(3):197–202.

**Ballard, R.** 1971. Interelationships between site factors and productivity of radiata pine at Riverhead Forest, New Zealand. Plant Soil 35:371–380.

**Ballard, R., and J. G. A. Fiskell.** 1974. Phosphorus retention in coastal plain forest soils: 1. Relationships to soil properties. Soil Sci. Soc. Amer. Proc. 38(2):250–255.

**Ballard, R., and W. L. Pritchett.** 1974. Phosphorus retention in coastal plain forest soils: II. Significance to forest fertilization. Soil Sci. Soc. Amer. Proc. 38(2):363–366.

**Ballard, R., and G. M. Will.** 1971. Distribution of aerially applied fertilizer in New Zealand forests. N.Z. J. For. Sci. 1:50–59.

**Barnes, R. L., and C. W. Ralston.** 1953. The effects of colloidal phosphate on height growth of slash pine plantations. Univ. Fla. Sch. For. Res. Note 1. 2 p.

**Barnes, R. L., and C. W. Ralston.** 1955. Soil factors related to the growth and yield of slash pine plantations. Fla. Agr. Exp. Sta. Bull. 559. 23 p.

**Barney, C. W.** 1951. Effect of soil temperature and light intensity on root growth of loblolly pine seedlings. Plant Physiol. 26:146–163.

**Bates, C. G.** 1918. Concerning site. J. For. 16:383–388.

**Baule, H.** 1973. World-wide forest fertilization: its present state, and prospects for the future. Potash Rev. 22(6):1–23.

**Baule, H., and C. Fricker.** 1970. The Fertilizer Treatments of Forest Trees. BLV Verlagsgesellschaft mbH, Munich. 260 p.

**Baver, L. D., W. H. Gardner, and W. R. Gardner.** 1972. Soil Physics. 4th ed. John Wiley and Sons, Inc., New York. 498 p.

**Beadle, N. C. W.** 1940. Soil temperatures during forest fires and their effect on the survival of vegetation. J. Ecol. 28:180–192.

**Beck, D. E.** 1971. Polymorphic site index curves for white pine in the southern Appalachians. USDA For. Serv. Res. Pap. SE-80. 8p.

**Beck, D. E., and K. B. Trousdell.** 1973. Site index: accuracy of prediction. USDA For. Serv. Res. Pap. SE-108. 7 p.

**Bednall, B. H.** 1968. The problem of lower volumes associated with second rotations in *Pinus radiata* plantations in South Australia. p. 3–12. *In* Proc. 9th British Commonw. For. Conf. New Delhi.

**Bengtson, G. W.** 1973. Fertilizer use in forestry: materials and methods of application. p. 97-153. *In* FAO-IUFRO Int. Symp. For. Fertilization. Paris.

**Bengtson, G. W., and J. J. Cornette.** 1973. Disposal of composted municipal waste in a plantation of young slash pine: effects on soil and trees. J. Environ. Quality 2(4):441–444.

**Bengtson, G. W., E. C. Sample, and S. E. Allen.** 1974. Response of slash pine seedlings to P sources of varying citrate solubility. Plant Soil 40:83–96.

**Bengtson, G. W., S. E. Allen, D. A. Mayo, and T. G. Zagner.** 1969. The use of fertilizers to speed pine establishment on reclaimed coalmine spoil in northeastern Alabama: I. Greenhouse experiments. p. 199-255. *In* R. J. Hutnik and G. Davis (ed.) Ecology and Reclamation of Devastated Land. Vol. 2. Gordon and Breach, New York.

**Benzian, B.** 1965. Experiments on nutrition problems in forest nurseries. For. Comm. Bull. (Britain) 39, London. 238 p.

**Beskow, G.** 1935. Soil freezing and frost heaving. Sveriges Geologiska Undersokning 26(3):1–242.

**Bilan, M. V.** 1967. Effect of low temperature on root elongation in loblolly pine seedlings. p. 74–82. *In* Proc. XIV IUFRO Congr., Sec. 23. Munich.

**Bilan, M. V.** 1968. Effects of physical soil environment on growth and development of root systems in southern pines. p. 15–19. *In* Forest Fertilization—theory and practice. Tennessee Valley Authority, Knoxville.

**Bilan, M. V.** 1971. Some aspects of tree root distribution. p. 69–80. *In* E. Hacskaylo (ed.) Mycorrhizae. USDA For. Serv. Misc. Pub. 1189.

**Bjorkman, E.** 1967. Manuring and resistance to diseases. p. 20–21. *In* A Colloquium: the Fertilization of Forest Trees. Int. Potash Inst., Berne, Switzerland.

**Black, C. A. (ed.)** 1965. Methods of Soil Analysis. Pt. 2. Amer. Soc. Agron., Madison. 1572 p.

**Boggie, R.** 1972. Effects of water-table height on root development of *Pinus contorta* on deep peat in Scotland. Oikos 23:304–312.

**Bonner, F. T.** 1968. Responses to soil moisture deficiency by seedlings of three hardwood species. USDA For. Serv. Res. Note 50–70. 3 p.

**Bouyoucos, G. J.** 1927. The hydrometer as a new method for the mechanical analysis of soils. Soil Sci. 23:343–353.

**Bouyoucos, G. J.** 1954. New type electrode for plaster of Paris moisture blocks. Soil Sci. 78:339–342.

**Bowen, G. D.** 1965. Mycorrhiza inoculation in nursery practice. Aust. For 29:231–237

**Boyle, J. R., and A. R. Ek.** 1972. An evaluation of some effects of bole and branch pulpwood harvesting on site macronutrients. Can. J. For. Res. 2:407–412.

**Brady, N. C.** 1974. The Nature and Properties of Soils. 8th ed. Macmillan Company, Inc., New York. 639 p.

**Braun-Blanquet, J.** 1932. Plant Sociology. McGraw-Hill Book Co., New York. 439 p.

**Bray, J. R., and E. Gorham.** 1964. Litter production in forests of the world. p. 101–157. *In* J. B. Cragg (ed.) Advance. Ecol. Res. 2. Academic Press, Inc., New York.

**Bremner, J. M., and L. A. Douglas.** 1971. Decomposition of urea phosphate in soils. Soil Sci. Soc. Amer. Proc. 35(4):575–578.

**Broadfoot, W. M.** 1969. Problems in relating soil to site index for southern hardwoods. For. Sci. 15(4):354–364.

**Broadfoot, W. M.** 1973. Water table depth and growth of young cottonwood. USDA For. Serv. Res. Note SO-167. 4 p.

**Broerman, F. S., and G. E. Gatherum.** 1967. Relationship of nitrogen and light intensity to growth, photosynthesis and respiration of green ash seedlings. Iowa State J. Sci. 42(2):137–148.

**Brouzes, L. J., and R. Knowles.** 1969. The effect of organic amendment, water content, and oxygen on the incorporation of $N_2$ by some agricultural and forest soils. Can. J. Microbiol. 15:899–905.

**Buffington, J. D.** 1967. Soil arthropod populations of the New Jersey pine barrens as affected by fire. Ann. Entomol. Soc. Amer. 60:530–535.

**Buol, S. W., F. D. Hole, and R. J. McCracken.** 1973. Soil Genesis and Classification. The Iowa State Univ. Press, Ames. 360 p.

**Burns, R. M., and E. A. Hebb.** 1972. Site Preparation and Reforestation of Droughty, Acid Sands. U.S. Dep. Agr., Agr. Handbk. 426. 61 p.

**Cajander, A. K.** 1926. The theory of forest types. Acta. For. Fenn. 29:1–108.

**Capel, J. C., and C. S. Coffman.** 1966. Growing cottonwood. For. Farmer 25(11):6–18.

**Carlisle, A., A. H. F. Brown, and E. J. White.** 1967. The nutrient content of tree stemflow and ground flora litter and leachates in a sessile oak (*Quercus petraa*) woodland. J. Ecol. 55:615–627.

**Carmean, W. H.** 1961. Soil survey refinements needed for accurate classification of black oak site quality in southeastern Ohio. Soil Sci. Soc. Amer. Proc. 25(5):394–397.

**Carmean, W. H.** 1975. Forest site quality evaluation in the United States. Advance. Agron. 27:209–269.

**Carter, M. C., and E. H. White.** 1971. The necessity for intensive cultural treatment in cottonwood plantations. Auburn Univer. Agr. Exp. Sta. Circ. 189. 11 p.

**Chavasse, C. G. R.** 1974. A review of land clearing for site preparation for intensive plantation forestry. p. 109–128. *In* IUFRO Symp. Stand Establishment. Wageningen, The Netherlands.

**Cochran, P. H.** 1969. Thermal properties and surface temperatures of seedbeds. USDA For. Serv., PNW For. Range Exp. Sta., Portland. 19 p.

**Coile, T. S.** 1935. Relation of site index for shortleaf pine to certain physical properties of the soil. J. For. 33:726–730.

**Coile, T. S.** 1952. Soil and the growth of forests. Advance. Agron. 4:329–398.

**Cole, D. W., and S. P. Gessel.** 1965. Movement of elements through a forest soil as influenced by tree removal and fertilizer additions. p. 95–104. *In* C. T. Youngberg (ed.) Forest-Soil Relationship in North America. Oregon State Univ. Press, Corvallis.

**Cole, D. W., S. P. Gessel, and S. F. Dice.** 1967. Distribution and cycling of nitrogen, phosphorus, potassium and calcium in the second-growth Douglas-fir ecosystem. p. 197–232. *In* H. E. Young (ed.) Symp. Primary Productivity and Mineral Cycling in Natural Ecosystems. Univ. Maine Press, Orono.

**Coleman, N. T., and G. W. Thomas.** 1967. The basic chemistry of soil acidity. p. 1–41. *In* R. W. Pearson and F. Adams (ed.) Soil Acidity and Liming. Spec. Pub. Ser. 12. Amer. Soc. Agron., Madison.

**Cooper, R. W.** 1957. Silvical characteristics of slash pine. USDA For. Serv., SE For. Exp. Sta. Pap. 81. 13 p.

**Cooper, R. W.** 1971. Current use and place of prescribed burning. p. 21–27. *In* Proc. Prescribed Burning Symp. USDA For. Serv., SE For. Exp. Sta., Asheville.

**Cooper, R. W.** 1975. Prescribed burning. J. For. 73:776–780.

**Cornwell, S. M., and E. L. Stone.** 1968. Availability of nitrogen to plants in acid coal mine spoils. Nature 217:768–769.

**Crocker, R. L., and J. Major.** 1955. Soils development in relation to vegetation and surface age at Glacier Bay, Alaska. J. Ecol. 43:427–448.

**Cromer, R. N.** 1967. The significance of the "ashbed effect" in *Pinus radiata* plantations. Appita 20(4):104–112.

**Czapowskyj, M. M.** 1970. Experimental planting of 14 tree species of Pennsylvania's anthracite strip-mine spoils. USDA For. Serv. Res Pap. NE-155. 18 p.

**Czapowskyj, M. M.** 1973. Establishing forest on surface-mined land as related to fertility and fertilization. p. 132–139. *In* Symp. For. Fertilization. USDA For. Serv. Gen. Tech. Rep. NE-3.

**Daly, G. T.** 1966. Nitrogen fixation by nodulated *Alnus rugosa*. Can. J. Bot. 44:1607–1621.

**Dasmann, R. F.** 1968. Environmental Conservation. John Wiley and Sons, Inc., New York. 375 p.

**Debano, L. F., L. D. Mann, and D. A. Hamilton.** 1970. Translocation of hydrophobic substances into soil by burning organic litter. Soil Sci. Soc. Amer. Proc. 34(1):130–133.

**Dement, J. A., and E. L. Stone.** 1968. Influence of soil and site on red pine plantations in New York. Cornell Univ. Agr. Exp. Sta. Bull. 1020. 25 p.

**Denison, W. C.** 1973. Life in tall trees. Sci. Amer. 228(6):75–80.

**Derr, H. J., and W. F. Mann, Jr.** 1970. Site preparation improves growth of planted pines. USDA For. Serv. Res. Note SO-106. 4 p.

**Dickerson, B. P.** 1976. Soil compaction after tree-length skidding in northern Mississippi. Soil Sci. Soc. Amer. J. 40(6):965–966.

**Dieterich, J. H.** 1971. Air-quality aspects of prescribed burning. p. 139–151. *In* Proc. Prescribed Burning Symp. USDA For. Serv., SE For. Exp. Sta., Asheville.

**Dorsser, J. C. van, and D. A. Rook.** 1972. Conditioning of radiata pine seedlings by undercutting and wrenching: Description of methods, equipment, and seedling response. N.Z. J. For. 17(1):61–73.

**Douglass, J. E., and W. T. Swank.** 1972. Streamflow modification through management of eastern forests. USDA For. Serv. Res. Pap. SE-94. 15 p.

**Durzan, D. J., and F. C. Steward.** 1967. The nitrogen metabolism of *Picea glauca* (Moench) Voss and *Pinus banksiana* Lamb. as influenced by mineral nutrition. Can. J. Bot. 45:697–710.

**Duvigneaud, P., and S. Denaeyer-DeSmet.** 1967. Biomass, productivity and mineral cycling in deciduous mixed forests in Belgium. p. 167–186. *In* H. E. Young (ed.) Symp. Primary Productivity and Mineral Cycling in Natural Ecosystems. Univ. Maine Press, Orono.

**Duvigneaud, P., and S. Denaeyer-DeSmet.** 1970. Biological cycling of minerals in temperate deciduous forests. p. 199–225. *In* D. E. Reichle (ed.) Analysis of Temperate Forest Ecosystems. Springer-Verlag, New York.

**Dyrness, C. T.** 1969. Hydrologic properties of soils on three small watersheds in the western Cascades of Oregon. USDA For. Serv. Res. Note PNW-111. 17 p.

**Dyrness, C. T.** 1976. Effect of wildfire on soil wettability in the high Cascades of Oregon. USDA For. Serv. Res. Pap. PNW-202. 18 p.

**Ebell, L. F.** 1972. Cone induction response of Douglas-fir to form of nitrogen fertilizer and time of treatment. Can. J. For. Res. 2:317–326.

**Eberhardt, P. J., and W. L. Pritchett.** 1971. Foliar applications of nitrogen to slash pine seedlings. Plant Soil 34:731–739.

**Ebermayer, E.** 1876. Die gesammte Lehre der Waldstreu mit Rucksicht auf die chemische Statik des Waldbaues. J. Springer, Berlin. 300 p.

**Edwards, C. A., D. E. Reichle, and D. A. Crossley.** 1970.  The role of soil invertebrates in turnover of organic matter and nutrients. p. 147–172. *In* D. E. Reichle (ed.) Analysis of Temperate Forest Ecosystems. Springer-Verlag, New York.

**Ehwald. E.** 1957.  Über den Nährstoffkreislauf des Waldes. Deutsche Akd. LandwWiss. Berlin. Sitz. ber. 6.1. 56 p.

**Ekberg, A., and R. Friberg.** 1972.  Gödselmedels fastilaggning i tradkronor. Inst. Skogsforbattring. Godsling Nr. 2. 4 p.

**Eyre. S. R.** 1963.  Vegetation and Soils. Aldine Publishing Co., Chicago. 324 p.

**Fayle, D. C. F.** 1975.  Extension and longitudinal growth during the development of red pine root systems. Can. J. For. Res. 5:109–121.

**Fellin, D. G., and P. C. Kennedy.** 1972.  Abundance of arthropods inhabiting duff and soil after prescribed burning on forest clearcuts in northern Idaho. USDA For. Serv. Res. Note INT-162. 8 p.

**Ffolliott, P. F., and D. B. Thorud.** 1974.  Vegetation management for increased water control in Arizona. Ariz. Agr. Exp. Sta. Tech. Bull. 215. 38 p.

**Fillip, S. M.** 1977.  How applicable is uneven-age management in northern forest types? p. 53–62. *In* Proc. Symp. Intensive Culture of Northern Forest Types. USDA For. Serv. Gen. Tech. Rep NE-29.

**Foil, R. R., and C. W. Ralston.** 1967.  The establishment and growth of loblolly pine seedlings on compacted soils. Soil Sci. Soc. Amer. Proc. 31(4):565–568.

**Fisher, R. F., and E. L. Stone.** 1969.  Increased availability of nitrogen and phosphorus in the root zone of conifers. Soil Sci. Soc. Amer. Proc. 33(6):955–961.

**Fisher, R. F., G. L. Rolfe, and R. P. Eastburn.** 1975.  Productivity and organic matter distribution in a pine plantation and an adjacent old field. Ill. Agr. Exp. Sta., For. Res. Rep. 75-1. 3 p.

**Fornes, R. H., J. V. Berglund, and A. L. Leaf.** 1970.  A comparison of the growth and nutrition of *Picea abies* (L.) Karst. and *Pinus resinosa* Ait. on a K-deficient site subjected to K fertilization. Plant Soil 33:345–360.

**Fourt, D. F., D. G. M. Donald, J. N. R. Jeffers, and W. O. Binns.** 1971.  Corsican pine (*Pinus nigra* var. *maritima* (Ait.) Melville) in southern Britain. Forestry 44(2):189–207.

**Fredriksen, R. L.** 1970.  Comparative chemical water quality—natural and disturbed streams following logging and slash burning. p. 125–137. *In* Proc. Symp. Forest Land Uses and Stream Environment. Oregon State Univ., Corvallis.

**Friberg, R.** 1974.  Results from fertilizer trials with applications at different times of the year. Inst. Skogsforbattring, Information NV.5. 3 p.

**Gadgil, R. L.** 1971a.  The nutritional role of *Lupinus arboreus* in coastal sand dune forestry: I. The potential influence of undamaged lupin plants on nitrogen uptake by *Pinus radiata*. Plant Soil 34:357–367.

**Gadgil, R. L.** 1971b.  The nutritional role of *Lupinus arboreus* in coastal sand dune forestry: III. Nitrogen distribution in the ecosystem before tree planting. Plant Soil 35:114–126.

**Gentle, S. W., F. R. Humphreys, and M. J. Lambert.** 1965.  An examination of a *Pinus radiata* phosphate fertilizer trial fifteen years after treatment. For. Sci. 11(3):315–324.

**Gerdemann, J. W., and J. M. Trappe.** 1975.  Taxonomy of Endogonaceae. p. 35–51. *In*

F. E. Sanders, B. Mosse, and P. B. Tinker (ed.) Endomycorrhizas. Academic Press, Inc., New York.

Gessel, S. P., and A. N. Balci. 1965. Amount and composition of forest floors under Washington coniferous forests. p. 11–23. *In* C. T. Youngberg (ed.) Forest-Soil Relationships in North America. Oregon State Univ. Press, Corvallis.

Gessel, S. P., and D. W. Cole. 1965. Influence of removal of forest cover on movement of water and associated elements through soils. J. Amer. Water Works Assoc. 57:1301–1310.

Gessel, S. P., and R. D. Walker. 1956. Height growth response of Douglas-fir to nitrogen fertilizer. Soil Sci. Soc. Amer. Proc. 20(1):97–100.

Gessel, S. P., D. W. Cole, and E. C. Steinbrenner. 1973. Nitrogen balances in forest ecosystems of the Pacific Northwest. Soil Biol. Biochem. 5:19–39.

Gilmore, A. R., W. A. Geyer, and W. R. Boggess. 1968. Microsite and height growth of yellow-poplar. For. Sci. 14(4):420–426.

Golley, F. B., J. T. McGinnis, R. G. Clements, G. I. Child, and M. J. Duever. 1975. Mineral Cycling in a Tropical Moist Forest Ecosystem. Univ. Georgia Press, Athens. 248 p.

Gooding, J. W., and W. H. Smith. 1972. Effects of fertilization on stem, wood properties, and pulping characteristics of slash pine (*Pinus elliottii* var. *elliottii* Engelm.). p. El-18. *In* Proc. Symp. Effects of Growth Acceleration on the Properties of Wood. USDA For. Serv., For. Prod. Lab., Madison.

Goor, C. P. van. 1974. Introduction to the theme. p. 27–33. *In* IUFRO Symp. Stand Establishment. Wageningen, The Netherlands.

Gosz, J. R., G. E. Likens, and F. H. Bormann. 1972. Nutrient content of litter fall on the Hubbard Brook Experimental Forest, New Hampshire. Ecology. 53:769–784.

Grano, C. X. 1970. Small hardwoods reduce growth of pine understory. USDA For. Serv. Res. Pap. SO-55. 11 p.

Gray, L. E. 1971. Physiology of vesicular-arbuscular mycorrhizae. p. 145–150. *In* E. Hacskaylo (ed.) Mycorrhizae. USDA For. Serv. Misc. Pub. 1189.

Grier, C. C. 1975. Wildfire effects on nutrient distribution and leaching in a coniferous ecosystem. Can. J. For. Res. 5:599–607.

Griffith, B. G., E. W. Hartwell, and T. E. Shaw. 1930. The evolution of soil as affected by the old field white pine-mixed hardwood succession in central New England. Harvard For. Bull. 15. 82 p.

Hacskaylo, E. 1971. Metabolite exchanges in ectomycorrhizae. p. 175–182. *In* E. Hacskaylo (ed.) Mycorrhizae. USDA For. Serv. Misc. Pub. 1189.

Hägglund, B. 1974. Treatment of young stands. p. 347–372. *In* IUFRO Symp. Stand Establishment. Wageningen, The Netherlands.

Hagner, S. 1971. The present standard of practical forest fertilization in different parts of the world. Proc. XV IUFRO Congr., Sec. 32., Gainesville, Fla. 16 p.

Hagner, S., and A. L. Leaf. 1973. Fertilization of established forest stands. p. 243–270. *In* FAO-IUFRO Int. Symp. For. Fertilization. Paris.

Haig, I. T. 1929. Colloidal content and related soil factors as indicators of site quality. Yale Univ. Sch. For. Bull. 24. 30 p.

Haines, L. W., and W. L. Pritchett. 1964. The effects of site preparation on the growth of slash pine. Soil Crop Sci. Soc. Fla. Proc. 24:27–34.

**Haines, L. W., and W. L. Pritchett.** 1965. The effects of site preparation on the availability of soil nutrients and on slash pine growth. Soil Crop Sci. Soc. Fla. Proc. 25:356–364.

**Haines, L. W., T. E. Maki, and S. G. Sanderford.** 1975. The effects of mechanical site preparation treatments on soil productivity and tree (*Pinus taeda* L. and *P. elliottii* Engelm. var. *elliottii*) growth. p. 379–395. *In* Forest Soils and Forest Land Management. Laval Univ. Press, Quebec.

**Hanna, P. R.** 1968. Topography and soil relations for white and black oak. USDA For. Serv. Res. Pap. NC-25. 7 p.

**Harley, J. L.** 1969. The Biology of Mycorrhizae. 2nd ed. Leonard Hill, London. 334 p.

**Harms, W. R.** 1969. Leaf-water deficit of tree seedlings in relation to soil moisture. For. Sci. 15(1):58–63.

**Hatch, A. B.** 1937. The physical basis for mycotrophy in *Pinus*. Black Rock For. Bull. 6. 168 p.

**Hatchell, G. E.** 1970. Soil compaction and loosening treatments affect loblolly pine growth in pots. USDA For. Serv. Res. Rep. SE-72. 9 p.

**Hayman, D. S.** 1975. The occurrence of mycorrhizae in crops as affected by soil fertility. p. 495–509. *In* F. E. Sanders, B. Mosse, and P. B. Tinker (ed.) Endomycorrhizas. Academic Press, Inc., New York.

**Hebb, E. A., and R. M. Burns.** 1975. Slash pine productivity and site preparation on Florida sandhill sites. USDA For. Serv. Res. Pap. SE-135. 8 p.

**Heiberg, S. O., and R. F. Chandler.** 1941. A revised nomenclature of forest humus layers for the northeastern United States. Soil Sci. 52:87–99.

**Heiberg, S. O., and D. P. White.** 1951. Potassium deficiency of reforested pine and spruce stands in northern New York. Soil Sci. Soc. Amer. Proc. 15:369–376.

**Heimburger, C. C.** 1934. Forest-type studies in the Adirondack region. New York Agr. Exp. Sta. Mem. 165 p.

**Heimburger, C. C.** 1941. Forest site classification and soil investigation on Lake Edward Forest Experiment Station Area. Can. Dep. Mines Res., Silv. Res. Note 66. 41 p.

**Helvey, J. D.** 1971. A summary of rainfall interception by certain conifers of North America. p. 103–113. *In* Biological Effects in the Hydrological Cycle. Purdue Univ., West Layfayette, Indiana.

**Hesselman, H.** 1926. Studier över barrskogens humustäche, dess egenskaper och beroende av skogsvarden. Statens Skogsförsöksant Meddel. 22:169–552.

**Hewlett, J. D.** 1972. An analysis of forest water problems in Georgia. Ga. For. Res. Council Rep. 30. 27 p.

**Hewlett, J. D., and A. R. Hibbert.** 1961. Increases in water yield after several types of forest cutting. Int. Assoc. Sci. Hydrol. 6:5–17.

**Heyward, F.** 1938. Soil temperatures during forest fires in the longleaf pine region. J. For. 36:478–491.

**Heyward, F., and R. M. Barnette.** 1934. Effect of frequent fires on chemical composition of forest soils in the longleaf pine region. Fla. Agr. Exp. Sta. Bull. 265. 39 p.

**Heyward, F., and R. M. Barnette.** 1936. Field characteristics and partial chemical analyses of the humus layer of longleaf pine forest soils. Fla. Agr. Exp. Sta. Bull. 302. 27 p.

**Heyward, F., and A. N. Tissot.** 1934. Some changes in the soil fauna associated with forest fires in the longleaf pine region. Fla. Agr. Expt. Sta. Tech. Bull. 265. 30 p.

**Hibbert, A. R.** 1967. Forest treatment effects on water yield. p. 527–543. *In* W. E. Sopper and H. W. Lull (ed.) Int. Symp. For. Hydrology. Pergamon Press, New York.

**Hicock, H. W., M. F. Morgan, H. J. Lutz, H. Bull, and H. A. Lunt.** 1931. The relation of forest composition and rate of growth to certain soil characters. Conn. Agr. Exp. Sta. Bull. 330:673–750.

**Hills, G. A.** 1952. The classification and evaluation of site for forestry. Ont. Dep. Lands For. Res. Rep. 24. 41 p.

**Hills, G. A.** 1960. Regional site research. For Chron. 36:401–423.

**Hills, G. A.** 1961. The ecological basis for land-use planning. Ontario Dep. Lands For. Res. Rep. 46. 204 p.

**Hodges, J. D., and P. L. Lorio.** 1969. Carbohydrate and nitrogen fractions of the inner bark of loblolly pines under moisture stress. Can. J. Bot. 47:1651–1657.

**Hodgkins, E. J.** 1960. Forest site evaluation in the southeast: an evaluation. p. 34–48. P. Y. Burnes (ed.) Southern Forest Soils. 8th Annu. For. Symp. Louisiana State Univ. Press, Baton Rouge.

**Hollis, C. A., W. H. Smith, R. A. Schmidt, and W. L. Pritchett.** 1975. Soil and tissue nutrients, soil drainage, fertilization and tree growth as related to fusiform rust incidence in slash pine. For. Sci. 21(2):141–148.

**Holmen, H.** 1969. Afforestation of peatlands. Skogs-o. Lantbr.-akad. Tidskr. 108(5):216–235.

**Holmen, H.** 1971. Forest fertilization in Sweden. Skogs-o. Lantbr.-akad. Tidskr. 110(3):156–162.

**Holmesgaard, E., H. Holstener-Jorgensen, and A. Yde-Anderson.** 1961. Soil formation, growth, and health of first and second generation spruce stands. I. North Zealand, Forstl. Forsogsv. Danm. 27:1–167.

**Hoover, M. D.** 1949. Hydrologic characteristics of South Carolina piedmont forest soils. Soil Sci. Soc. Amer. Proc. 14:353–358.

**Hoover, M. D.** 1952. Water and timber management. J. Soil Water Conserv. 7:75–78.

**Hoover, M. D., and H. A. Lunt.** 1952. A key for the classification of forest humus types. Soil Sci. Soc. Amer. Proc. 16(4):368–370.

**Hornbeck, J. W., R. S. Pierce, and C. A. Federer.** 1970. Streamflow changes after forest clearing in New England. Water Resour. Res. 6:1124–1132.

**Hortenstine, C. C., and D. F. Rothwell.** 1968. An evaluation of garbage compost as a source of plant nutrients for oats and radishes. Compost Sci. 9(2):23–25.

**Horton, K. W.** 1958. Rooting habits of lodgepole pine. Can. Dep. North. Affairs and Natur. Resour., For. Res. Div. Tech. Note 67. 26 p.

**Hoyle, M. C.** 1971. Effects of the chemical environment on yellow birch, root development and top growth. Plant Soil 35:623–633.

**Huikari, O.** 1973. Results of fertilization experiments on peatlands drained for forestry. Finnish For. Res. Inst. Rep. 1.

**Humphreys, F. R., and W. L. Pritchett.** 1971. Phosphorus adsorption and movement in some sandy forest soils. Soil Sci. Soc. Amer. Proc. 35(3):495–500.

**Huppuch, C. D.** 1960. The effects of site preparation on survival and growth of syca-more cuttings. USDA For. Serv. Res. Note SE-140. 2 p.

**Ilvessalo, Y.** 1929. Notes on some forest (site) types in North America. J. Ecol. 34:1–111.

**Ingestad, T.** 1963. Microelement nutrition of pine, spruce, and birch seedlings in nutrient solutions. Rep. For. Res. Inst. Sweden 51(7):150.

**Jackson, D. S.** 1965. Species siting: climate, soil, and productivity. N.Z. J. For. 10(1):90–102.

**Jackson, M. L.** 1958. Soil Chemical Analysis. Prentice-Hall, Inc., Englewood Cliffs, N.J. 498 p.

**Johnson, R. D.** 1964. Water relations in *Pinus radiata* under plantation conditions. Aust. J. Bot. 12:111–124.

**Jorgensen, J. R., and C. S. Hodges, Jr.** 1971. Effect of prescribed burning on the microbial characteristics of soil. p. 107–114. *In* Proc. Prescribed Burning Symp. USDA For. Serv., SE For. Exp. Sta. Asheville.

**Jorgensen, J. R., and C. G. Wells.** 1971. Apparent nitrogen fixation in soil influenced by prescribed burning. Soil Sci. Soc. Amer. Proc. 35(5):806–810.

**Kalela, E.** 1957. Über Veranderungen in den Wurzelverhältnissen der Kiefernfestände in Laufe der Vegetationsperiod. Acta For. Fenn. 65:3–41.

**Katznelson, H., J. W. Rovatt, and E. A. Peterson.** 1962. The rhizosphere effect of mycorrhizal and non-mycorrhizal roots of yellow birch seedlings. Can. J. Bot. 40:378–382.

**Kaufman, C. M.** 1968. Growth of horizontal roots, height, and diameter of planted slash pine. For. Sci. 14(3):265–274.

**Kaufman, C. M., W. L. Pritchett, and R. E. Choate.** 1977. Growth of slash pine (*Pinus elliottii* Engel. var. *elliottii*) on drained flatwoods. Fla. Agr. Exp. Sta. Bull. 792. 30 p.

**Kayll, A. J.** 1974. Use of fire in land management. p. 483–504. *In* T. T. Kozlowski and C. E. Ahlgren (ed.) Fire and Ecosystems. Academic Press, Inc., New York.

**Keeves, A.** 1966. Some evidence of loss of productivity with successive rotations of *Pinus radiata* in the south-east of South Australia. Aust. For. 30:51–63.

**Kellison, R. C.** 1975. Genetic improvement of forest trees. p. 13–16. *In* S. Prahacs (ed.) New Horizons for the Chemical Engineer in Pulp and Paper Technology. AIChE Symp. Series. 72. 157.

**Kellogg, C. E.** 1964. Potentials for food production. p. 57–69. *In* Yearbk. of Agr. U.S. Dep. Agr., Washington, D.C.

**Kessell, S. L.** 1927. Soil organisms: the dependence of certain pine species on a biological soil factor. Empire For. J. 6:70–74.

**Kittridge, J.** 1948. Interception and stemflow. p. 99–114. *In* Forest Influences. McGraw-Hill Book Co., New York.

**Klausing, O.** 1956. Untersuchungen uber den Mineralunsatz in Buchenwalden auf Granit und Diorit. Forstw. Cbl. 75:18–32.

**Klawitter, R. A.** 1970a. Does bedding promote pine survival and growth on ditched wet sands? USDA For. Serv. Res. Note NE-109. 4 p.

**Klawitter, R. A.** 1970b. Water regulation on forest land. J. For. 68:338–342.

**Klock, G. O.** 1971. Streamflow nitrogen loss following erosion control fertilization. USDA For. Serv. Res. Note PNW-169. 9 p.

**Kohmann, K.** 1972. Root ecological investigations on pine (*Pinus silvestris*) II. The root system's reaction to fertilization. Meddr. Norske Skogfors Ves. 30:392–396.

**Kormanik, P. P., W. C. Bryan, and R. C. Schultz.** 1977. The role of mycorrhizae in plant growth and development. p. 1–10. *In* H. M. Vines (ed.) Physiology of Root-Microorganisms Associations. Proc. Symp. South. Sec. Amer. Soc. Plant Physiol. Atlanta.

**Köstler, J. N., E. Brückner, and H. Bibelriether.** 1968. Die Wurzeln der Waldbäume. Paul Parey, Hamburg. 284 p.

**Kozlowski, T. T.** 1968. Soil water and tree growth. p. 30–57. *In* N. E. Linnartz (ed.) 17th Annu. For. Symp. The Ecology of Southern Forest. Louisiana State Univ. Press, Baton Rouge.

**Kozlowski, T. T.** 1971. Growth and Development of Trees. Vol. II. Academic Press, Inc., New York. 514 p.

**Kramer, P. J.** 1969. Plant and Soil Water Relationships: A Modern Synthesis. McGraw-Hill Book Co., New York. 482 p.

**Kreutzer, K.** 1972. The influence of litter utilization on the nitrogen cycle in Scots pine (*Pinus sylvestris* L.) stands. Forstw. Cbl. 91:263–270.

**Krugman, S. L., and E. L. Stone.** 1966. The effect of cold nights on the root-regenerating potential of ponderosa pine seedlings. For. Sci. 12(4):451–459.

**Kubiěna, W. L.** 1953. The Soils of Europe. Thomas Murby and Co., London. 318 p.

**Langdon, O. G.** 1971. Effects of prescribed burning on timber species of the southeastern coastal plain. p. 34–44. *In* Proc. Prescribed Burning Symp. USDA For. Serv., SE For. Exp. Sta., Asheville.

**Leaf, A. L.** 1968. K, Mg, S deficiencies in forest trees. p. 88–122. *In* Forest Fertilization—theory and practice. Tennessee Valley Authority, Knoxville.

**Leaf, A. L., R. E. Leonard, J. V. Berglund, A. R. Eschner, P. H. Cochran, J. B. Hart, G. M. Marion, and R. A. Cunningham.** 1970. Growth and development of *Pinus resinosa* plantations subjected to irrigation-fertilization treatments. p. 97–118. *In* C. T. Youngberg and C. B. Davey (ed.) Tree Growth and Forest Soils. Oregon State Univ. Press, Corvallis.

**Lehotsky, K.** 1972. Sand dune fixation in Michigan—thirty years later. J. For. 70:155–160.

**Levno, A., and J. Rothacher.** 1969. Increases in maximum stream temperatures after slash burning in a small experimental watershed. USDA For. Serv. Res. Note. PNW-110. 7 p.

**Levy, G.** 1972. Premiers resultats concernant deux experiences d'assainissement du sol sur plantations de resineux. Ann. Sci. For. 29(4):427–450.

**Likens, G. E., and F. H. Bormann.** 1972. Nutrient cycling in ecosystems. p. 26–67. *In* J. Wiens (ed.) Ecosystems: Structure and Function. Oregon State Univ. Press, Corvallis.

**Likens, G. E., F. H. Bormann, N. M. Johnson, D. W. Fisher, and R. S. Pierce.** 1970. Effects of forest cutting and herbicide treatment on nutrient budgets in the Hubbard Brook watershed-ecosystem. Ecol. Monogr. 40:23–47.

**Loehr, R. C.** 1968. Pollution implications of animal wastes—a forward oriented review. Robert S. Kerr Water Res. Ctr., USDI Water Pollution Control Admin. Ada, Okla.

**Lohrey, R. E.** 1974. Site preparation improves survival and growth of direct-seeded pines. USDA For. Serv. Res. Note SO-185. 4 p.

**Lopushinsky, W.** 1969. Stomatal closure in conifer seedlings in response to leaf moisture stress. Bot. Gaz. 130(4):258–263.

**Lopushinsky, W.** 1975. Water relations and photosynthesis in lodgepole pine. p. 135–153. *In* D. M. Baumgartner (ed.) Management of Lodgepole Pine Ecosystems, Symp. Proc. Washington State Univ., Pullman.

**Lorio, P. L., V. K. Howe, and C. N. Martin.** 1972. Loblolly pine rooting varies with microrelief on wet sites. Ecology 53:1134–1140.

**Low, A. J., and G. van Tol.** 1974. Initial spacing in relation to stand establishment. p. 296–315. *In* IUFRO Symp. Stand Establishment. Wageningen, The Netherlands.

**Lutz, H. J., and R. F. Chandler.** 1946. Forest Soils, John Wiley and Sons, Inc., New York. 514 p.

**Lyford, W. H.** 1943. The palatability of freshly fallen forest tree leaves to millipedes. Ecology 24:252–261.

**Lyford, W. H.** 1952. Characteristics of some podzolic soils of the northeastern United States. Soil Sci. Soc. Amer. Proc. 16(3):231–234.

**Lyford, W. H.** 1963. Importance of ants to brown podzolic soil genesis in New England. Harvard For. Pap. 7. 18 p.

**Lyford, W. H.** 1964. Water-table fluctuations in periodically wet soils of central New England. Harvard For. Pap. 8. 15 p.

**Lyford, W. H.** 1973. Forest soil microtopography. p. 47–58. *In* D. L. Dindal (ed.) Proc. 1st Soil Microcommunities Conf. U.S. Atomic Energy Comm., Syracuse.

**Lyford, W. H.** 1975. Rhizography of non-woody roots of trees in the forest floor. p. 179–196. *In* J. G. Torrey and D. T. Clarkson (ed.) The Development and Function of Roots. Academic Press, Inc., New York.

**Lyford, W. H., and D. W. MacLean.** 1966. Mound and pit microrelief in relation to soil disturbance and tree distribution in New Brunswick, Canada. Harvard For. Pap. 15. 18 p.

**Lyford, W. H., and B. F. Wilson.** 1966. Controlled growth of forest tree roots: techniques and application. Harvard For. Pap. 16. 12 p.

**Lyr, H., and G. Hoffman.** 1967. Growth rates and growth periodicity of tree roots. p. 181–236. *In* J.A.A. Romberger and P. Mikola (ed.) International Review of Forestry Research. Vol. 2. Academic Press, Inc., New York.

**Mader, D. L.** 1953. Physical and chemical characteristics of the major types of forest humus found in the United States and Canada. Soil Sci. Soc. Amer. Proc. 17(2):155–158.

**Mader, D. L.** 1964. Where are we in soil-site classification? p. 23–32. *In* Applications of Soils Information in Forestry. New York State Coll. Agr. (Ithaca) Misc. Pub.

**Maftoun, M., and W. L. Pritchett.** 1970. Effects of added nitrogen on the availability of phosphorus to slash pine on two lower coastal plain soils. Soil Sci. Amer. Proc. 34(4):685–690.

**Madgwick, H. A. I.** 1964. Variations in the chemical composition of red pine (*Pinus resinosa* Ait.) leaves: a comparison of well-grown and poorly-grown trees. Forestry 67(1):87–94.

**Mann, W. F., and T. R. Dell.** 1971. Yields of 17-year-old loblolly pine planted on a cutover site at various spacings. USDA For. Serv. Res. Pap. SO-70. 8 p.

**Mann, W. F., and J. M. McGilvray.** 1974. Response of slash pine to bedding and phosphorus application in southeastern flatwoods. USDA For. Serv. Res. Pap. SO-99. 9 p.

**Marx, D. H.** 1966. The role of ectotrophic mycorrhizal fungi in the resistance of pine roots to infection by *Phytophthora cinnamomi* Rands. Ph.D. dissertation. North Carolina State Univ., Raleigh. 179 p.

**Marx, D. H.** 1972. Ectomycorrhizae as biological deterrents to pathogenic root infections. Annu. Rev. Phytopath. 10:429–454.

**Marx, D. H.** 1977. The role of mycorrhizae in forest production. p. 151–161. *In* TAPPI Conf. Pap. Atlanta, Ga.

**Marx, D. H., W. C. Bryan, and C. B. Davey.** 1970. Influence of temperature on aseptic synthesis of ectomycorrhizae by *Thelephora terrestris* and *Pisolithus tinctorius* on loblolly pine. For. Sci. 16(4):424–431.

**Marx, D. H., A. B. Hatch, and J. F. Mendicino.** 1977. High soil fertility decrease sucrose content and susceptibility of loblolly pine roots to ectomycorrhizal infection by *Pisolithus tinctorius*. Can. J. Bot. 55:1569–1574.

**Mather, J. R.** 1953. The disposal of industrial effluent by woods irrigation. p. 434–448. *In* Proc. Purdue Ind. Waste Conf. 8. W. Lafayette, Ind.

**Matsui, Z.** 1956. The ecological features in foggy districts. p. 103–123. *In* Hokkaido (Japan) For. Exp. Sta. Spec. Rep. 205.

**Matthews, S. W.** 1973. This changing earth. Nat. Geogr. Mag. 143(1):1–37.

**May, J. T., C. C. Parks, and H. F. Perkins.** 1969. Establishment of grasses and tree vegetation on spoil from kaolin strip mining. p. 137–147. *In* Ecology and Revegetation of Devastated Land. Vol. 2. Gordon and Breach Sci. Pub., New York.

**Mayer-Krapoll, H.** 1956. The use of commercial fertilizers–particularly nitrogen–in forestry. Ruhr-Stickstoff, Aktiengesellschaft, Bochum. (Transl. and Pub.) Allied Chemical and Dye Corp., New York. 111 p.

**McColl, J. G., and D. F. Grigal.** 1975. Forest fire: Effects on phosphorus movement to lakes. Science 188:1109–1111.

**McFee, W. W., and E. L. Stone.** 1965. Quantity, distribution, and variability of organic matter and nutrients in a forest podzol in New York. Soil Sci. Soc. Amer. Proc. 29(4):432–436.

**McFee, W. W., and E. L. Stone.** 1968. Ammonium and nitrate as nitrogen sources for *Pinus radiata* and *Picea glauca*. Soil Sci. Soc. Amer. Proc. 32(6):879–884.

**McGee, C. E.** 1977. Planted yellow-poplar grows well after essential site preparation. Tree planters' Notes 28(1):5–7.

**McGregor, W. H. D.** 1957. Fertilizer increases growth rate of slash pine. USDA For. Serv., SE For. Exp. Sta. Res. Note 101. 2 p.

**McKee, W. H., and R. A. Sommers.** 1971. Slash pines respond equally to nitrogen applications in fall and spring. USDA For. Ser. Res. Note SO-128. 4 p.

**McLean, H. R., and R. F. Ward.** 1976. Is "smoke-free" burning possible? Fire Manage. 4:10–13.

**McLemore, B. F.** 1975. Cone and seed characteristics of fertilized and unfertilized longleaf pines. USDA For. Serv. Res. Pap. SO-109. 10 p.

**McMinn, R. G.** 1963. Characteristics of Douglas-fir root systems. Can. J. Bot. 41:105–122.

**McQuilkin, W. E.** 1935. Root development of pitch pine, with some comparative observations on short-leaf pine. J. Agr. Res. 51:983–1016.

**Mead, D. J., and W. L. Pritchett.** 1971. A comparison of tree responses to fertilizers in field and pot experiments. Soil Sci. Soc. Amer. Proc. 35(2):346–349.

**Mead, D. J., and W. L. Pritchett.** 1974. Variation of N, P, K, Ca, Mg, Mn, Zn, and Al in slash pine foliage. Commun. in Soil Sci. Plant Anal. 5(4):291–301.

**Mead, D. J., and W. L. Pritchett.** 1975. Fertilizer movement in a slash pine ecosystem II. N distribution after two growing seasons. Plant Soil 43:467–478.

**Megahan, W. F.** 1972. Logging, erosion, sedimentation—are they the dirty words? J. For. 70:403–407.

**Melin, E.** 1963. Some effects of forest tree roots on mycorrhizal Basidiomycetes. p. 125–145. *In* Symbiotic Associations. Proc. 13th Symp. Soc. Gen. Microbiol. Cambridge. Univ. Press, London.

**Metz, L. J., and D. L. Dindal.** 1975. Collembola populations and prescribed burning. Environ. Entomol. 4(4):583–587.

**Metz, L. J., and M. R. Farrier.** 1971. Prescribed burning and soil mesofauna on the Santee Experimental Forest. p. 100–106. *In* Proc. Prescribed Burning Symp. USDA For. Serv., SE For. Exp. Sta., Asheville.

**Metz, L. J., C. G. Wells, and P. P. Kormanik.** 1970. Comparing the forest floor and surface soil beneath four pine species in the Virginia piedmont. USDA For. Serv. Res. Pap. SE-55. 8 p.

**Miehlich, G. von.** 1971. Effect of spruce mono-culture on pore size distribution, pH, organic matter, and nutrient content of a loess-pseudo gley. Forstw. Cbl. 90:301–318.

**Mikola, P.** 1973. Application of mycorrhizal symbiosis in forestry practice. p. 383–411. *In* G. C. Marks and T. T. Kozlowski (ed.) Ectomycorrhizae: their Ecology and Physiology. Academic Press, Inc., New York.

**Miller, D. E., and W. H. Gardner.** 1962. Water infiltration into stratified soil. Soil Sci. Soc. Amer. Proc. 26(2):115–119.

**Miller, R. E., and L. V. Pienaar.** 1973. Seven year response of 35-year-old Douglas-fir to nitrogen fertilizer. USDA For. Serv. Res. Pap. PNW-165. 24 p.

**Miller, R. E., and R. L. Williamson.** 1974. Dominant Douglas-fir respond to fertilizing and thinning in southwest Oregon. USDA For. Serv. Res. Note PNW-216. 7 p.

**Mina, V. N.** 1955. Cycle of N and ash elements in mixed oakwood of the forest steppe. Pochvovedenie 6:32–44.

**Minore, D., and C. E. Smith.** 1971. Occurrence and growth of four northwestern tree species over shallow water tables. USDA For. Serv. Res. Note PNW. 160. 9 p.

**Minore, D., C. E. Smith, and R. F. Wollard.** 1969. Effects of high soil density on seedling root growth of seven northwestern tree species. USDA For. Serv. Res. Note PNW-112. 6 p.

**Mitchell, R. G., and H. G. Paul.** 1974. Field fertilization of Douglas-fir and its effect on *Adelges cooleyi* population. Environ. Entomol. 3(3):501–504.

**Möller, G.** 1973. Prognoskurvor för godslingseffekt i tall och gran. Inst. Skogsförbättring. Gödsling Nr. 2. 7 p.

**Möller, G.** 1974. Practical and economical aspects of forest fertilization. Phosphorus Agr. 62:33–48.

**Moore, D. G.** 1975. Effects of forest fertilization with urea on stream quality—Quilcene Ranger District, Washington. USDA For. Serv. Res. Note PNW-241. 9 p.

**Monteith, J. L.** 1965. Evaporation and environment. Symp. Soc. Exptl. Biol. 19:205.

**Moser, M.** 1967. Ectotrophic nutrition at timberline. Midd. d. Forst. Bundesversuchsanst. Wien. 75:357–380.

**Müller, P. E.** 1879. Studier över skovjörd, som bidrag til skovdyrkningens theori. Tidsskr. Skovbr. 3:1–124.

**Munevar, F., and A. G. Wollum.** 1977. Effects of the addition of phosphorus and inorganic nitrogen on carbon and nitrogen mineralization in Andepts from Columbia. Soil Sci. Soc. Amer. J. 41(3):540–545.

**Mustanoja, K. J., and A. L. Leaf.** 1965. Forest fertilization research, 1957–1964. Bot. Rev. 31:151–246.

**Mutatkar, V. K., and W. L. Pritchett.** 1967. Effects of added aluminum on some soil microbial processes and on the growth of oats (*Avena sativa*) in Arrendondo fine sand. Soil Sci. 103:39–46.

**Nehring, K.** 1934. Über die schwankungen der reaktionsverhaltnisse im boden. Zeitschr. f. Pflanzenernahrung, Dungung u. Bodenkundi, Teil A. 36:257–270.

**Nicolson, T. H.** 1967. Vesicular-arbuscular mycorrhiza—a universal plant symbiosis. Sci. Prog. (Oxford) 55:561–581.

**Nolan, C. N., and W. L. Pritchett.** 1960. Certain factors affecting the leaching of potassium from sandy soils. Soil Crop Sci. Soc. Fla. Proc. 20:139–145.

**Neumann, E. E.** 1966. Nutrient balance and increment of a fertilized pole-stage pine stand on a degraded sandy site of medium quality. Arch. Forster. 15(10):1115–1138.

**Nömmik, H.** 1973. The effect of pellet size on the ammonia loss from urea applied to forest soil. Plant Soil 39:309–318.

**Norris, L. A.** 1970. Degrading of herbicides in the forest floor. p. 397–411. *In* C. T. Youngberg and C. B. Davey (ed.) Tree Growth and Forest Soils. Oregon State Univ. Press, Corvallis.

**Norris, L. A.** 1975. Behavior and impact of some herbicides in the forest. p. 159–176. *In* Herbicides in Forestry. Proc. John S. Wright For. Conf. Purdue Univ., W. Lafayette, Ind.

**Norris, L. A., and D. G. Moore.** 1971. The entry and fate of forest chemicals in streams. p. 138–158. *In* J. T. Krygier and J. D. Hall (ed.) Proc. Symp. Land Uses and Stream Environment. Oregon State Univ., Corvallis.

**Nye, P. H.** 1961. Organic material and nutrient cycles under moist tropical forests. Plant Soil 13:333–346.

**Olson, J. S.** 1963. Energy storage and the balance of producers and decomposers in ecological systems. Ecology 44:322–331.

**Oswald, H.** 1974. Biological and ecological aspects of the treatment of young stands. p. 320–340. *In* IUFRO Symp. Stand Establishment. Wageningen, The Netherlands.

**Overrein, L. N.** 1968. Lysimeter studies on tracer nitrogen in forest soils. I. Nitrogen losses by leaching and volatilization after addition of urea-$N^{15}$. Soil Sci. 106:280–290.

**Overrein, L. N.** 1970. Immobilization and mineralization of tracer nitrogen in forest raw humus: II. Effect of temperature and incubation time on the interchange of urea-, ammonium-, and nitrate-$N^{15}$. Plant Soil 32:207–220.

**Ovington, J. D.** 1954. Studies of the development of woodland conditions under different trees. II. The forest floor. J. Ecol. 42:71–80.

**Ovington, J. D.** 1962. Quantitative ecology and the woodland ecosystem concept. p. 103–192. *In* J. B. Cragg (ed.) Advances in Ecological Research. Vol. 1. Academic Press, Inc., New York.

**Ovington, J. D.** 1968. Some factors affecting nutrient distribution within ecosystems. UNESCO, Natur. Resour. 5:95–105.

Ovington, J. D., and H. A. I. Madgwick. 1959. The growth and composition of natural stands of birch: II. The uptake of mineral nutrients. Plant Soil 10:389–400.

Paavilainen, E. 1972. Reaction of Scots pine to various nitrogen fertilizers on drained peatland. Commun. Inst. For. Fenn. 77. 3. 46 p.

Page, J. M., and M. L. Gustafson. 1969. Equipment for forest fertilization. Amer. Soc. Automotive Eng. (New York) Pub. 690553. 10 p.

Patric, J. H. 1961. The San Dimas large lysimeters. J. Soil Water Conserv. 16:13.

Patric, J. H., and K. G. Reinhart. 1971. Hydrologic effects of deforesting two mountain watersheds in West Virginia. Water Resour. Res. 7:1182–1188.

Patt, J., D. Carmeli, and T. Zafrir. 1966. Influence of soil physical conditions on root development and on productivity of citrus trees. Soil Sci. 102:82–84.

Pearse, A. S. 1943. Effects of burning-over and raking-off litter on certain soil animals in the Duke Forest. Amer. Midl. Nat. 29:406–424.

Pearson, G. A. 1931. Forest types in the Southwest as determined by climate and soil. U.S. Dep. Agr. Tech. Bull. 247. 144 p.

Penman, H. L. 1963. Vegetation and hydrology. Commonw. Bur. Soil Sci. Tech. Commun. 53. 124 p.

Pereira, H. C. 1973. Land Use and Water Resources. Cambridge Univ. Press, London. 246 p.

Peterson, J. R., T. M. McCalla, and G. E. Smith. 1971. Human and animal wastes as fertilizers. p. 557–596. In Fertilizer Technology and Use. II. Soil Sci. Soc. Amer., Madison.

Phares, R. E. 1971. Fertilizer tests with potted red oak seedlings. USDA For. Serv. Res. Note NC-114. 4 p.

Pharis, R. P., R. L. Barnes, and A. W. Naylor. 1964. Effects of nitrogen level, calcium level, and nitrogen source upon the growth and composition of Pinus taeda L. Physiol. Plant. 17:560–572.

Phipps, H. M. 1973. Growth response of some tree species to plastic greenhouse culture. J. For. 71:28–30.

Plass, W. T. 1975. An evaluation of trees and shrubs for planting surface-mine spoils. USDA For. Serv. Res. Pap. NE-317. 8 p.

Pomeroy, K. B. 1949. Germination and initial establishment of loblolly pine under various surface soil conditions. J. For. 47:541–543.

Post, B. W. 1974. Soil preparation in stand establishment. p. 141–165. In IUFRO Symp. Stand Establishment. Wageningen, The Netherlands.

Priester, D. S., and W. R. Harms. 1971. Microbial population in two swamp soils of South Carolina. USDA For. Serv. Res. Note SE-150. 6 p.

Pritchett, W. L. 1968. Progress in the development of techniques and standards for soil and foliar diagnosis of phosphorus deficiency in slash pine. p. 81–87. In Forest Fertilization—theory and practice. Tennessee Valley Authority, Knoxville.

Pritchett, W. L. 1972. The effect of nitrogen and phosphorus fertilizers on the growth and composition of loblolly and slash pine seedlings in pots. Soil Crop Sci. Soc. Fla. Proc. 32:161–165.

Pritchett, W. L., and R. E. Goddard. 1967. Differential responses of slash pine progeny lines to some cultural practices. Soil Sci. Soc. Amer. Proc. 31(2):280–284.

Pritchett, W. L., and J. W. Gooding. 1975. Fertilizer recommendations for pines in the U.S. Southeastern coastal plain. Fla. Agr. Exp. Sta. Bull. 774. 21 p.

**Pritchett, W. L., and J. L. Gray.** 1974. Is forest fertilization feasible? For. Farmer 33(8):6–14.

**Pritchett, W. L., and W. H. Smith.** 1970. Fertilizing slash pine on sandy soils of the lower coastal plain. p. 19–41. *In* C. T. Youngberg and C. B. Davey (ed.) Tree Growth and Forest Soils. Oregon State Univ. Press, Corvallis.

**Pritchett, W. L., and W. H. Smith.** 1972. Fertilizer responses in young pine plantations. Soil Sci. Soc. Amer. Proc. 36(4):660–663.

**Pritchett, W. L., and W. H. Smith.** 1974. Management of wet savanna soils for pine production. Fla. Agr. Exp. Sta. Tech. Bull. 762. 22 p.

**Pritchett, W. L., and W. H. Smith.** 1975. Forest fertilization in the U.S. southeast. p. 467–476. *In* B. Bernier and C. H. Winget (ed.) Forest Soils and Forest Land Management. Laval Univ. Press, Quebec.

**Pritchett, W. L., and K. R. Swinford.** 1961. Response of slash pine to colloidal phosphate fertilization. Soil Sci. Soc. Amer. Proc. 25(5):397–400.

**Pritchett, W. L., and C. G. Wells.** 1978. Harvesting and site preparation increases nutrient mobilization. p. 98–110. *In* T. Tippin (ed.) Proc. Symp. Principle of Main. Prod. on Prepared Sites. USDA For. Serv., SE State Priv. Forest. Atlanta.

**Pruitt, W. O.** 1971. Factors affecting potential evapotranspiration. p 82–102. *In* E. J. Monke (ed.) Biological Effects in the Hydrological Cycle. Purdue Univ. Agr. Exp. Sta., W. Lafayette, Ind.

**Pyatt, D. G.** 1970. Soil groups of upland forests. For. Comm. For. Rec. (London) 71. 51 p.

**Ralston, C. W.** 1964. Evaluation of forest site productivity. p. 171–201. *In* J. A. Romberger and P. Mikola (ed.) Int. Rev. For. Res. Vol. 1. Academic Press, Inc., New York.

**Ralston, C. W., and G. E. Hatchell.** 1971. Effects of prescribed burning on the physical properties of soil. p. 68–86. *In* Proc. Prescribed Burning Symp. USDA For. Serv., SE For. Exp. Sta., Asheville.

**Ramann, E.** 1893. Forstliche Bodenkunde und Standortslehre. Julius Springer, Berlin. 479 p.

**Raupach, M.** 1967. Soil and fertilizer requirements for forests of *Pinus radiata*. Advance. Agron. 19:307–353.

**Rehfuess, K. E., and A. Schmidt.** 1971. Effects of lupine establishment and nitrochalk on state of nutrition and increment in older pine stands of the Oberpfalz. Forstw. Cbl. 90: 237–259.

**Remezov, N. P.** 1959. The method of studying the biological cycle of elements in forests. Pochvovedenie 1:71–79.

**Remezov, N. P., and P. S. Pogrebnyak.** 1969. Forest Soil Science. U.S. Dept. Agr., Nat. Sci. Found., Washington, D.C. 261 p.

**Rennie, P. J.** 1955. The uptake of nutrients by mature forest growth. Plant Soil. 7(1):49–95.

**Rennie, P. J.** 1971. The role of mechanization in forest site preparation. XV IUFRO Cong., Div. 3. Rep., Gainesville, Fla. 102 p.

**Rice, R. M., J. S. Rothacher, and W. F. Megahan.** 1972. Erosional consequences of timber harvesting: an appraisal. p. 321–329. *In* Watersheds in Transition. Amer. Water Resour. Assoc. Proc. Ser. 14, Urbana, Ill.

**Richards, B. N.** 1967. Introduction of the rain-forest species *Araucaria cunninghamii* Ait. to a dry sclerophyll forest environment. Plant Soil 27:201–216.

**Richards, B. N.** 1968. Effect of soil fertility on the distribution of plant communities as shown by pot cultures and field trials. Commonw. For. Rev. 47:200–210.

**Richards, B. N., and D. I. Bevege.** 1972. Principles and practices of foliar analysis as a basis for crop-logging in pine plantations: I. Basic considerations. Plant Soil. 36:109–119.

**Richards, B. N., and G. K. Voigt.** 1965. Nitrogen accretion in coniferous forest ecosystems. p. 105–116. *In* C. T. Youngberg (ed.) Forest-Soil Relationships in North America. Oregon State Univ. Press, Corvallis.

**Richards, L. A.** 1949. Methods of measuring soil moisture tension. Soil Sci. 68:95–112.

**Richards, L. A., and S. J. Richards.** 1957. Soil moisture, p. 49–60. *In* 1957 Yearbk. of Agr. U.S. Dep. Agr., Washington, D.C.

**Richenderfer, J. L., W. E. Sopper, and L. T. Kardos.** 1975. Spray-irrigation of treated municipal sewage effluent and its effect on chemical properties of forest soils. USDA For. Serv. Gen. Tech. Rep. NE-17. 24 p.

**Robertson, W. K., W. H. Smith, and D. M. Post.** 1975. Effect of nitrogen and placed phosphorus and dolomitic limestone in an Aeric Haplaquod on slash pine growth and composition. Soil Crop Sci. Soc. Fla. Proc. 34:58–60.

**Rodin, L. E., and N. I. Bazilevich.** 1967. Production and Mineral Cycling in Terrestrial Vegetation. Oliver and Boyd, London. 288 p.

**Romans, J. C. C.** 1970. Podzolisation in a zonal and altitudinal context in Scotland. p. 88–101. *In* Rep. Welsh Soils Discuss. Grp. II. Macaulay Inst., Aberdeen.

**Romell, L. G.** 1922. Luftväxlingen i marken som ekologisk faktor. Statens Skogsförsöksanst. Meddel. 19:125–359.

**Romell, L. G.** 1935. Ecological problems of the humus layer in the forest. Cornell Univ. Mem. 170. 28 p.

**Romell, L. G., and S. O. Heiberg.** 1931. Types of humus layer in the forests of northeastern U.S. Ecology. 12:567–608.

**Rousseau, L. Z.** 1932. Le rendement des peuplements et la coverture vivante. Ass. Ing. For. Prov. Quebec.

**Rowe, J. S.** 1972. Les regions forestieres du Canada. Ministere de l'Environnement. Ottawa. 30 p.

**Rutter, A. J.** 1968. Water consumption by forests. p. 23–84. *In* T. T. Kozlowski (ed.) Water Deficits and Plant Growth. Academic Press, Inc., New York.

**Sallenave, H.** 1969. The cultivation of maritime pine in southwest France. Phosphorus Agr. 54:17–26.

**Sartz, R. S.** 1973. Snow and frost depths on north and south slopes. USDA For. Serv. Res. Note NC-157. 2 p.

**Savage, S. M., J. Osborn, J. Letey, and C. Heaton.** 1972. Substances contributing to fire-induced water repellency in soils. Soil Sci. Soc. Amer. Proc. 36(4):674–678.

**Schlapfer, T. A.** 1972. Soil resource guide—Southern Region. USDA For. Serv., So. Region-8. 48 p.

**Schmidtling, R. C.** 1973. Intensive culture increases growth without affecting wood quality of young southern pines. Can. J. For. Res. 3:565–573.

**Schmidtling, R. C.** 1975. Fertilizer timing and formulation affect flowering in a loblolly pine seed orchard. p. 153–160. *In* Proc. 13th South. For. Tree Improvement. Conf., Raleigh.

**Schofield, R. K.** 1935. The pF of the water in soil. p. 37–48. *In* Trans. 3rd Int. Congr. Soil Sci. Vol II. Thomas Murby and Co., London.

**Schramm, J. R.** 1966. Plant colonization studies in black wastes from anthracite mining in Pennsylvania. Trans. Amer. Philos. Soc. (Philadelphia) 56(1). 194 p.

**Schultz, R. P.** 1971. Stimulation of flower and seed production in a young slash pine orchard. USDA For. Serv. Res. Pap. SE-91. 10 p.

**Schultz, R. P.** 1972. Root development of intensively cultivated slash pine. Soil Sci. Soc. Amer. Proc. 36(1):158–162.

**Schultz, R. P.** 1973. Site treatment and planting method alter root development of slash pine. USDA For. Serv. Res. Pap. SE-109. 11 p.

**Schultz, R. P.** 1976. Environmental changes after site preparation and slash pine planting on a flatwood site. USDA For. Serv. Res. Pap. SE-156. 20 p.

**Schultz, R. P., and L. P. Wilhite.** 1969. Differential response of slash pine families to drought. USDA For. Serv. Res. Note SE-104. 3 p.

**Schultz, R. P., and L. P. Wilhite.** 1974. Changes in a flatwood site following intensive preparation. For. Sci. 20(3):230–237.

**Shoulders, E.** 1967. Growth of slash and longleaf pines after cultivation, fertilization, and thinning. USDA For. Serv. Res. Note SO-59. 4 p.

**Silver, W. S., and T. Maque.** 1970. Assessment of nitrogen fixation in terrestrial environments in field conditions. Nature 227:378–379.

**Singer, M. J., and R. H. Rust.** 1975. Phosphorus in surface runoff from a deciduous forest. J. Environ. Quality 4(3):307–311.

**Sisam, J. W. B.** 1938. Site as a factor in silviculture, its determination with special reference to the use of plant indicators. Can. Dep. Mines Resour. Silvi. Res. Note. 54 p.

**Smalley, G. W.** 1964. Topography, soils, and the height of planted yellow-poplar. J. Ala. Acad. Sci. 35:39–44.

**Smith, C. E., and R. F. Woollard.** 1969. Effects of high soil density on seedling root growth of seven northwestern tree species. USDA For. Serv. Res. Note PNW-122. 6 p.

**Smith, D. M.** 1972. The continuing evolution of silviculture practice. J. For. 70:89–92.

**Smith, W. H., and J. O. Evans.** 1977. Special opportunities and problems in using forest soils for organic waste application. p. 429–454. *In* Soils for Management of Organic Wastes and Waste Waters. Amer. Soc. Agron., Madison.

**Smith, W. H., H. G. Underwood, and J. T. Hays.** 1971. Ureaforms in the fertilization of young pines. J. Agr. Food Chem. 19:816–821.

**Snowdon, P.** 1973. Boron deficiency in relation to growth of *Pinus radiata* D. Don. Masters Thesis. Aust. Nat. Univ., Canberra.

**Soil Survey Staff.** 1975. Soil Taxonomy—A basic system of soil classification for making and interpreting soil surveys. U.S. Dep. Agr., Agr. Handbk. 436. 754 p.

**Sopper, W. E., and L. T. Kardos.** 1972. Effects of municipal waste disposal on the forest ecosystem. J. For. 70:540–545.

**Spurr, S. H.** 1952. Forest Inventory. The Ronald Press Co., New York. 476 p.

**Spurr, S. H.** 1955. Soils in relation to site-index curves. Soc. Amer. For. Proc. 80–85.

**Spurr, S. H., and B. Y. Barnes.** 1973.   Forest Ecology. The Ronald Press, New York. 571 p.

**Stanford, G., C. B. England, and A. W. Taylor.** 1970.   Fertilizer use and water quality. U.S. Dep. Agr. Res. Serv. Rep. ARS 41–168. 19 p.

**Stanhill, G.** 1970.   The water flux in temperate forests: precipitation and evapotranspiration. p. 242–256. *In* D. E. Reichle (ed.) Analysis of Temperate Forest Ecosystems. Springer-Verlag, New York.

**Stein, W. I., J. L. Edwards, and R. W. Tinus.** 1975.   Outlook for container-grown seedling use in reforestation. J. For. 73:337–341.

**Steinbrenner, E. C.** 1975.   Mapping forest soils on Weyerhaeuser lands in the Pacific Northwest. p. 513–525. *In* B. Bernier and C. H. Winget (ed.) Forest Soils and Forest Land Management. Laval Univ. Press, Quebec.

**Steinbrenner, E. C., and J. H. Rediske.** 1964.   Growth of ponderosa pine and Douglas-fir in controlled environment. Weyerhaeuser For. Pap. 1.

**Stephens, C. G.** 1961.   The soil landscape of Australia. CSIRO Soils Pub. 18. 43 p.

**Stephens, F. R.** 1965.   Relation of Douglas-fir productivity to some zonal soils in the northwestern Cascades of Oregon. p. 245–260. *In* C. T. Youngberg (ed.) Forest-Soil Relationships in North America. Oregon State Univ. Press, Corvallis.

**Stephens, G. R., and D. E. Hill.** 1972.   Waste to wood. Can. Poultryman. Feb:44–46.

**Stoate, T. N.** 1950.   Nutrition of the pine. For. Timber Bur. (Aust.) Bull. 30. 61 p.

**Stoeckeler, J. H.** 1960.   Soil factors affecting the growth of quaking aspen forests in the Lake States. Minn. Agr. Exp. Sta. Tech. Bull. 323.

**Stone, E. L.** 1968.   Micronutrient nutrition of forest trees: a review. p. 132–175. *In* Forest Fertilization—theory and practice. Tennessee Valley Authority, Knoxville.

**Stone, E. L.** 1973.   The impact of timber harvest on soil and water. p. 427–467. *In* Rep. President's Advisory on Timber and the Environment. USDA For. Serv., Washington, D.C.

**Stone, E. L.** 1975.   Soil and man's use of forest land. p. 1–9. *In* B. Bernier and C. Winget (ed.) Forest Soils and Forest Land Management. Laval Univ. Press, Quebec.

**Stone, E. L., and A. L. Leaf.** 1969.   Potassium deficiency and response in young conifer forests in eastern North America. p. 217–229. *In* Colloquium on Forest Fertilization. Int. Potash Inst., Jyvaskyla, Finland.

**Stone, E. L., and G. M. Will.** 1965.   Nitrogen deficiency of second generation radiata pine in New Zealand. p. 117–139. *In* C. T. Youngberg (ed.) Forest Soil Relationships in North America. Oregon State Univ. Press, Corvallis.

**Stout, B. B.** 1956.   Studies on the root systems of deciduous trees. Black Rock For. Bull. 15. 45 p.

**Strand, R. F., H. W. Anderson, and R. T. Bergland.** 1973.   Forest fertilization in the Northwest and Canada. For. Ind. 100:68–70.

**Striffler, W. D., and E. W. Mogren.** 1971.   Erosion, soil properties, and revegetation following a severe burn in the Colorado Rockies. p. 25–36. *In* C. W. Slaughter, R. J. Barney, and G. M. Hansen (ed.) Fire in the Northern Environment—a Symposium. USDA For. Serv., PNW For. Range Exp. Sta., Portland.

**Swan, H. S. D.** 1972.   Foliar nutrient concentrations in red pine as indicators of tree nutrient status and fertilizer requirement. Pulp Pap. Res. Inst. Can., Woodlands Pap. 41. 19 p.

**Swank, W. T., and J. E. Douglass.** 1974. Streamflow greatly reduced by converting deciduous hardwood stands to pine. Science 85:857–859.

**Swank, W. T., N. B. Goebel, and J. D. Helvey.** 1972. Interception loss in loblolly pine stands of the South Carolina piedmont. J. Soil Water Conserv. 26:160–163.

**Swanston, D. N., and C. T. Dyrness.** 1973. Stability of steep land. J. For. 71:264–269.

**Sweeney, J. R., and H. H. Biswell.** 1961. Quantitative studies of the removal of litter and duff by fire under controlled conditions. Ecology. 42:572–575.

**Swift, L. W., and J. B. Messer.** 1971. Forest cuttings raise temperatures of small streams in the southern Appalachians. J. Soil. Water. Conserv. 26:111–116.

**Switzer, G. L., and L. E. Nelson.** 1972. Nutrient accumulation and cycling in loblolly pine (*Pinus taeda* L.) plantations ecosystems: the first twenty years. Soil Sci. Soc. Amer. Proc. 36(1):143–147.

**Switzer, G. L., L. E. Nelson, and W. H. Smith.** 1968. The mineral cycle in forest stands. p. 1–9. *In* Forest Fertilization—theory and practice. Tennessee Valley Authority, Knoxville.

**Tamm, C. O.** 1958. The atmosphere. Encycl. Plant Physiol. 4:233–242.

**Tamm, C. O.** 1964. Determination of nutrient requirements of forest stands. Int. Rev. For. Res. 1:115–170.

**Tamm, C. O.** 1968. The evolution of forest fertilization in European silviculture. p. 242–247. *In* Forest Fertilization—theory and practice. Tennessee Valley Authority, Knoxville.

**Tamm, C. O.** 1973. Effects of fertilizers on the environment. p. 299–317. *In* FAO-IUFRO Int. Symp. For. Fertilization. Paris.

**Tamm, C. O., H. Holmen, P. Popovic, and G. Wiklander.** 1974. Leaching of plant nutrients from soils as a consequence of forestry operations. Ambio. 3:211–221.

**Tarrant, R. F., H. J. Gratkowski, and W. E. Waters.** 1973. The future role of chemicals in forestry. USDA For. Serv. Gen. Tech. Rep. PNW-6. 10 p.

**Tarrant, R. F., K. C. Lu, W. B. Bollen, and J. F. Franklin.** 1969. Nitrogen enrichment of two forest ecosystems by red alder. USDA For. Serv. Res. Pap. PNW-76. 8 p.

**Terman, G. L.** 1971. Phosphate fertilizer sources: agronomic effectiveness in relation to chemical and physical properties. Fertilizer Soc. London Proc. 123, London. 39 p.

**Theodorou, C., and G. D. Bowen.** 1969. The influence of pH and nitrate on mycorrhizal associations of *Pinus radiata* D. Don. Aust. J. Bot. 17:59–67.

**Theodorou, C., and G. D. Bowen.** 1970. Mycorrhizal responses of radiata pine in experiments with different fungi. Aust. For. 34:183–191.

**Theodorou, C., and G. D. Bowen.** 1971. Influence of temperature on the mycorrhizal associations of *Pinus radiata* D. Don. Aust. J. Bot. 19:13–20.

**Tiedemann, A. R., G. O. Klock, L. L. Mason, and D. E. Sears.** 1976. Shrub plantings for erosion control in eastern Washington—progress and research of needs. USDA For. Serv. Res. Note PNW-279. 11 p.

**Tisdale, S. L., and W. L. Nelson.** 1966. Soil Fertility and Fertilizers. The Macmillan Company, New York. 694 p.

**Torrey, J. G.** 1978. Nitrogen fixation by actinomycete-nodulated angiosperms. Bioscience 28 (9):586–592.

**Toumey, J. W., and C. F. Korstian.** 1947. Foundations of Silviculture Upon an Ecological Basis. John Wiley and Sons, Inc., New York. 468 p.

**Troedsson, T., and W. H. Lyford.** 1973.   Biological disturbance and small-scale spatial variations in a forested soil near Garpenberg, Sweden. Studia Forestalia Auecica. 109. 23 p.

**Trousdell, K. B., and M. D. Hoover.** 1955.   A change in ground-water level after clearcutting of loblolly pine in the coastal plain. J. For. 53:493–498.

**Trousdell, K. B., D. E. Beck, and F. T. Lloyd.** 1974.   Site index for loblolly pine in the Atlantic Coastal Plain of the Carolinas and Virginia. USDA For. Serv. Res. Pap. SE-115. 11 p.

**Tubbs, C. H.** 1973.   Allelopathic relationships between yellow birch and sugar maple seedlings. For. Sci. 19(2):139–145.

**Ursic, S. J.** 1970.   Hydrologic effects of prescribed burning and deadening upland hardwoods in northern Mississippi. USDA For. Serv. Res. Pap. SO-54. 15 p.

**Ursic, S. J.** 1975.   Harvesting southern forests: a threat to water quality? p. 145–151. *In* Non-Point Sources of Water Pollution. Proc. South. Reg. Conf. Virginia Polytechnic Inst. State Univ., Blacksburg.

**Vail, J. W., M. S. Parry, and W. E. Carlton.** 1961.   Boron deficiency dieback in pines. Plant Soil. 14:393–398.

**Vallee, G., and G. L. Lowry.** 1970.   Forest soil-site studies II. The use of forest vegetation for evaluating site fertility of black spruce. Pulp. Pap. Res. Inst. Can., Woodl. Pap. 16. 32 p.

**Van Lear, D. H., and J. F. Hosner.** 1967.   Correlation of site index and soil mapping units. J. For. 65:22–24.

**Verry, E. S.** 1972.   Effect of an aspen clearcutting on water yield and quality in northern Minnesota. p. 276–284. *In* Symp. on Watersheds in Transition. Amer. Water Resour. Assoc., Urbana, Ill.

**Viro, P. J.** 1966. Manuring of young plantations. Commun. Inst. For. Fenn. 61.4. 30 p.

**Viro, P. J.** 1967.   One tree plots in measuring mature stands. Proc. XIV IUFRO Congr. Munich.

**Viro, P. J.** 1969.   Prescribed burning in forestry. Commun. Inst. For. Fenn. 67.7. 49 p.

**Viro, P. J.** 1970.   Time and effect of forest fertilization. Commun. Inst. For. Fenn. 70. 5. 17 p.

**Viro, P. J.** 1974.   Effects of forest fire on soil. p. 7–45. *In* T. T. Kozlowski and C. E. Algren (ed.) Fire and Ecosystems. Academic Press, Inc., New York.

**Voigt, G. K.** 1965.   Nitrogen recovery from decomposing tree leaf tissue and forest humus. Soil Sci. Soc. Amer. Proc. 29(6):756–759.

**Voigt, G. K.** 1971.   Mycorrhizae and nutrient mobilization. p. 122–131. *In* E. Hacskaylo (ed.) Mycorrhizae. USDA For. Ser. Misc. Pub. 1189.

**Voigt, G. K., and G. L. Steucek.** 1969.   Nitrogen distribution and accretion in an alder ecosystem. Soil Sci. Soc. Amer. Proc. 33(6):946–949.

**Volk, G. M.** 1970.   Gaseous loss of ammonia from prilled urea applied to slash pine. Soil Sci. Soc. Amer. Proc. 34(3)513–516.

**Vosso, J. A.** 1971.   Field inoculation with mycorrhizae fungi. p. 187–196. *In* E. Hacskaylo (ed.) Mycorrhizae. USDA For. Ser. Misc. Pub. 1189.

**Wadleigh, C. H.** 1964.   Fitting modern agriculture to water supply. p 8–14. *In* Research on Water. Spec. Pub. Ser. 4. Amer. Soc. Agron., Madison.

**489**

**Wakeley, P. C., and J. Marrero.** 1958. Five-year intercept as site index in southern pine plantations. J. For. 56:332–336.

**Waksman, S. A.** 1936. Humus: Origin, Chemical Composition, and Importance in Nature. Williams & Wilkins Co., Baltimore. 494 p.

**Waksman, S. A.** 1952. Soil Microbiology. John Wiley and Sons, Inc., New York. 356 p.

**Walter, H.** 1973. Vegetation of the Earth. Springer-Verlag, New York. 237 p.

**Waring, H. D.** 1973. Early fertilization for maximum production. p. 215–241. *In* FAO-IUFRO Int. Symp. For. Fertilization. Paris.

**Watkins, S. H., R. F. Strand, D. S. DeBell, and J. Esch.** 1972. Factors influencing ammonia losses from urea applied to Northwestern forest soils. Soil Sci. Soc. Amer. Proc. 36 (2): 354–357.

**Watson, G. A.** 1973. Soil and plant nutrient studies in rubber cultivation. p. 331–350. *In* FAO-IUFRO Int. Symp. For. Fertilization. Paris.

**Watson, R.** 1917. Site determinations, classification, and application. J. For. 15:553–565.

**Weetman, G. F.** 1962. Mor humus: a problem in a black spruce stand at Iroquois Falls, Ontario. Pulp Pap. Res. Inst. Can., Woodl. Res. Index 130. 18 p.

**Weetman, G. F.** 1977. Present methods and technology available for intensive management and extent of present use. p. 31–42. *In* Proc. Symp. Intensive Culture of Northern Forest Types. USDA For. Serv. Gen. Tech. Rep. NE-29.

**Weetman, G. F., and B. Webber.** 1972. The influence of wood harvesting on the nutrient status of two spruce stands. Can. J. For. Res. 2:351–369.

**Weidemann, E.** 1923. Zuwacksruckgang und Wuchstockingen der Fichte in den mittleren und den unteren Hohenlagen der Sachsischen Staatsforsten. Tharandt. (Translation 302, U.S. Dep. Agr., 1936).

**Wells, C. G.** 1971. Effects of prescribed burning on soil chemical properties and nutrient availability. p. 86–99. *In* Proc. Prescribed Burning Symp. USDA For. Serv., SE For. Expt. Sta., Asheville.

**Wells, C. G., and D. M. Crutchfield.** 1969. Foliar analysis for predicting response to phosphorus fertilization on wet sites. USDA For. Serv. Res. Note SE-129. 4 p.

**Wells, C. G., and J. R. Jorgensen.** 1975. Nutrient cycling in loblolly pine plantations. p. 137–158. *In* B. Bernier and C. H. Winget (ed.) Forest Soils and Forest Land Management. Laval Univ. Press, Quebec.

**Wells, C. G., and L. J. Metz.** 1963. Variation in nutrient content of loblolly pine needles with season, age, soil, and position on the crown. Soil Sci. Soc. Amer. Proc. 27(1):90–93.

**Wells, C. G., J. R. Jorgensen, and C. E. Burnette.** 1975. Biomass and mineral elements in a thinned loblolly pine plantation at age 16. USDA For. Serv. Res. Pap. Se-126. 9 p.

**Wells, C. G., D. Whigham, and H. Lieth.** 1972. Investigation of mineral nutrient cycling in upland Piedmont forest. J. Elisha Mitchell Sci. Soc. 88:66–78.

**Werner, R. A.** 1970. Persistence of Bidrin in two forest soils. USDA For. Serv. Res. Note SE-139. 5 p.

**Westveld, M.** 1952. A method of evaluating forest site quality from soil, forest cover, and indicator plants. USDA For. Serv., NE For. Exp. Sta. Pap. 48. 12 p.

**White, D. P.** 1973. Effects of fertilization on quality of wood and other forest products. p. 271–297. *In* FAO-IUFRO Int. Symp. For. Fertilization. Paris.

**White, D. P., and A. L. Leaf.** 1956.   Forest fertilization. World For. Ser. Bull. 2. New York State Univ., Coll. of For., Syracuse, N.Y. 305 p.

**White, E. H., and D. D. Hook.** 1975.   Establishment and regeneration of silage plantings. Iowa State J. Res. 49(3):287–296.

**White, E. H., and W. L. Pritchett.** 1970.   Water-table control and fertilization for pine production in the flatwoods. Fla. Agr. Exp. Sta. Tech. Bull. 743. 41 p.

**White, E. H., W. L. Pritchett, and W. K. Robertson.** 1971.   Slash pine root biomass and nutrient concentrations. p. 165–176. *In* H. E. Young (ed.) Forest Biomass Studies. Proc. XV IUFRO Congr. Gainesville, Fla.

**White, L. D., L. D. Harris, J. E. Johnston, and D. G. Milchunas.** 1975.   Impact of site preparation on flatwoods wildlife habitat. Proc. SE Assoc. Game Fish Comm. 29:347–353.

**Whittaker, R. H., and G. M. Woodwell.** 1972.   Evolution of natural communities. p. 137–159. *In* J. A. Weins (ed.) Ecosystem Structure and Function. Proc. 31st Annu. Biol. Colloquium, Oregon State Univ. Press, Corvallis.

**Whyte, A. G. D.** 1973.   Productivity of first and second crops of *Pinus radiata* on the Moutere gravel soils of Nelson. N.Z. J. For. 18:87–103.

**Wilcox, H. E.** 1971.   Morphology of ectendomycorrhizae in *Pinus resinosa*. p. 54–80. *In* E. Hacskaylo (ed.) Mycorrhizae. USDA For. Serv. Misc. Pub. 1189.

**Wilde, S. A.** 1958.   Forest Soils. Ronald Press Company, New York. 537 p.

**Wilde, S. A.** 1966.   A new systematic terminology of forest humus layers. Soil Sci. 101:403–407.

**Wilde, S. A., and G. K. Voigt.** 1967.   The effects of different methods of tree planting on survival and growth of pine plantations on clay soils. J. For. 65:99–101.

**Wilde, S. A., J. G. Iyer, C. Tanzer, W. L. Trautmann, and K. G. Watterson.** 1965.   Growth of Wisconsin coniferous plantations in relation to soils. Wisc. Agr. Exp. Sta. Res. Bull. 262. 81 p.

**Wilkinson, S. R., and A. J. Ohlrogge.** 1964.   Mechanism for nitrogen-increased shoot/root ratios. Nature 204:902–904.

**Will, G. M.** 1961.   Mineral requirements of radiata pine seedlings. N.Z. J. Agr. Res. 4(3–4):309–327.

**Will, G. M.** 1966.   Root growth and dry-matter production in a high-producing stand of *Pinus radiata*. N.Z. For. Res. Note 44. 15 p.

**Will, G. M.** 1968.   The uptake, cycling, and removal of mineral nutrients by crops of *Pinus radiata*. Proc. N.Z. Ecol. Soc. 15:20–24.

**Will, G. M., and P. J. Knight.** 1968.   Pumice soils as a medium for tree growth. III. Pot trial evaluation of nutrient supply. N.Z. J. For. 13(1):50–65.

**Wolaver, T. G.** 1972.   The distribution of natural and anthropogenic elements and compounds in precipitation across the U.S.: theory and quantitative models. Masters Thesis. School of Public Health, Univ. North Carolina, Chapel Hill. 118 p.

**Wollum, A. G.** 1973.   Characterization of the forest floor in stands along a moisture gradient in southern New Mexico. Soil Sci. Soc. Amer. Proc. 37(4):637–640.

**Wollum, A. G., and C. B. Davey.** 1975.   Nitrogen accumulation, transformation, and transport in forest soils. p. 67–107. *In* B. Bernier and C. H. Winget (ed.) Forest Soils and Forest Land Management. Laval Univ. Press, Quebec.

**Wolters, G. L., and R. C. Schmidtling.** 1975.   Browse and herbage in intensively managed pine plantations. J. Wildl. Manage. 39(3):557–562.

**Wood, H. B.** 1977. Hydrologic differences between selected forested and agricultural soils in Hawaii. Soil Sci. Soc. Amer. J. 41(1):132–136.

**Woods, R. V.** 1976. Early silviculture for upgrading productivity on marginal *Pinus radiata* sites in the southeast region of South Australia. Wood For. Dep. Bull. 24. 90 p.

**Woodwell, G. M., and R. H. Whittaker.** 1967. Primary production and the cation budget of the Brookhaven forest. p. 151–166. *In* H. E. Young (ed.) Symp. on Primary Productivity and Mineral Cycling in Natural Ecosystems. Univ. Maine Press, Orono.

**Wooldridge, D. D.** 1970. Chemical and physical properties of forest litter layers in central Washington. p. 327–337. *In* C. T. Youngberg and C. B. Davey (ed.) Tree Growth and Forest Soils. Oregon State Univ. Press, Corvallis.

**Worst, R. H.** 1964. A study of the effects of site preparation and spacing on planted slash pine in the coastal plains of southeast Georgia. J. For. 62:556–557.

**Wright, E., and R. F. Tarrant.** 1957. Microbial soil properties after logging and slash burning. USDA For. Serv., Pacific NW For. Range Exp. Sta. Res. Note 157. 5 p.

**Xydias, G. K., and A. L. Leaf.** 1964. Weevil infestation in relation to fertilization of white pine. For. Sci. 10(4):428–431.

**Yawney, H. W., and R. S. Walters.** 1973. Special forest crops and fertilization. p. 122–131. *In* Symp. For. Fertilization. USDA For. Serv. Gen. Tech. Rep. NE-3.

**Young. H. E.** 1967. Symp. on Primary Productivity and Mineral Cycling in Natural Ecosystems. Univ. Maine Press, Orono. 245 p.

**Young, C. E., Jr., and R. H. Brendemuehl.** 1973. Response of slash pine to drainage and rainfall. USDA For. Serv. Res. Note SE-186. 8 p.

**Youngberg, C. T.** 1959. The influence of soil conditions following tractor logging on the growth of planted Douglas-fir seedlings. Soil Sci. Soc. Amer. J. 23(1):76–78.

**Youngberg, C. T., and A. G. Wollum.** 1970. Nonleguminous symbiotic nitrogen fixation. p. 383–395. *In* C. T. Youngberg and C. B. Davey (ed.) Tree Growth and Forest Soils. Oregon State Univ. Press, Corvallis.

**Zahner, R.** 1968. Water deficits and growth of trees. p. 191–254. *In* Water Deficits and Plant Growth. Vol. 2. Academic Press, Inc., New York.

**Zak, B.** 1964. Role of mycorrhizae in root disease. Annu. Rev. Phytopath. 2:377–393.

**Zimmerman, M. H., and C. L. Brown.** 1971. Tree Structure and Function. Springer-Verlag, New York. 336 p.

**Zinke, P. J.** 1967. Forest interception studies in the United States. p. 137–161. *In* W. E. Sopper and H. W. Hull (ed.) Proc. Int. Symp. For. Hydrology. Pergamon Press, London.

**Zon. R.** 1913. Quality classes and forest types. Soc. Amer. For. Proc. 8:100–104.

**Zöttl, H. W.** 1960. Dyrnamik der stickstoffmineralisation im organischen waldboden-material: I. Beziehung zwischen Bruttomineralisation und nettomineralisation. Plant Soil. 13:166–182.

**Zöttl, H. W.** 1973. Diagnosis of nutritional disturbances in forest stands. p. 75–95. *In* FAO-IUFRO Int. Symp. For. Fertilization. Paris.

# INDEX